THE LISTENERS

THE LISTENERS

A History of **WIRETAPPING**

in the **UNITED STATES**

BRIAN HOCHMAN

Harvard University Press

Cambridge, Massachusetts

London, England

2022

Library of Congress Cataloging-in-Publication Data

Names: Hochman, Brian, 1980– author.
Title: The listeners : a history of wiretapping in the United States / Brian Hochman.
Description: Cambridge, Massachusetts : Harvard University Press, 2022. |
Includes bibliographical references and index.
Identifiers: LCCN 2021032833 | ISBN 9780674249288 (hardcover)
Subjects: LCSH: Wiretapping—United States—History. | Electronic surveillance—
United States—History.
Classification: LCC HV7936.T4 H63 2022 | DDC 363.25/2—dc23
LC record available at https://lccn.loc.gov/2021032833

To Lena:

I love you.

CONTENTS

Introduction

The Ballad of D. C. Williams

D. C. WILLIAMS WORKED THE TRADING DESK at a small investment firm in Placerville, California, about forty-five miles east of Sacramento. He spent his days doing what most everyone in the financial sector does for a living. He studied trends and he made deals. He bought low and he sold high. He kept a close eye, all the while, on the regular movements of the market.

Williams also had a knack for electronics, and his specialized technical skills helped him pick up a second job that involved wiretapping. For this job, Williams sometimes went by the names H. Franklin or D. C. Hannahs. It depended on what sort of work was required. Over the span of a short few months, Williams gained notoriety across the state of California by devising a scheme that allowed him to do both jobs—trading stocks and tapping wires—at the same time. It made him a wealthy man.

The scheme was as effective as it was clever. Williams would tap into the communications of manufacturing firms and mining companies in Sacramento and nearby San Francisco, hoping to intercept news of a price quote, a patent application, or an impending sale—anything that a corporate entity would consider confidential. Williams would then relay the news to a syndicate of stockbrokers posted in locations as far-flung as New York and Virginia. The brokers made financial moves based on the intercepted information, returning a cut of the profits to Placerville.

The genius of the scheme wasn't simply that it allowed Williams to eavesdrop on privileged communications. It also capitalized on the time it takes to send an electronic signal across a region as vast as that of the continental United States—a short period of time, but a period of time nonetheless. After bribing a few well-connected officials, Williams found a way to transmit the contents of his wiretaps while slowing the speed with which the original corporate messages reached their intended destinations.

His brokers could buy and sell stocks moments before anyone else, taking advantage of illegal tips while appearing to go along with the rhythms of the market.

The arrangement proved lucrative. Williams's correspondence, later produced as evidence in court, revealed that the members of his syndicate had made a small fortune while the wiretapping scam was up and running. But everything came crashing down when an anonymous tip put the authorities onto Williams's trail. After a brief investigation, detectives in Placerville arrested Williams in the act of tapping the corporate network. He was soon tried, convicted, and sent to prison under an obscure California statute prohibiting the interception of electronic messages. Reporters covering the case pronounced it a "new chapter in crime," a reminder that advances in communications almost always produce advances in eavesdropping.

The year—and here's the twist to the story—was 1864.

D. C. Williams was the first American ever jailed for tapping a wire.[1]

. . .

I first stumbled onto D. C. Williams's case while putting the finishing touches on another research project. His story was buried in the columns of a nineteenth-century newspaper. Reading the account induced a curious sort of historical vertigo. Williams had been tapping telegraph messages, not telephone calls or digital conversations. The obscure California statute under which he was prosecuted was written in 1862. That means wiretapping was enough of a concern for the state legislature to outlaw the practice just one year after the completion of the transcontinental telegraph, at the very moment when the dream of a national communications network became a reality.

All the same, Williams's story reads like something we might encounter in our news feeds today. In the elegance of its design and the sophistication of its execution, the Placerville wiretapping scheme recalls the database hacks and cybersecurity breaches that now make headlines with almost metronomic regularity. For some readers, Williams may even sound like a character straight out of *Flash Boys* (2015), journalist Michael Lewis's award-winning exposé of digital trading fraud on Wall Street.[2]

Yet Williams was a creature of his own era, not ours. He dealt in dots and dashes, not ones and zeroes. Above all else, the timelessness of his story illustrates a basic rule of media studies: the past is never far from the present—or at least it isn't as far from the present as the ceaseless march of innovation would make it seem.[3] In truth, technological progress doesn't so much erase history as write over it, like a palimpsest, whose obscured layers comprise the basis of the visible surface of things. Scratch the surface but slightly and you realize that what we see before us is merely the product of historical accumulation.

Williams's 1864 case is one of those layers of the palimpsest, overwritten and obscured by the present. It should remind us that technological innovations come with both promises and pitfalls, and that the social realities they produce adhere to patterns long in the making. It should remind us that some of the most urgent challenges we face in our own surveillance society—the ubiquity of data, the insecurity of networks, the attempts of third parties of all sorts to monitor and monetize private information—are as old as electronic communications themselves. It should remind us that wiretapping has always been a part of the wires. We've been here before.

. . .

D. C. Williams was hardly the first to tap a wire. Proponents of electronic communication expressed concerns about the security of telegraph networks early on. In the United States, the initial designs for a functioning telegraph system took into account the potential for eavesdropping and sabotage. In 1837—three years before obtaining a patent for the telegraph, and seven years before demonstrating its workings to Congress—Samuel Morse proposed hiding the lines of telegraph networks underground, rather than stringing them overhead. The idea was to prevent "mischievously disposed persons" from cutting into the wires and "injur[ing] the circuit."[4] Morse took the issue so seriously that some of his earliest experiments with telegraph signaling failed because he buried his lines without first considering the basic principles of electrical insulation.

There is little in the early history of the telegraph to suggest that Morse's fears were warranted. But along with the proliferation of telegraph ciphers in the 1840s and 1850s, designed to prevent operators and outsiders alike from reading confidential messages, Morse's proposal for an underground telegraph suggests that the earliest visions of a networked society were haunted by its operational vulnerabilities. In hindsight, Morse's experiments were a prophecy of a future to come. The wiretap realized it soon enough.

In all likelihood, the birth of wiretapping in America coincides with the start of the Civil War, the world's first armed conflict in which the use of electronic communications proved decisive. As early as 1861, telegraph operators in the Union and Confederate armies were employing wiretaps to intercept enemy dispatches and transmit disinformation. By the time Williams was convicted in 1864, the exploits of Civil War wiretappers were already the stuff of legend. Using creative techniques to gather and send electronic intelligence, often under impossibly difficult conditions, military signal operators like William Forster, Charles Gaston, and George "Lightning" Ellsworth received fawning national news coverage for their wiretapping missions. Working on both sides of the conflict, they appeared to raise the practice of telegraph surveillance to the level of art.

Wiretapping would proliferate within the criminal class in the post-bellum years, as the telegraph expanded its reach. D. C. Williams was among the first of many crooks who attempted to take advantage of the nation's ever-increasing dependence on electronic information. Throughout the second half of the nineteenth century, unscrupulous speculators tapped lines to defraud stock exchanges while gamblers worked with corrupt telegraph professionals to place surefire bets at horse-racing poolrooms. Confidence artists exploited fears about the insecurity of electronic networks, too. In cities and towns across America, smooth-talking criminals perfected what came to be known as the "wireless wiretap game," swindling ordinary men and women who were all too eager to believe that a well-placed telegraph tap could provide returns on a commodities investment or a bookmaking wager.

It was only later that the wiretap emerged as a tool of criminal detection—and much later still that the courts decided it was acceptable for agents of the state to wield its powers. Yet the specter of government surveillance nevertheless shaped the history of America's electronic networks from the very beginning. During the late nineteenth century, state and federal officials seldom tapped telegraph lines to investigate corruption and crime. (It was much easier to subpoena copies of telegraph messages, which common carriers typically retained on file for up to a year.) But the first American telephone exchanges would provide fertile ground for law enforcement wiretaps. Police detectives were regularly tapping telephone lines in cities like New York as early as 1895, almost as soon as local telephone companies established citywide networks. Corporate communications providers were in on the act. By the turn of the twentieth century, law enforcement agencies could rely on telephone carriers to cough up customer information, establish centralized listening posts, and provide other forms of technical support to assist in the proceedings of criminal investigations.

Wiretapping thus has a long history—much longer than today's battles over privacy and surveillance would appear to suggest. In the wake of Edward Snowden's sensational disclosures about the scope of warrantless eavesdropping in the United States, much of the national debate remains focused on the unique challenges to privacy that the digital age presents: on new technologies of communication and connection; on new techniques for monitoring them; and on the newfound ease with which a range of institutions—from government agencies like the National Security Agency (NSA) and the Federal Bureau of Investigation (FBI) to corporate conglomerates like AT&T, Verizon, and Google—can monitor the things we say and the data we produce. But our inability to see beyond the present obscures the fact that the work of surveillance in the digital age follows patterns that the analog past has already established.[5] Our historical amnesia also prevents us from appreciating that electronic surveillance didn't always seem so pervasive and routine, and that calls to resist technology's incursions neither came from the political margins nor appeared so futile.

For more than a century, the proper place of wiretapping in American life was the subject of intense debate. The popular arguments about its evils were in fact strong enough to keep the wiretap's legal and political status unsettled. As late as the mid-1960s, a vocal majority of Americans believed that no one—not even sworn agents of the law, under strict controls—had the right to eavesdrop on private conversations. Compare that to recent public opinion polls about electronic surveillance, which reveal that most of us have resigned ourselves to the idea that our communications can and should be monitored in one form or another.[6] Wiretapping was once a "dirty business," as Supreme Court Justice Oliver Wendell Holmes Jr. famously characterized it more than ninety years ago.[7] Now it's a standard investigative tactic, indispensable in the detection of crime and essential to the protection of national security. How did we get from there to here?

This book answers that question, showing how the wiretap has evolved from a specialized intelligence-gathering tool to a mundane fact of American life. In the pages that follow, I consider the origins of wiretapping in military campaigns and criminal confidence games, and I explore the U.S. government's uses of audio surveillance in the nation's self-declared "wars" on alcohol, communism, terrorism, and crime. I track the evolution of wiretapping and eavesdropping technologies, which have long outpaced legal frameworks for guaranteeing privacy. And I reexamine some of American history's great wiretapping scandals—every generation has one, it seems—tracing their influence on popular debates about national security, crime control, and the rights and liberties of individual citizens. My central argument is that surveillance is, and always has been, a constitutive element of our communications ecosystem.[8] The American ideal of electronic privacy has never existed in practice.

This might seem a somewhat grim conclusion to draw from more than 150 years of history. But it doesn't downplay the formative role that opposition to wiretapping has played in the recognition of privacy as a shared right, and it doesn't foreclose the possibility that we can imagine alternatives to our surveillance society. In the end, I regard my argument about the relationship between surveillance and communications as a provocation, intended to draw attention to registers of historical debate that have been lost to time. If you listen closely enough to earlier generations of Americans who argued about wiretapping—its uses and abuses, its benefits and evils, its necessities and excesses—you start to hear a story that isn't just about privacy, a legal concept that the technology of the wiretap in fact predates by more than three decades.[9] Beneath it, on a much lower frequency, the story of wiretapping turns out to be a story about what it means to live in a networked society, a story that returns time and again to some of the most enduring questions Americans have asked about the rise of modern communications. Do technologies connect us, or do they make us more isolated? Do they make us more powerful, or do they render us powerless? Should we marvel at the utopian promises of innovation, or should we fear a more dystopian future to come—a future where every advance in technology robs us of privacy, connection, and power?

From the mid-nineteenth century to the present, ordinary Americans have raised questions like these when confronted with men and women like D. C. Williams. The questions are still unanswered, and will remain so as long as we have reasons to ask them.

. . .

This isn't exactly the book I expected to write. When I started working on this project several years ago, I envisioned myself playing the role of historical detective. Inspired by the D. C. Williams case, and equipped with little more than an educated guess about what the sources would tell me, I saw myself embarking on an arduous mission of retrieval and recovery—one that would probe the recesses of the past and expose their long-guarded secrets. I imagined filing scores of Freedom of Information Act (FOIA) requests, cobbling together a narrative in the face of the usual documentary obstacles: omissions, code names, redactions. I envisioned myself talking with former government spooks—hard-boiled and chain-smoking, the sorts of men and women who agree to meet with people like me in crowded parks or strip-mall coffee shops—and I fancied them affording me glimpses of a dark world occluded from public view. Living in Washington, D.C., as I have for more than a decade, seemed to provide a number of obvious tactical advantages. Where else to learn about eavesdropping than the nation's capital, in the shadows of the government agencies whose clandestine surveillance powers have shaped the contours of American life for so long?

I now realize how naive I was. So many of my assumptions were fed by the stories we tell ourselves about electronic surveillance today. Two stories, master narratives of sorts, dominate news reports, legislative debates, and congressional hearings, and are reinforced in the domain of culture. The first is a story about *secrecy:* a narrative that encourages us to think of the world of wiretapping and eavesdropping as at once both pervasive and hidden, far-reaching and invisible, everywhere and nowhere. I suspect that it's this narrative, nurtured by my love for investigative journalism and the paranoid thrillers of 1970s cinema, that drove my grandiose ideas about suppressed truths, redacted documents, and the power of FOIA. It assumes that unseen forces have hidden the past from us. Only a painstaking historical eye or a well-placed source can bring its truths to light.

The second story, closely related to the first, concerns *governmentality:* a narrative that encourages us to regard wiretaps and bugs as technologies of power that the state brings to bear on the citizens it seeks to control.[10] This is the narrative that made me think I'd find everything I needed to know in my own backyard. It assumes that wiretapping is the sole province of government agencies like the FBI and the NSA, and that sooner or later all of us are forced to submit to their creeping intrusions. In recent years, scholars and activists have begun to explore just how much corporate capitalism has had a stake in the development of electronic surveillance, and just how disproportionately communities of color have borne the brunt of its incursions into private life.[11] But it isn't a stretch to say that when we talk about wiretapping today, we still tend to talk about what goes on in the name of government, behind closed doors, and more often than not at the expense of individual liberties that all of us share equally.

These are true stories, needless to say. The realities of the world in which we live have made them more urgent than ever. At times I take them up directly in the pages that follow—exploring, for example, the importance of government wiretapping scandals to changing conceptions of national security power, and the role of racial politics in the normalization of law enforcement eavesdropping. But the master narratives of secrecy and governmentality shape our expectations about what the past can tell us. They determine what we see and what we don't. One of the most powerful lessons from my research is that earlier generations of Americans talked about the twin threats of wiretapping and electronic eavesdropping differently than we do today. They told *other stories,* which has meant that the terms of the debate over eavesdropping have in fact varied widely over time. It's easy to miss them if we only look through the lens of the present. Our propensity to do so is one of many reasons the American public so often appears to discover wiretapping anew, expressing fresh outrage and despair with every revelation, despite the fact that the practice has been around as long as the wires themselves.

Consider a few of the more glaring examples of historical discontinuity that I discuss in these pages. Today most Americans express apprehensions about the scope of government wiretapping for national security, a logical result of the U.S. intelligence community's activities in the years following the terrorist attacks of September 11, 2001. But the wiretaps that domestic law enforcement agencies regularly monitor in the service of criminal investigations—a far greater number, according to the U.S. government's annual *Wiretap Report*—receive little in the way of attention or scrutiny.

At face value, this disparity may not seem surprising. But the disorienting fact of history is that the currents of public concern over the dangers of wiretapping at one time ran in the opposite direction. For much of the twentieth century, wiretapping for national security was generally regarded as a necessary evil, a mild concession that individual citizens needed to make in order to combat dissent and subversion. By contrast, most Americans disapproved of the prospect of wiretapping for crime control. As late as the 1960s, the idea of giving police officers the power to tap telephones under judicial supervision, now considered harmless, was widely condemned as an abuse of government power. Somewhat remarkably, the passage of the 1968 law that at long last sanctioned law enforcement eavesdropping—the law that legalized the use of wiretaps and bugs in the investigation of common federal crimes like gambling and narcotics trafficking—was one of the earliest political events that led ordinary citizens to decry the "end of privacy" in America. It wasn't the last time the nation would reach that dubious milestone.[12]

The instability of the American public's notions about the proper balance of electronic surveillance power has also led to some confounding shifts in the nation's collective understanding of what constitutes wiretap abuse. Today many consider the government as the primary perpetrator of any and all eavesdropping excesses. For better or for worse, the state tends to play the role of "Big Brother" in media representation and political debate. By contrast, lawmakers and activists of the middle decades of the twentieth century tended to direct their outrage toward a much more diffuse range of actors and institutions. Their most pressing concern was the routine use of phone taps and bugs by individuals and groups who operated in "unofficial" capacities, outside the halls of government: private detectives who tapped for hire to resolve marital disputes; labor spies who listened to telephone calls to disrupt union activity; and corporate security experts who used the tools of the eavesdropping trade to gather competitive intelligence. These were the "listeners" whom the nation feared for much of the wiretap's history, Big Brother's little siblings. Even telephone company technicians provided cause for alarm because they monitored lines to perform service checks. That fact seems almost quaint in hindsight.

Concerns with the eavesdropping abuses of the *federal* government are themselves skewed toward the present, beholden as they are to a sea change in electronic surveillance law that occurred in the late 1960s. In the century that preceded that legislative transformation, the government's right to tap remained unsettled. If federal eavesdropping powers existed—and it's crucial to understand that it wasn't always clear that they did—the reliance on wiretaps and bugs at agencies like the FBI and the

Central Intelligence Agency (CIA) flew in the face of the states, which took conflicting stances on the issue from the beginning. Some states allowed their law enforcement agencies to tap telegraph and telephone lines. Others had statutes on the books forbidding the practice. Many state jurisdictions lacked wiretapping laws altogether.

Notwithstanding the messiness of the past, the existing body of academic scholarship on wiretapping has tended to reflect the biases of the present, reinforcing the canonical narratives of secrecy and governmentality despite superficial differences in subject matter, approach, and political temperament. On one side runs a tradition of writing on code-breaking and spycraft, which casts wiretapping as a weapon that the U.S. military and intelligence bureaucracies have deployed in the global theaters of war.[13] On the other side runs a tradition of writing on policing and politics, which casts wiretapping as a tool that law enforcement agencies like the FBI have used to squash dissent.[14]

The scholars who have shaped these dual historiographical traditions—one oriented toward the foreign, the other toward the domestic—all appear to share the underlying sense that the origins of our surveillance society lie somewhere buried in the depths of the state. This is easy to see if you peruse the footnotes to any recent monograph on the NSA or the FBI. There you'll see what perhaps amounts to the most obvious traces of the secrecy and governmentality narratives: citation after citation of government documents, court cases, and first-person interviews, the products of herculean efforts of investigation and reconstruction. Similar assumptions about the origins of electronic surveillance also animate a much broader tradition of writing on electronic privacy, a multifaceted concept that scholars have almost always approached by exploring the official realms of law and policy, but seldom venturing beyond them.[15]

These traditions of scholarship have enabled my own research. But I follow a slightly different set of storylines in this book—revising, in the process, many of the foundational assumptions we share about electronic surveillance. The history of wiretapping in America—my own version, at least—proceeds along two interlocking tracks, both of which cover much more prosaic ground than the narratives of secrecy and governmentality seem to allow. The first track concerns technologies and laws—how wiretapping devices developed, how Americans employed them, and how officials of all political stripes attempted to control and exploit their capabilities. The second track concerns culture and representation—how ordinary citizens discovered, experienced, imagined, and debated wiretapping over time.

By bringing these two tracks together, this book shows just how far, and just how visibly, electronic surveillance has worked its way into the fabric of American life since the middle decades of the nineteenth century. Wiretapping has functioned as a vital medium in the exercise of state power, and its workings have at times remained hidden from the American public. But the practice also infiltrated many other social domains along the way: from the stock market to the gambling hall, from the bedroom to the boardroom, and from the Supreme Court to the silver screen. The dirty business is everywhere if you know where to look for it.

. . .

There are several core arguments about the history of wiretapping in America that inform the chapters that follow. They provide a more or less constant backdrop to these pages: their lessons help to shape the stories I tell and the connections I make. It's worth highlighting them here briefly.

The first lesson, by now somewhat obvious, is that wiretapping has never exactly been secret. Over the last 150 years, the world of the eavesdropper has proved anything but invisible, so much so that it sometimes seems strange that we ever came to think otherwise. Legal and political debates about wiretapping have preoccupied American citizens since the 1860s. Moreover, stories of tapped phones and bugged rooms have provided perennial fodder for fiction, television, journalism, and film. From the pulp wire thrillers of the nineteenth century to *The Wire* today, wiretapping has left a distinctive imprint on the course of American culture. Its intrusion into the most mundane corners of social life is one of many reasons the existence of the U.S. "surveillance state" has seemed so natural, and so intractable, for so long.

The second lesson of this book is that wiretapping wasn't always the exclusive province of the surveillance state itself. The U.S. government came late to the game of electronic surveillance, it turns out. From the 1890s onward, and accelerating around World War I, state and federal law enforcement agencies routinely installed telephone taps to gather evidence and combat espionage. Yet crooks and con artists employed wiretaps in criminal schemes long before that, establishing an association between wiretapping and criminality that would prove difficult to shed.

Whatever the nation's laws may or may not have said about the legality of the practice, the strength of the association between wiretapping and criminality destined the government's early experiments with electronic surveillance to proceed on shaky ground. It would take decades to convince ordinary Americans that the government had the right to tap under judicial controls. And while the issue wreaked havoc in the courts and in Congress, electronic surveillance established firm roots in the private sector, evolving into a multimillion-dollar industry that served both domestic and international markets. Employed throughout the twentieth century by electrical engineers, private detectives, and security contractors, wiretaps and other eavesdropping technologies helped corporate interests gather competitive intelligence and ensure quality service. They also played an important role in the history of divorce litigation. The state's claim on eavesdropping was thus never monolithic or exclusive. For more than a century—until the late 1960s, and even well after—taps and bugs flourished outside of government channels.

The instability of the U.S. government's historical relationship to electronic surveillance also points to the third and final lesson of this book: that wiretapping is political, but not in the way we normally think. To be sure, government agencies and corporate interests have exploited electronic surveillance as a method of imposing conformity and quashing dissent. They still do, perhaps more than we know. But anticommunism and antiradicalism, the two most notorious ideological underpinnings of state surveillance in America, played a much less central role in the institutionalization of government wiretapping than the standard historical accounts have suggested.[16] As I demonstrate in the pages that follow, many of the most significant legislative expansions of the state's wiretap authority in the twentieth century were in fact first articulated as cutting-edge "crime control" measures, designed to assist the routine work of law enforcement. The emergence—and eventual triumph—of a racialized politics of law and order during the 1960s, in particular, proved instrumental in widening the permissible boundaries of government wiretapping in the United States. The world of normalized electronic surveillance in which we live today thus has roots in an imagined imperative to police the nation's streets, in communities of color especially.

Politics, moreover, seldom moves in one direction. One of the most surprising things I learned while writing this book is just how widespread popular opposition to electronic surveillance really was for most of this country's history. Until the late 1960s, when fears about social disorder led the nation to acquiesce to more punitive approaches to law enforcement, most Americans strongly disapproved of wiretapping. Well after the courts made warranted wiretaps legal—until the terrorist attacks of September 11, 2001, really—lingering concerns about the dangers of the practice were pervasive on both sides of the ideological spectrum. At times they made strange political bedfellows. On the left, Americans protested audio surveillance as an affront to constitutional rights and civil liberties; on the right, the issue seemed to encapsulate the persistence of government overreach. Whatever its basis, opposition to wiretapping was once a mainstream political stance in America, one not limited to a paranoid fringe. Combined with the usual lags between technological change and regulatory action, the consensus surrounding the evils of the wiretap had the effect of delaying the codification of coherent surveillance policies over time. On occasion, the American public even held the men who abused wiretaps and bugs accountable (they were almost always men) and effected meaningful change.

Political resistance to wiretapping once prevailed in this country. Taking the long view, it's even possible to say that the fights Americans waged against the practice enjoyed modest success in achieving their aims. We'd do well to remember the resolve that ordinary citizens once demonstrated in challenging technology's intrusions into private life. Today we fight the same battles on a much wider scale.

. . .

The reach of the wiretap and the bug extends well beyond the confines of the United States, of course. A book on the global history of eavesdropping would look very different from this one.[17] On the one hand, countries with totalitarian governments—most famously Russia in the era of the Soviet Union and East Germany under communist rule—have made extensive investments in the use of audio surveillance. In the process, they have terrorized their subjects, chilled dissent, and given state functionaries the leverage they need to extort, harass, imprison, and worse.[18] On the other hand, countries with more democratic legal systems and communications infrastructures either codified coherent wiretap laws well before the United States (as did Australia and the United Kingdom) or developed wiretap policies through much less tumultuous political processes (as did Canada).[19] National boundaries have seldom held firm in the history of eavesdropping to begin with. Since the early 1940s, for example, the United States has worked hand in hand with Australia, Canada, New Zealand, and the United Kingdom to establish technical surveillance networks, share electronic intelligence, and protect democratic interests around the world. The Five Eyes alliance, so called, still exerts a powerful influence on geopolitical affairs today.

All the same, the United States remains a worthy case study for two main reasons. First, Americans were in all likelihood the first to pioneer the art of wiretapping, and for more than a century the United States served as the world's primary exporter of audio surveillance technologies. (America's position at the vanguard of the global electronic surveillance market has changed only slightly in recent years.)[20] Second, official wiretapping policies evolved more chaotically here than elsewhere—a product not only of the unique relationship between the federal government and the states, whose competing interests delayed the institutionalization of coherent wiretap laws for almost a century, but also of the nation's formidable tradition of civil libertarian activism, which over time found ways to thwart the government's attempts to assert its right to tap.

The *culture* of wiretapping in the United States also has its own distinct character, a reality that's perhaps easier to describe than explain. The list of U.S. citizens who have had their telephones tapped is as decorated as it is long, encompassing something like a who's-who compendium of modern America. Anyone worth talking to, it seems, has been on the wrong end of a wiretap at one time or another. To follow their names and stories is to follow nothing less than a shadow history of American culture, unruly in its logic and pervasive in its reach.

The widespread use of wiretaps and bugs in the investigation of syndicated crime, both on and off the record, made gangsters and other public enemies the earliest and most visible targets of electronic surveillance. Charles "Lucky" Luciano, Al Capone, Dutch Schultz, and John Dillinger all had telephones that were tapped. Legend has it that Schultz became so accustomed to the arrangement that he greeted the government agents who monitored his line when he called to order his breakfast every morning.[21] Los Angeles mob boss Mickey Cohen was the subject of multiple eavesdropping operations. Later he turned the tables and began employing his own in-house audio surveillance expert, an electronics technician who had made his name in southern California by tapping the home telephone of comedian Mickey Rooney, and who would later pen a best-selling autobiography after renouncing the sin of wiretapping at a Billy Graham tent revival. Cohen agreed to write the book's preface; Graham later turned it into a feature film. The title, predictably enough, was *Wiretapper* (1955), starring Hollywood bit players Bill Williams and Georgia Lee.

Martin Luther King Jr., Malcolm X, Muhammad Ali, Bobby Seale, and Stokely Carmichael were all subjected to wiretaps, among other persistent forms of government harassment. So were Joan Baez, Roy Cohn, Jane Fonda, Barry Goldwater, Abby Hoffman, William Safire, and Benjamin Spock, as well as political organizations ranging from the National Lawyers Guild, the United Auto Workers Union, the Nation of Islam, and the Ku Klux Klan.[22] The FBI tapped Jimmy Hoffa's telephones throughout the early 1960s in an effort to put the notorious Teamsters president behind bars. Faced with a litany of criminal charges, Hoffa eventually appealed to the Supreme Court on the grounds that the FBI's eavesdropping operation had violated his right to attorney-client privilege. The justices who deliberated on the case of *Hoffa v. United States* (1967) were unaware that many of their own telephones were tapped at the time, too.[23]

In 1969, responding to leaked news reports about the U.S. government's secret bombing campaign in Cambodia, Secretary of State Henry Kissinger ordered wiretaps on President Richard Nixon's entire National Security Council and three investigative journalists. ("We must crush these people! We must destroy them," Kissinger advised his boss before dialing J. Edgar Hoover's office and arranging the eavesdropping operation.)[24] Kissinger would eventually find himself on the other end of a FBI wiretap, while Nixon of course went on to bug himself—a decision that unintentionally ended up serving as the nail in his political coffin. Two decades later, a government employee named Linda Tripp secretly recorded her telephone conversations with Monica Lewinsky, the White House intern then involved in a romantic relationship with the president of the United States. Many regard the discovery of the Tripp tapes as the spark that set off the explosive series of events that led to the impeachment of Bill Clinton.

Wiretaps haven't merely served as a reliable engine of political intrigue. In recent years they have also created scandals in the worlds of religion, sports, and entertainment. In the early 2000s, Church of Scientology head David Miscavige allegedly arranged to tap the phones of Hollywood actress Nicole Kidman, then the wife of celebrity Scientology spokesperson

Tom Cruise. The Miscavige-Kidman affair is the latest in a long line of eavesdropping cases involving religious and spiritual leaders. Their ranks are as diverse as the archbishop of New York City, whose phones were tapped throughout the early 1910s, and Bhagwan Shri Rajneesh, whose new-age commune in Rajneeshpuram, Oregon, was wired for sound seven decades later, in the early 1980s.

In 2014, Donald Sterling, the former owner of the Los Angeles Clippers, made a series of racist remarks to his mistress in conversations that took place on a recorded cell phone line. When the tapes of Sterling's calls were leaked to the press, the resulting firestorm led to Sterling's ouster and the largest sale of a sports franchise in American history. Other figures involved in professional and collegiate athletics whose conversations have recently been recorded (or who have recorded conversations themselves) include erstwhile NBA point guard and coach Mark Jackson, now a broadcast commentator for ESPN and ABC; former New England Patriots stars Aaron Hernandez and Antonio Brown; longtime Major League slugger Matt Williams; and dozens of basketball coaches, recruiters, and athletics office personnel employed by the University of Arizona, Auburn University, the University of Kansas, Louisiana State University, the University of Louisville, Oklahoma State University, and the University of Southern California, among several high-profile schools involved in an ongoing NCAA "pay-for-play" investigation.

And then there's Donald Trump, whose story returns to recorded conversations time and again. Trump's former attorney Michael Cohen was known for his prolific use of audio recordings. Throughout his twelve years of work for the Trump organization, Cohen secretly amassed hundreds of hours of taped exchanges for purposes we can only hazard to guess. Trump is also rumored to have wired some of his real estate properties for sound, including his famed resort at Mar-A-Lago, the so-called Winter White House.[25] The former president's longstanding proximity to electronic surveillance may well provide an explanation for the moment when he took to Twitter to threaten former FBI director James Comey with recordings of their disputed meetings in the Oval Office. "Lordy I hope there are tapes," Comey famously quipped. At this writing, none have emerged.

Some of these incidents are discussed in these pages. Others are not. Much like the investigative injunction to "follow the money"—a catchphrase popularized in the wake of Watergate, an affair that itself began with an audio surveillance operation—the impulse to follow the wiretaps is tempting, promising as it does the ability to recover a unified truth accessible to history. But fashioning a comprehensive account of wiretapping in America is an impossible task, and not just because the practice has always lingered in the uncertain gray areas between fact and fiction, on-the-record data and off-the-record hearsay. The number of wiretaps is simply too large. Their cultural reach is too vast. Even though historians and journalists have spent decades in government archives working to reconstruct the past that has been hidden from us, and even though this book builds directly on their efforts, it's imperative to recognize that no single history of wiretapping and electronic eavesdropping is complete. There's always another tap to uncover.

To combat the problem of sprawl, the ensuing chapters explore a series of public flash points that illuminate the chaotic course of the wiretap's history in America. They run the gamut from court cases and congressional hearings to pulp novels, investigative reports, and Hollywood films. Archival sources have figured prominently in my research as well. The paper trail for this project ended up taking me far beyond the confines of the nation's capital, leading me to locations as diverse as Saint Louis, San Antonio, Chicago, and Seattle. The tiny town of Arkadelphia, Arkansas, somehow cropped up on my research itinerary—a product of the fact that the political architect of America's first comprehensive federal wiretap law willed his personal papers to a small bible college there. But in general, the shape of this book reflects my interest in balancing the public with the secret, the published with the unpublished, contrary to the disposition of so much of the extant scholarship.

Wherever possible, this book is centered on people. In part, this is to counteract the long-standing tendency in studies of surveillance to grant extraordinary agency to agencies, so to speak. In our own age of mass surveillance, when numbers, clicks, and cookies appear much more critical to the work of eavesdropping than human voices and their individual sources, both the tappers and the tapped tend to remain nameless and faceless. Mammoth government institutions and sprawling corporate firms—known to us only through a sea of acronyms: FBI, NSA, AT&T—seem to do all of the work, while their targets are forced to either submit to abstraction or turn to the cold realm of the law for recourse. To my mind, it's inadequate to approach the history of wiretapping and electronic eavesdropping from this angle alone. Above all else, doing so ignores what I regard as the *banality* of electronic surveillance in America: its reliance on everyday technologies of communication; its penetration into mundane social contexts; and its dependence on individuals who make difficult choices and live real lives on both sides of every wiretap. Staying true to that core belief has governed the types of stories I tell in these pages, however partial they may be.

．　．　．

The Listeners is divided into three parts. Parts One and Two cover the period between 1860 and 1968, when most Americans associated wiretapping with criminals and spies. Part Three explores the period between 1968 and 2001, when wiretapping evolved into a routine police tactic—a result of the ascendancy of "tough on crime" law-and-order politics. Each chapter proceeds in chronological order, although I occasionally opt to follow narrative threads that travel both backward and forward in time. My hope is that doing so will offer added context to the cases and debates that have mattered over the course of the wiretap's long history in the United States.

Chapters 1 and 2 move from the second half of the nineteenth century to the early decades of the twentieth, an extended period in which wiretapping seemed more like a trick of the lawless than a tool of the law. The earliest wiretappers intercepted telegraph messages during the American Civil War. Although Union and Confederate signal operators were celebrated as heroes both during and after the conflict, the surveillance tactics they perfected soon fell into the hands of the common crook. In the postbellum decades, tapping wires emerged as an easy way to rig economic activities that relied on the convenience of the telegraph to function. Some rogue telegraph professionals intercepted messages to get ahead in the stock market, extending the tradition of wiretap fraud that D. C. Williams helped to inaugurate. Others used wiretaps to place illegitimate bets on the results of horse races. Reports on the use of wiretaps in criminal conspiracies, often sensationalized, fueled pervasive fears about the insecurity of America's early electronic networks. Fiction often trumped fact in this period. Although criminal wiretapping schemes remained rare, con artists began exploiting the public's familiarity with their workings in an effort to defraud ordinary citizens, and writers of literary fiction began using wiretaps as stock plot devices in the pages of pulp novels. By the early 1900s, news outlets across the country were warning of an epidemic of "wireless wiretap" scams, which tricked unsuspecting individuals to invest in telegraph tapping schemes that didn't actually exist.

This first half-century of history, running roughly from the late 1860s to the early 1910s, solidified wiretapping's reputation as a dirty business— the domain of criminals, cheats, and con artists. The kneejerk association between wiretapping and crime was one reason telecommunications providers were able to convince state legislatures around the country to implement wiretapping prohibitions long before the federal government established an official position on the issue of electronic surveillance. By the end of World War I, fourteen state jurisdictions had adopted California's early model, ratifying statutes that made tapping lines a punishable offense.[26] Fifteen more followed suit in the coming decade.[27] Yet an irony of this period of early legislative activity is that wiretapping had already begun to infiltrate the ranks of American law enforcement, off the record and under fuzzy legal pretext. Throughout World War I and the "Red Scare" years, government agents at the Bureau of Investigation— rechristened as the Federal Bureau of Investigation (FBI) in 1935—tapped telephones in their frenzied hunt for communist spies and political subversives. At the same time, America's war on alcohol afforded the wiretap a new role as a routine medium of criminal investigation. Because bootlegging syndicates often relied on the convenience of the telephone to coordinate their operations, federal Prohibition agents began wiretapping to gather intelligence on their inner workings.

Wiretapping still remained a dirty business during this period. Most Americans believed that the use of wiretaps in criminal investigations was dishonest and immoral, if not unconstitutional. In a surprise development, the U.S. Supreme Court sanctioned law enforcement wiretapping in *Olmstead v. United States* (1928), a watershed case involving one of the largest bootlegging syndicates in the country. But the ruling drew fierce opposition, particularly among Americans who opposed the excesses of Prohibition enforcement. Several legislative proposals that would have reversed *Olmstead* made their way to the floor of Congress in the early 1930s. All of them failed.

The tide seemed to turn when Congress passed the Federal Communications Act (FCA) in 1934. Section 605 of the FCA made it illegal to "intercept . . . and divulge" communications by wire, a stance that the Supreme Court would uphold in three major rulings on wiretapping in the late 1930s. Yet the vague language of the statute blunted the law's force as a wiretap deterrent. Some officials saw the "intercept . . . and divulge" clause as an outright ban on wiretapping. Others regarded it as a mere prohibition against the use of wiretap evidence in court. The clash of perspectives, never resolved, opened the door to wiretap abuse among law enforcement agencies and private citizens alike.

Chapters 3 and 4 focus on the confusion of the decades that followed the passage of the Federal Communications Act, exploring two scandalous wiretapping cases that helped to lay the groundwork for national wiretap reform. The first case involved a Soviet spy named Judith Coplon, whose arrest and subsequent trials were in part the result of a dragnet FBI wiretapping operation. The second case involved a private investigator whose midtown Manhattan apartment contained a secret listening post connected to more than 100,000 telephone lines across New York City. Both cases hastened a series of state and federal hearings on threats to communications privacy. One indirect result of the push for reform was the U.S. Supreme Court's decision in *Benanti v. United States* (1957), which marked an initial attempt to bring order to the chaos of state and federal wiretap law.

Part Two spans the late 1950s to mid 1960s, when debates over wiretapping in the United States reached a fever pitch. Chapter 5 focuses on the publication of Samuel Dash, Richard Schwartz, and Robert Knowlton's *The Eavesdroppers* (1959), the first comprehensive study of wiretapping to appear in the United States. The product of sixteen months of investigative research in twelve different cities, *The Eavesdroppers* uncovered a century's worth of wiretap abuse on the part of state authorities and private investigators. Yet the book elicited conflicting responses in light of the Supreme Court's decision in *Benanti v. United States* two years earlier. The result of a failed New York Police Department drug investigation, the *Benanti* ruling suggested that all official forms of law enforcement wiretapping—municipal, state, and federal—fell under the jurisdiction of the 1934 Federal Communications Act. At least in theory, this meant that police agencies around the country could no longer rely on recorded telephone conversations to gather evidence and convict criminal suspects. As the law enforcement community scrambled to bring its surveillance protocols in line with the *Benanti* decision, *The Eavesdroppers* began to look less like a neutral investigative report and more like an opportunistic political attack. Well into the 1960s the book continued to serve as a lightning rod for debates about the ability of a wide range of third parties—not just the federal government—to listen in on American citizens.

Chapter 6 explores the shadowy world of private-sector eavesdropping that emerged in exactly the same period. A quirk of the *Benanti* ruling was that it appeared to outlaw wiretapping (eavesdropping via telephone) but not bugging (eavesdropping via microphone). Where one door closed, another opened: on the heels of the invention of the electronic transistor, which helped to miniaturize clandestine listening devices, the loopholes in the *Benanti* decision led to the explosive growth of the private-sector eavesdropping industry in America. As I demonstrate, the rise of the hidden microphone during this period is perhaps best captured in the popular visibility of two professional eavesdroppers who ended up having an outsized role in national debate: Jim Vaus, a mob wiretapper who left the criminal underworld and found faith, becoming a Christian evangelist, and Harold K. Lipset, a San Francisco private eye known for his creative use of bugging technologies. The response to their exploits reveals just how dramatically public opinion of electronic surveillance shifted over the course of the 1960s, when miniature listening devices began to seem everywhere and nowhere at once.

Part Three of *The Listeners* considers how wiretapping and eavesdropping became synonymous with government surveillance. For the majority of the twentieth century, professional eavesdroppers like Vaus and Lipset did as much wiretap work in the fields of divorce litigation and corporate espionage as law enforcement agents did in the fields of crime control and national security. It wasn't until the 1970s that wiretaps and bugs began to seem the exclusive province of the state—a product of a momentous shift in the politics of electronic surveillance that occurred over the course of a turbulent twelve-month period in 1967–1968. In 1967, the U.S. Supreme Court's landmark decisions in *Berger v. New York* and *Katz v. United States* sanctioned government wiretapping under judicial oversight. Congress responded the following year with the passage of Title III of the Omnibus Crime Control and Safe Streets Act (1968), a controversial set of provisions that put the Supreme Court's new formula for warranted wiretapping into law, barring all other forms of the activity in the process. The result was nothing short of revolutionary: after more than four decades of uncertainty and debate, the wiretap was finally approved as a tool of state surveillance.

We still live under the regime of court-authorized government eavesdropping that *Berger, Katz,* and Title III helped to create. But it's important to realize that history almost turned out differently. In order to place the Title III revolution in proper context, Chapter 7 follows the arc of the political career of Missouri senator Edward V. Long, perhaps the most vocal advocate for electronic privacy to emerge in the period. In the mid-1960s, Long chaired a Senate subcommittee that exposed illegal wiretapping at the Internal Revenue Service, one among a host of federal agencies that engaged in the practice. Responding to widespread concerns about government invasions of privacy, in 1967 Long worked with a sympathetic White House to craft a proposal for the Right of Privacy Act, a legislative measure that would have permanently wiped out wiretapping in America.

Somewhat surprisingly, at least in hindsight, a majority of Americans supported Long's bill. As late as June 1967, the Right of Privacy Act appeared ripe for passage. But the social unrest of the late 1960s—war protests, political assassinations, and, especially, race riots—altered its political fortunes. In a matter of months Long's proposal to end government wiretapping foundered, and a faction of conservative lawmakers rushed to codify electronic surveillance as a tactic for preserving law and order. The passage of Title III was the result. As I demonstrate in Chapter 8, by the time of the Watergate and Church Committee scandals of the mid-1970s, at the very moment when most Americans became aware of government surveillance abuse at the highest levels of political office, wiretaps and bugs had paradoxically become normalized in the enforcement of criminal law. Such an outcome would have been unthinkable only a decade prior.

Chapters 9 and 10 examine the two decades between the breakup of the Bell Telephone system in the early 1980s and the passage of the U.S. PATRIOT Act in 2001, a period of explosive innovation and change in the telecommunications industry. For U.S. law enforcement agencies— particularly agencies charged with fighting the nation's War on Drugs— the new telecommunications technologies of the period posed thorny logistical problems. Many of them proved resistant to established methods of eavesdropping. Faced with a political mandate to increase wiretapping, on the one hand, and an increasingly untappable national communications network, on the other, law enforcement officials were forced to lobby U.S. telecom companies to provide expanded technical assistance in wiretapping operations. The telecommunications industry pushed back, wary of consumer backlash in an increasingly competitive marketplace. By 1990, the two sides began looking to Congress to broker a solution.

After years of back-channeling and negotiation, compromise eventually came in the form of the Communications Assistance for Law Enforcement Act (CALEA), a controversial 1994 law that required telecommunications companies to build surveillance-friendly networks.

The technological solutions that CALEA offered to U.S. law enforcement agencies were far from politically neutral, however. As I show in Chapter 10, CALEA institutionalized workarounds for government wiretapping that reflected the long-standing racial biases of American drug enforcement. The public spats over "backdoor" police surveillance that followed CALEA's passage also found their way into the narrative arc of David Simon's acclaimed television series *The Wire* (2002–2008). Long praised for its exacting portrayal of American city life, *The Wire* provides an ironic coda to the debates over race, technology, and government surveillance that emerged in CALEA's wake, as the electronic age gave way to the digital.

The book concludes with a brief consideration of the history of wiretapping and electronic eavesdropping in America after September 11, 2001. The Epilogue explores the technological and political backstory behind the apparent decline of wiretapping—and the concomitant rise of "dataveillance"—in the twenty-first century.

.   .   .

Before proceeding, a brief note on terminology:

Throughout this book I use the word *wiretapping* to describe the act of intercepting or recording messages or voice conversations transmitted over electronic communications networks. This limited, technically specific definition of the term accords to its historical uses over the course of the nineteenth and twentieth centuries. In that extended period, wiretapping was an activity that arose in conjunction with the dominant communications media of the day: the telegraph and the telephone. Monitoring these media usually, although not always, involved a physical connection to live wires.

In the late twentieth century, the telegraph and the telephone were joined by the computer and the Internet, our great gateways to the digital age. At the exact same time, the rise of wireless and fiber-optic networks fundamentally altered the nation's communications infrastructure. Among other things, the technical shift from analog to digital meant that eavesdroppers could no longer connect to live wires to do their work. Wiretapping soon became something of a metaphorical investigative activity. The individuals charged with monitoring the nation's electronic communications—state and federal investigators, mostly—could only eavesdrop through backdoor listening posts, established under law by the nation's telecommunications companies. As a measure of how far things had come since the age of the telegraph and telephone, the legal definitions of what it means to "intercept" and "record" electronic communications changed during this period as well.[28]

Today the average American tends to use the word *wiretapping* somewhat promiscuously, often deploying it in reference to a broad range of activities and technologies that involve the recording of personal information. With a great deal of variation and imprecision, "wiretapping" applies to the interception of radio transmissions, to the surreptitious recording of conversations by microphone (often called *bugging*), and to the mining of stored electronic data (e-mail, text messages) and metadata (location information, call records). Household devices such as the Amazon "Alexa" and Google Home are sometimes referred to as "personal wiretaps" because they have been known to record ambient conversations without the explicit consent of their users.

The term *electronic eavesdropping* more accurately encompasses these activities and technologies, and I take pains to distinguish them from the uses and abuses of the wiretap in the pages that follow. Using the terms "wiretapping" and "electronic eavesdropping" interchangeably betrays a misleading lack of technical specificity. The wiretap is a tool of electronic eavesdropping, but not all electronic eavesdropping technologies and operations involve wiretaps. Substituting the two terms is also anachronistic. It wasn't until 1968, more than a century after D. C. Williams went to prison for tapping corporate telegraph messages in California, that wiretaps and electronic eavesdropping devices were in fact governed by a unified statutory framework in federal law.

Wiretapping and electronic eavesdropping are themselves specialized forms of *electronic surveillance*, a term that now encompasses activities as diverse as video recording via closed-circuit television systems and aerial drone cameras; geolocation tracking via license plate readers, cell-site towers, and communications satellite connections; biometric monitoring via fingerprint databases and facial recognition software; and digital data mining via malware, cookies, and other computational media formats that track online behavior—to name but a few of the most visible and embattled variations.[29] To earlier generations for whom wiretaps and bugs were the only familiar forms of electronic surveillance, many of these activities would have seemed like the stuff of science fiction. But they are now our reality. We can only imagine what the future will bring. A history of electronic surveillance in the twenty-first century will almost certainly look different from the book you now hold in your hand.

I point out all of these distinctions to alleviate any terminological confusion that might arise in the pages that follow—and to signal my own interest in staying faithful to the language of the historical actors that lie at the center of my story. Wiretapping, for them, was but one species in the family of activities known as electronic eavesdropping; electronic eavesdropping was but one family of activities in the vast kingdom of electronic surveillance. I follow certain narrative threads, narrow though they may seem by the standards of the digital age, to show how far the kingdom has come over time: to mark our distance from, and connections to, the analog past that came before us.

I leave it to the reader to decide whether that past can help us find a way out of our current predicament.

Part I

DIRTY BUSINESS

Chapter 1

Stolen Signals and Whispering Wires

THE EIGHTEENTH AMENDMENT TO THE UNITED STATES CONSTITUTION prohibited the manufacture, transportation, and sale of alcoholic beverages across America. It also created a thriving black market for booze, and an entrepreneurial class of criminals who rallied to capitalize on the thirsts of a nation gone dry. Few enjoyed more success than Roy Olmstead, a former police officer known throughout the Pacific Northwest as the "King of the Bootleggers." During the early years of Prohibition, Olmstead smuggled more than 200 cases of liquor into the city of Seattle every day, a volume of traffic that annually netted his organization as much as $2 million. By most accounts, the story of wiretapping in America begins when federal officials decided to make a last-ditch effort to stop him.

A bootlegging operation as large as Olmstead's had many moving parts.[1] Running it required careful coordination. The liquor Olmstead was peddling originated in England and arrived in the Pacific Northwest via Canada. To transport wholesale shipments into the country—more than 4,000 cases at a time, smuggled across the northern border—Olmstead chartered three Canadian freighters that sailed from Vancouver to a small island just beyond American waters. Speedboats met the freighters there to collect their cargo. Then they whizzed off to rendezvous with Olmstead's men on the coast of Puget Sound. On shore, a fleet of trucks conveyed each shipment to an underground storage facility on a ten-acre ranch outside of Seattle. A second fleet of trucks, disguised as grocery wagons, moved smaller batches of liquor to four storefront distribution centers scattered across the city.

Olmstead sent orders to the storefronts every morning. By midday his drivers were heading out on their rounds. They used four different automobiles to make deliveries: a Cadillac and a Packard (for orders headed to wealthy clients and downtown speakeasies), and two Fords (for orders headed to everyone else). From the rumrunner's chartered vessel to the customer's front door, the entire process relied on the labor of more than fifty men. Olmstead also kept dozens of city employees on his payroll, for protection.

As with almost every complex business venture, legitimate or not, in the age of electronic communications, the Olmstead organization relied on the telephone to facilitate its operations. Olmstead's skippers arranged their shipments by phone, and Olmstead's drivers confirmed their arrivals by phone. Clerks and salesmen talked to Olmstead's customers by phone. Packers confirmed the day's orders by phone, and dispatchers relayed instructions to delivery-men by phone. Patrol officers at the Seattle Police Department, their pockets fat with graft, used the telephone to warn Olmstead's underlings of impending government raids. Olmstead himself relied on the phone to manage everyone he employed. Never above the fray, he attended to every detail.

At the center of Olmstead's bootlegging empire were three telephones— ELliott-6785, -6786, and -6787—located in a spacious office on the tenth floor of the Henry Building, in the heart of downtown Seattle. A former taxi dispatcher named John McLean manned this notorious exchange, answering every call while seated at a roll-top desk. If you dialed one of these three numbers during normal business hours, you were certain to hear McLean's harried voice on the other end of the line. Every evening around 8:00 p.m., McLean pulled the plugs and the phones went silent.

Through ELliott-6785, -6786, and -6787, Olmstead's bootlegging operation had run like clockwork. But everything changed on the afternoon of June 25, 1924, when Richard Fryant—a former wire technician for the New York Telephone Company, recently transplanted to Seattle—tapped one of the lines.[2]

A hard-line "dry" faction in the Seattle mayor's office had hired Fryant and a private detective named Henry Behneman to investigate corruption among city officials, many of whom had accepted bribes from Olmstead. Monitoring telephones seemed like the most reliable way to identify the crooked parties involved. To achieve their mission, Fryant and Behneman had bluffed their way past the Henry Building's security guard by posing as phone company employees on a routine service check. Once inside, they descended into a dark, airless basement and located a telephone junction box connected to every handset in the building. Fryant attached an extension line to the wires leading up to Olmstead's central dispatch. Sitting on overturned milk crates, passing a makeshift earpiece back and forth, the two men took turns listening to the steady stream of calls. They hastily scribbled notes on what they heard.

Within days the eavesdropping operation began to snowball. When the director of the Seattle Prohibition Bureau, William Whitney, got wind of the wiretap in the basement of the Henry Building, he saw a rare opportunity not just to root out municipal corruption but to bring down the King of the Bootleggers himself. Whitney soon put Fryant on the Prohibition Bureau payroll, renting him a more secure listening post in a vacant office on the Henry Building's ninth floor, directly below Olmstead's headquarters. Whitney also fired Behneman, a detective with a reputation for playing both sides of the law, and replaced him with someone he trusted to keep the investigation secret: his own wife, Clara. Clara Whitney had an important skill that Fryant lacked. She was an expert in stenographic shorthand.

In the months that followed, Richard Fryant and Clara Whitney—the wiretapper and the stenographer—listened to every call coming into and going out of Olmstead's central dispatch. William Whitney occasionally joined them, although he never bothered to secure a search warrant for the operation. Day by day, conversation by conversation, the eavesdroppers pieced together the inner workings of the Olmstead syndicate: when the wholesale shipments arrived, and where the retail was stored; which customers were purchasing liquor, and how their orders were delivered; which members of the police department were getting a cut, and which city officials were looking the other way. At night, while the telephones on the tenth floor of the Henry Building sat unplugged, Clara Whitney performed the arduous task of translating her shorthand rendering of the day's wiretaps into a typewritten transcript. Her husband would then glance over the log of calls, notate a few conversations for clarity, and file away the records for safekeeping. In a move that later became the subject of legal controversy, he chose to discard the original stenographic notes on which the transcripts were based.

The "black book" of wiretaps, as the collected typewritten transcripts came to be known, was to serve as the centerpiece of an ironclad federal case against Olmstead and his men. The text grew longer and longer as the summer of 1924 dragged on. The bigger the bootlegging enterprise, it turned out, the more telephones in its service. Soon Whitney ordered Fryant to tap four more lines: ELliott-1737, GArfield-3039, GArfield-4680, and KEnwood-7130. In late July, Fryant found himself clambering up a telephone pole to install a tap on BEacon-4390, the connection at Olmstead's stylish suburban home. By that time, the King of the Bootleggers was well aware of the government's investigation into his affairs. He had long since instructed his subordinates to speak in code when talking on the telephone. Sometimes he took to the lines to broadcast false information about shipments and deliveries, hoping to throw Whitney and the wiretappers off the scent. On more than one occasion, Olmstead's lieutenants warned unsuspecting callers that federal agents might be listening in.

Still, the telephone was too convenient, too crucial for business. Olmstead never stopped using it. By the time the Prohibition Bureau disconnected Fryant's wiretaps in November 1924, the black book ran to more than 775 typewritten pages. Whitney stored copies of its contents in safe deposit boxes scattered around the city, in case Olmstead tried to use his political connections to have them destroyed. In January 1925, the black book was the primary piece of evidence that convinced a grand jury to indict Olmstead and ninety associates for conspiring to import and sell alcoholic beverages. The ensuing trial was thought to be the largest in the history of Prohibition. Local newspapers dubbed it the "Case of the Whispering Wires."

From the outset, Olmstead's attorneys raised a stubborn series of questions about the Prohibition Bureau's investigation. They believed that the wiretaps had violated Olmstead's constitutional rights. The Fourth Amendment was written to protect American citizens against unreasonable searches and seizures. Wasn't eavesdropping on a telephone call without a warrant tantamount to an illegal search? And hadn't the government seized something important—Olmstead's conversations, now recorded in the black book—while listening over the wires? Fifth Amendment protections against self-incrimination also seemed relevant to Olmstead's case. Wasn't referring to the black book at trial—in effect, if not in intent—a way of compelling Olmstead to testify against himself?

There were procedural questions, too. Washington was one of twenty-eight states in the country where the practice of wiretapping was already prohibited by law.[3] From the very beginning, Fryant and the Whitneys were aware that eavesdropping on Olmstead's calls was a punishable offense. In a federal trial, was it appropriate for the government to rely on evidence obtained in direct violation of a state statute?

Olmstead's attorneys harped on all of these questions when the Case of the Whispering Wires went to trial, but the evidence against Olmstead and his confederates was overwhelming. Olmstead was convicted and sentenced to four years in prison. He appealed on the basis of his defense team's argument about the evils of wiretapping, and his case made it all the way to the U.S. Supreme Court three years later. News outlets around the country immediately recognized the stakes of the legal battle.

"If wires may be tapped for one reason or on the strength of anyone's suspicion, they may be tapped for all reasons and suspicions and for none," wrote a columnist in the *Seattle Times* the day after the Supreme Court agreed to hear Olmstead's case. "Every law-abiding and decent telephone patron is concerned in this matter. If they are to be subject to eavesdropping annoyance, they will at least wish to know it so that they may govern themselves accordingly."[4] The editorial staff at the *Washington Post* echoed many of the same sentiments, warning that a ruling against Olmstead's claims of government misconduct might endanger the privacy of telephone communications and even encourage civilian lawlessness: "If it is in violation of a citizen's rights to enter his house through a door or window, how can it be lawful to enter over a wire? The furtive act of wire tapping is dishonorable at best, and doubly offensive when perpetrated by officers of the law. Certainly the inviolability of a man's house against unlawful search should extend to his means of communication. Moreover, a decision legalizing wire tapping for the purpose of enforcing the law might encourage blackmailers and other criminals to tap the wires."[5]

The Supreme Court handed down a ruling in the case of *Olmstead v. United States* on June 4, 1928. In a narrowly contested 5-4 decision, the Court ended up siding with the government, arguing that wiretapping was constitutionally permissible. Common sense seemed to dictate the majority's logic. Because the Prohibition Bureau hadn't forced Olmstead to talk to his employees on the telephone (the bootleggers, in the words of the opinion, were "continually and voluntarily transacting business" with "no evidence of compulsion"), the Fifth Amendment didn't apply in the case.[6] And because Fryant and the Whitneys had neither trespassed on Olmstead's private property nor confiscated anything tangible when tapping his lines, the case fell outside the purview of the Fourth Amendment as well. Writing for the majority, Chief Justice (and former President) William Howard Taft was unyielding in his judgment that wiretapping could not be construed as an illegal search. "The [Fourth] Amendment does not forbid what was done here," Taft concluded. "There was no searching. There was no seizure. The evidence was secured by the use of the sense of hearing, and that only. There was no entry of the houses or offices of the defendants."[7] Tapping a telephone line thus didn't appear to violate the Constitution.

The Case of the Whispering Wires was closed, at least for the time being.

. . .

Today *Olmstead v. United States* is remembered less for the reasoning of the majority than for the dissenting opinions it prompted from the bench.[8] Justice Louis Brandeis—the influential "co-inventor" of the legal right to privacy, which he and the attorney Samuel Warren first formulated in an 1890 article for the *Harvard Law Review*—penned an extended dissent that would go on to serve as a touchstone for Fourth Amendment jurisprudence.[9]

Inspired by a legal brief filed on Olmstead's behalf by representatives of the telephone industry, Brandeis argued that wiretapping violates a citizen's "right to be let alone—the most comprehensive of rights and the right most favored by civilized men."[10] He reasoned that eavesdropping on telephone calls invades the privacy not just of the primary caller under investigation, but of any other innocent person who happens to be talking on the other end of the line. Brandeis warned that the government would someday acquire "subtler and more far-reaching means of invading privacy" than wiretapping.[11] He looked to the ideals implicit in the Fourth Amendment as a bulwark against the intrusions of the state that seemed certain to come.

Justice Oliver Wendell Holmes Jr. joined Brandeis in the *Olmstead* minority. In a blistering addendum to Brandeis's dissent, Holmes denounced wiretapping as a "dirty business" and argued that criminal evidence so obtained had no place in a court of law.[12] Holmes had first affirmed the majority's opinion in conference. He changed his mind when one of his clerks alerted him to the fact that wiretapping was a criminal offense in the state of Washington.[13]

Opponents of wiretapping made frequent use of the dissenting opinions to *Olmstead v. United States* in the decades that followed. While

34

Brandeis's argument would inspire several generations of civil libertarian activism and electronic privacy law, eventually providing the basis for the Supreme Court's decision to overturn *Olmstead* nearly forty years later, versions of Holmes's "dirty business" characterization became a ubiquitous refrain in the political arena. Wiretapping wasn't just unconstitutional, many Americans alleged throughout the first half of the twentieth century. It was "indecent," "immoral," "odious," "dishonorable," and "despicable."[14] Such sentiments were especially strong in the era of Prohibition. Mabel Walker Willebrandt, the U.S. assistant attorney general in charge of prosecuting federal Prohibition cases, refused to argue before the Supreme Court in *Olmstead* on the grounds that she "thoroughly disapproved" of the "wire-tapping tactics" then common among segments of the federal law enforcement community.[15] Even Chief Justice Taft, in a nifty piece of rhetorical misdirection, revealed that he shared many of Holmes's reservations about wiretapping when he acknowledged that eavesdropping on telephone calls was something "other than nice ethical conduct."[16] In other words, Taft believed, just as much as Holmes and Willebrandt did, that wiretapping was dirty. But he also believed that criminals like Olmstead were doing things that were much dirtier. It made sense to permit the former to prevent the latter.[17]

"Dirty business" is a curious turn of phrase for a sitting justice of the U.S. Supreme Court to use in a case like Olmstead's. Why did it make sense for Holmes to rely on it, particularly when such rhetoric opened his dissenting opinion to charges of kneejerk moralism?[18] Why did his characterization of wiretapping resonate for so many, and for so long? Why did tapping Roy Olmstead's telephone seem so bad?

Answering these questions takes us beyond commonsense American ideas about the right to privacy, and beyond long-standing legal prohibitions against eavesdropping.[19] It also takes us beyond the immediate purview of the *Olmstead* case, which is often regarded as the starting point for the debates about electronic surveillance that still rage today.

It turns out that wiretapping was a dirty business almost from the very start. Throughout the second half of the nineteenth century—and well into the early decades of the twentieth—the act of tapping telegraph and telephone lines was generally considered to be the province of common crooks and confidence artists. The criminals were the ones who wiretapped, the story went, not sanctioned and upstanding agents of the law. When Holmes resorted to the phrase "dirty business" in the *Olmstead* dissent, he was deliberately yanking on that chain of cultural associations.

His moralistic language invoked a long history of criminal activity that generations of Americans had already come to associate with the practice of wiretapping: from stock market cheats and betting parlor swindles, to slick confidence games that suckered ordinary men and women out of their life savings.

In the years leading up to the *Olmstead* ruling, a loose coalition of state police officials, private investigators, and federal policymakers began working to shake the wiretap free from its criminal associations. They hoped to transform electronic surveillance into a legitimate police tactic for the modern American age. Despite what the Supreme Court ruled from on high in 1928, their efforts never quite succeeded.

. . .

The "dirty business" that Justice Holmes saw the Seattle Prohibition Bureau conducting in the Olmstead case was a long way from the storied origins of wiretapping. The earliest wiretappers were military men. They perfected the tricks of their trade during the American Civil War, the world's first armed conflict in which the use of electronic communications proved decisive.[20] In both contemporaneous news accounts and retrospective wartime memoirs, military wiretappers were celebrated for their courage, ingenuity, and bravado. In their time, unlike Olmstead's, the men who tapped wires were heroes.

The telegraph had been acclaimed as an annihilator of time and distance from the moment of its invention in the 1840s, and the necessities of wartime only magnified the technology's possibilities. Both armies, Union and Confederate, were quick to recognize the strategic value of telegraphic communication. Telegraphy facilitated coordination among troop divisions and improved the management of supply lines. It accelerated the delivery of military intelligence. And it provided ranking officers and political officials, often posted in command centers hundreds of miles away, with near-instantaneous updates from the battlefield. The traffic in telegraph messages became so vital to the unfolding conflict, the story goes, that President Abraham Lincoln ordered a cot delivered to the telegraph room of the War Department so he could be present when dispatches from his generals arrived overnight.[21] Secretary of War Edwin Stanton, whose office adjoined the room where Lincoln sometimes slept, often referred to the telegraph as his "right arm."[22] Union and Confederate troops appear to have shared similar feelings about the technology's tactical worth. There was an oft-repeated Yankee ditty about the signal operators who dashed off telegraph messages from the field: "Let the telegraphers through / they'll get us into a fight / and out of it too."[23]

Given the outsized role that the telegraph came to play in the war between the states, it was only natural that both armies would try to find ways to read each other's messages. It was here that the wiretap was born: as a novel method of acquiring intelligence and sowing disinformation, perhaps the earliest known example of electronic espionage in human history.

Historical records offer us few clues as to the exact date of its birth. But retrospective accounts written by the soldiers who worked in the military telegraph service, published in the Civil War's aftermath, suggest that Union and Confederate operators began tapping wires almost as soon as military telegraph lines were up and running. From the beginning, wiretapping missions proved exceedingly risky—they usually required sending small groups of soldiers into well-traveled areas beyond enemy lines. On the Union side, the historical ledger reveals just how dangerous tapping a wire could be. In September 1863, a Yankee scout named William Forster tapped the Confederate telegraph line that ran alongside the Charleston-Savannah railroad, relying on the help of two runaway slaves who were familiar with the geography of the region. Forster's mission proved fatal. After Confederate soldiers caught him in the act and pursued him through a swamp, he ended up dying as a prisoner of war. The slaves who helped him managed to escape.[24] That same year, two signal operators in the service of Union general William S. Rosecrans tapped the wires that followed the tracks of the Chattanooga railroad, just outside of Knoxville, Tennessee. Over the course of one eventful week, the two clerks learned not only where Confederate troops were marching in other areas of the war's western theater, but also that a group of spies had discovered their presence on the line. They quickly disconnected their tap and fled for their lives, surviving thirty-three days in Confederate territory before falling in with another Union division. One of them was later captured on an unrelated wiretapping assignment in Ohio.[25]

Confederate wiretapping missions garnered more public attention throughout the war, perhaps because the Union army relied on a more robust electronic communications network. Reports of successful rebel wiretaps were published in northern and southern newspapers as early as 1862, and word of their ingenuity even traveled across the Atlantic.[26] (The

British began using wiretaps in colonial military campaigns in the 1880s. British soldiers initially referred to the act of telegraph tapping as "wire milking." Wiretappers were themselves known, somewhat improbably, as "milkmen.")[27] Confederate brigadier general John H. Morgan's chief signal operator, George "Lightning" Ellsworth—a soldier whose gift for wiretapping would later become the stuff of legend—made headlines throughout the summer of 1862 after he tapped a Union telegraph line at several different points between Louisville, Kentucky, and Nashville, Tennessee. For more than a month, according to one news account, Ellsworth gave Morgan the "pleasure of learning from the enemy's own mouth all that it concerned him to know, and what he had best do or avoid doing."[28] Charles Gaston, a signal clerk for Confederate general Robert E. Lee, likewise tapped Ulysses S. Grant's Fort Monroe telegraph line for an extended period in 1864. The operation proved fruitless until Gaston intercepted a reckless message announcing the delivery of a supply of beef to Union soldiers stationed in South Carolina. Confederate troops ended up hijacking the rations before they reached their intended destination.[29]

Concerns about the prevalence of wiretapping during wartime led to the widespread adoption of military telegraph ciphers, formulated to prevent opposing operators from reading confidential messages. Cipher codes were a conventional feature of telegraphic communications from the beginning. The year after Samuel Morse successfully demonstrated the workings of his signature invention to Congress in 1844 ("What hath God wrought?" Morse famously signaled), his attorney and publicist devised a commercial telegraph code to ensure that "neither the attendants of the Telegraph, nor any other person, except the parties who are in concert, will be able to read, or decipher, the communications that are transmitted."[30] Cipher systems proliferated in the coming decades, mostly among businessmen attempting to shorten their messages in order to save on the cost of transmittal. But codes took on redoubled importance in wartime. To protect the secrecy of each military telegram, one signal serviceman recalled, telegraph operators on both sides of the conflict "strove hard to clothe . . . despatches [sic] in a strange, uncouth garb."[31] For some military officers, the task of encoding and decoding confidential telegraph messages became a full-time job.

Telegraph ciphers served another strategic purpose in the early age of the wiretap: ensuring a message's authenticity. In an effort to create confusion among enemy ranks, Civil War telegraphers frequently tapped lines to send bogus communications.[32] By all accounts, Union and Confederate signal operators installed wiretaps to transmit disinformation as often as they did to intercept enemy dispatches and gather intelligence. Telegraph ciphers provided an indispensable safeguard against counterfeit communications. In the chaotic new arena of information warfare, ciphers could ensure that an inbound signal was coming from a trusted source. Messages sent in an outdated cipher—or "in the clear," as the saying goes—betrayed the fact that the opposing side had a tap on the line.[33]

By the end of the Civil War, military telegraph wires crisscrossed more than 15,000 miles of northern and southern territory.[34] Because vast swaths of the Union and Confederate networks had been erected in haste and left unprotected, tapping them was a relatively simple affair—a mere matter of cutting into a telegraph wire and attaching a fine copper extension line to the exposed cable. Signal clerks who were assigned to monitor enemy communications for extended periods of time typically covered their wiretap lines with silk to protect against wear and tear.[35] They also buried them under dirt and leaves to prevent discovery.[36] Notwithstanding the inherent dangers involved, anyone with the right tools and a cursory knowledge of telegraph signaling could install a wiretap, although some seem to have been more skilled than others. George "Lightning" Ellsworth was reputed to have had such a talent for wiretapping that he could place both ends of a severed telegraph line against his tongue and "read" the incoming electrical signals as they pulsed through his mouth.[37]

Few would have had to go that far. Early in the war, military telegraph operators developed handheld signaling devices—so-called "pocket sounders," battery-powered and portable—that could be attached to telegraph wires to send and receive messages on the fly. One of the earliest visual representations of a wiretap, a sketch published in the March 1865 issue of *Frank Leslie's Illustrated Newspaper*, suggests the power and convenience of this unique electronic instrument (see Figure 1.1).[38] The image depicts a nameless signal service operator—identified in the accompanying news article as a clerk to Union Army colonel Benjamin Grierson—tapping a Confederate line along the tracks of the Mississippi Central Railroad. The clerk strikes a somber, almost meditative pose as he sits beneath the wires, recording rebel messages in a small notebook while a pocket sounder clicks away in his left hand. The size of the instrument would have made such multitasking possible, even under duress. Most were no bigger than a cigarette case, fitting easily into the palm of a grown man's hand (see Figure 1.2). By the turn of the twentieth century,

Figure 1.1. "Telegraph Operator Tapping Rebel Telegraph Line Near Egypt, on the Mississippi Central Railroad." This sketch is one of the earliest visual representations of a wiretap ever published in the United States. Note the size of the pocket signaling device in the telegraph operator's left hand. *Frank Leslie's Illustrated Newspaper,* March 18, 1865, 404.

Figure 1.2a,b. Pocket telegraph sounder (ca. 1861–1865). Division of Work and Industry, National Museum of American History, Smithsonian Institution.

portable telegraph sounders had shrunk to the size of pocket watches. Versions of the device were widely available on the open market.

. . .

The retrospective accounts of the military telegraph service that appeared in the decades following the Civil War often romanticized the practice of wiretapping as ingenious and heroic. The most frequently cited chronicle, written by former Union telegraph clerk William Plum, praised signal operators for their "intrepid conduct" and noble "self-denial" during their wiretapping missions.[39] George Ellsworth, who published several colorful installments of his wartime memoirs in the *New Orleans Times-Democrat* in 1882, took pains to underscore the inventiveness and success of his eavesdropping schemes for the Confederacy. He depicted the wiretap as a routine "Telegraphic strategy" that southern signal clerks had raised to the level of art in order to "throw federal forces off our track Countermand their orders etc. [*sic*]."[40] Ellsworth would come to be hailed as a pioneer in the field of guerilla warfare.[41]

Other memoirs of the military telegraph service lamented that the signal operators who regularly risked their lives in the line of duty hadn't received their historical due. "The operators of the military-telegraph service performed work of the most vital import to the army in particular and to the country in general," remarked Union field general A. W. Greely in 1912. According to Greely, who took pains in his memoirs to glorify telegraph tapping as an act of "desperate daring," Civil War telegraphers "fully merited the gratitude of the Nation for their efficiency, fidelity, and patriotism, yet their services have never been practically recognized by the Government or appreciated by the people."[42] There was more to such statements than a mere hope for historical acclaim. More than 300 signal operators—around one in twelve—perished in the field over the course of the conflict.[43] Because the majority of them were never assigned an official military rank, their families seldom received the commissions that were owed to them after the war.[44]

Plum, Ellsworth, and Greely were all writing about wiretapping in the late nineteenth and early twentieth centuries. As with so many chronicles of the Civil War published in this period, the tales they told were steeped in the politics of national reconciliation.[45] Heroic wiretappers were captured behind enemy lines. Cunning signal servicemen tapped the telegraph to deceive the opposition. Virtuoso cipher clerks worked day and night to decode intercepted messages. Such romantic characterizations of the labor of Union and Confederate operators helped to rebrand the war as an honorable struggle between brothers and men, erasing its contested origins in the battle over the institution of slavery. In a telling rhetorical pattern, several postbellum accounts of the military telegraph service would extol the genial "sentiments of fraternity" that signal clerks expressed toward their counterparts on the opposite side of the battlefield.[46] The telegraph tap, in this narrative, was a novel gambit in the gentleman's game of information warfare, an art that helped advance a noble conflict that had long since come to a just conclusion.

Romantic stories of the heroism of the Civil War telegraph service were also intended to salvage the reputations of military wiretappers themselves. This was because the techniques of technical surveillance they developed during the war had suddenly fallen from grace in the public eye.

After the Civil War, the practice of wiretapping lost its connotations of technological ingenuity and soldierly sacrifice. The shift was the product of a shocking new brand of crime: petty crooks and con artists had begun to employ wiretaps in lucrative electronic schemes, exploiting a host of economic activities that depended on the flow of telegraph communications. Incidents of criminal wiretapping were by no means common in the postbellum decades. But sensationalized accounts of their successes made wiretaps seem pervasive enough to raise troublesome questions about the security of the nation's electronic networks. As early as 1874, a rash of criminal wiretapping cases in New York City and Washington, D.C., led a reporter for the *New York Times* to warn that "there are gentlemen prowling about . . . to whom the telegraph reveals its actual secrets, who make a somewhat mischievous use of their power of reading a language [Morse code] which to most of the world is an unknown tongue."[47]

These were the new wiretappers. They ranged from amateur tinkerers with a flair for electronics to corrupt signal operators with paying jobs in the telegraph industry. Allan Pinkerton, the legendary private eye who founded the nation's leading detective agency in this period, used the moniker "lightning stealers" to refer to the specialized class of criminals who began exploiting wiretaps in this way.[48] Pinkerton knew them well. He had a hand in bringing at least one to justice early in his career.[49]

As we saw in the case of D. C. Williams in the Introduction, the stock market was the new wiretapper's earliest arena of criminal activity, an unwelcome byproduct of the telegraph's sudden importance to the workings of the American economy. Businesses were some of the earliest adopters of telegraphy in the 1840s and 1850s, and after the Civil War the traffic in telegrams gradually became central to day-to-day financial operations across the country, as telegraph networks expanded their reach. Two main technological innovations had strengthened the relationship between telegraphy and commerce. The first was the telegraph ticker, invented in 1867. These mechanical devices transmitted commercial price quotations in real time, forerunners to the modern-day stock tickers that scroll on the sides of skyscrapers in New York City. Ticker-tape instruments accelerated the flow of information between exchange floors and financial brokers, and they enabled accurate and continuous commercial trading without the burdensome intervention of human operators.[50] The second technological advancement was the quadruplex, a device invented by Thomas Edison in 1874 and shortly thereafter acquired by Western Union, the nation's largest telegraph carrier. The quadruplex dramatically expanded the bandwidth of commercial telegraph services. True to its name, the technology increased the telegraph's carrying capacity to accommodate four signals at once, an improvement that allowed Western Union to begin providing exclusive services to press associations, financial houses, and other commercial venues, contrary to a long-standing industry policy against the leasing of private lines.[51]

As the historian David Hochfelder has shown, the ticker and the quadruplex together transformed the telegraph into the "backbone of modern American finance capitalism."[52] Yet the U.S. economy's new dependence on electronic information at times left markets vulnerable to criminals with knowledge of the telegraph system. The telegraph wires leading to and from commercial houses were easy to tap—as in the 1867 case of George Crowdrey, a former Civil War signal clerk who embarked on an elaborate plot to divert corporate cables and supply the intercepted intelligence to a group of Chicago businessmen.[53] Telegraph operators could also be bribed to divulge the contents of corporate messages—as in the 1868 case of William Roche, an operator at the Franklin Telegraph Company in New York who furnished a crooked speculator with a trove of confidential bank telegrams regarding the price of gold. In sworn testimony before the Tombs Police Court, Roche admitted that several of his colleagues at Franklin Telegraph had long been "doing the same thing."[54] He had no idea why he was singled out for indictment.

Much like the telegraphic disinformation campaigns waged throughout the Civil War, the commercial wiretapping conspiracies of the second half of the nineteenth century often involved the transmission of counterfeit messages. The most notorious such case occurred in September 1899, when a group of corrupt commodity speculators created the "wildest panic ever witnessed" at the New Orleans Stock Exchange by tapping the lines of the Western Union Commercial News Bureau and disseminating inflated cotton prices.[55] The fabricated jump in the commodity's value wreaked havoc on the trading floor. Trades in New Orleans were suspended for several days, and exchanges in Atlanta, Boston, Charleston, Little Rock, and Savannah were also disrupted.[56] The speculators made off with more than $170,000 before Western Union could determine the source of the fraudulent cables.

Commercial wiretapping schemes were seldom so extensive. Far more common were petty eavesdropping operations conducted through informal financial establishments known as "bucket shops," where ordinary citizens could wager on fluctuating commodity prices.[57] Bucket shops began cropping up in cities and towns across the United States during the 1870s. And much like the licensed stock exchanges whose economic functions they simulated, bucket-shop operations depended almost entirely on the ticker and the quadruplex: the continuous flow of pricing information, transmitted by telegraph, was what enabled them to do business. Unscrupulous speculators regarded the arrangement as ripe for exploitation, especially because the transactions conducted through bucket shops already seemed to toe the line of the law. Dozens of news outlets in the late nineteenth century filed reports about speculators tapping bucket-shop lines in order to make foolproof bets on the price of cotton, grain, corn, and oil.[58] As late as 1908, the *Los Angeles Times* found that commodities brokers were using pocket wiretap instruments—technological descendants of the portable sounders military operators employed throughout the Civil War—on the floors of bucket shops across Nevada and California.[59] Bucket-shop proprietors were also known to have tapped the telegraph lines leading to legitimate stock exchanges in an effort to secure free price quotations.[60]

Savvy criminals also used wiretaps to rig another field of economic activity that had grown dependent on electronic information: gambling. The historical ironies here are rich. The earliest warranted wiretap investigations in the United States, installed under court order during the 1970s, were explicitly targeted at large gambling syndicates, many of which used the telephone system to collect and communicate wagers across state lines. A century earlier, by contrast, the wiretap was often employed as a gambler's tool, particularly after betting parlors began receiving the results of horse races via private telegraph services. Here again, real-time news traveled over the wires to semi-legal establishments dedicated to financial speculation: "poolrooms," they were called. Here again, anyone who tapped a bookmaker's line could place a surefire bet without the house knowing. Much like the stock exchange and the bucket shop, the poolroom's reliance on the telegraph had rendered its routine operations susceptible to criminal tampering.

Methods of employing wiretaps to defraud horse-race poolrooms—a practice that came to be known as "past-posting"—varied widely. In its most common form, the scheme involves a wiretapper cutting into the line of communication between a poolroom and a racetrack: first severing the poolroom's private telegraph wire (thus preventing any messages from reaching the bookmaker), and then affixing an extension line to the original connection (thus securing a monopoly on up-to-the-minute racing information). With an exclusive connection established, the wiretapper can relay the results of each race to a middleman, whose only job is to place a bet on the winning horse. Once the wiretapper "releases" the racing outcome to the poolroom, usually after a short delay, the middleman collects his winnings, flees the scene, and the cycle begins anew. Versions of the poolroom wiretapping swindle were well known by the 1880s and 1890s. "The trick worked on the poolroom men is almost as old as telegraphy and is so simple that the wonder is that it is not attempted more frequently," remarked one writer in the wake of a past-posting conspiracy uncovered in Los Angeles in 1902 (see Figure 1.3).[61] Another variation on the scheme involved tapping poolroom lines and transmitting fabricated racing results.

One of the earliest known past-posting scams conducted in the United States involved the Jerome Park racetrack in the Bronx, New York. The case also happens to be one of the most extensive ever recorded. In October 1883, a group of crooked telegraph clerks installed a tap on the Jerome Park line and proceeded to send bogus racing information to poolrooms as far-flung as Baltimore, Boston, Philadelphia, and St. Louis. The conspirators netted more than $100,000 over the course of one day. In the wake of the affair, a representative from Western Union conceded that wiretaps were a basic fact of life in an age when real-time racing news traveled around the country by telegraph: "I know of no way to guard against tapping the wires, or to prevent an operator from sending a false report of the race. . . . This whole matter is simply one of the in-

SECTIONAL VIEW OF BLACK & CO.'S POOL ROOM AND BUILDING. THE UPPER LEFT-HAND CORNER SHOWS HOW TAPPERS WORKED WITH THE OPENING IN THE BRICK. THE ROOM ON THE RIGHT IS WHERE THEY SLEPT, AND THE LOWER SECTION SHOWS THE POOL ROOM IN OPERATION, WITH BLACK'S OPERATOR ON THE LEFT.

Figure 1.3. The *Los Angeles Times* published this rough sketch of a past-posting scheme to illustrate how two telegraph operators defrauded the Black & Fitzgerald poolroom in Los Angeles in September 1902. One of the perpetrators, Charles L. Matfeldt, was eventually caught and sentenced to prison under California's 1862 wiretap law. "Electric Crooks: One of the Most Skillfully Concocted Crimes Ever Done to a Turn in Los Angeles," *Los Angeles Times,* September 15, 1902, 5.

cidents of life on the race course."[62] A poolroom owner who suffered losses in a similar scam run out of Long Branch, New Jersey, grumbled that "such attempts, in one way or another, are made every year. . . . A wire is so easily tapped that the [telegraph] companies are wholly unable to furnish absolute protection."[63]

All the same, past-posting operations weren't without their risks. Experienced poolroom operators had the ability to recognize the signaling

patterns of the telegraph clerks who regularly transmitted their racing results. Because a telegrapher's "key style" was as identifiable as his voice
or handwriting, any deviation from the norm could alert a poolroom operator to the presence of a wiretap on the line.[64] This partly explains why
turn-of-the-century news outlets reported so often on failed incidents of
poolroom fraud, rather than on successful past-posting ventures. (Except
in the most brazen of cases, successful wiretapping scams typically went
undetected anyway.) In January 1893, a veteran poolroom operator
helped New York City police arrest a group of five men—a stockbroker,
a plumber, and three former Western Union linemen—who had rented a
room at the Spingler House, near West 14th Street and University Place,
and placed a tap on the private line of a poolroom two blocks away.[65]
Seven years later, another "gang of wire tappers," all Western Union employees, fleeced several bookmaking establishments in Chicago, Omaha,
and St. Louis, evading authorities for months until they arrived in New
York City. When a poolroom operator in midtown Manhattan noticed
suspicious signaling patterns on his line, the wiretappers fled the area before placing any bets. Investigators later found more than $700 worth of
signaling equipment left behind in a nearby hotel room.[66] In the coming
years, poolroom operators in Chicago, Los Angeles, and Washington,
D.C., would all spare their employers tens of thousands of dollars in losses
by detecting wiretaps.[67] The case in the nation's capital, reported in November 1902, holds special significance: it was the first recorded attempt
to work a past-posting wiretap on racing results communicated by telephone rather than telegraph.

Despite the sensational headlines that incidents like these generated,
the number of messages tapped on the private telegraph lines leased to
poolrooms, bucket shops, and stock exchanges was negligible compared
to the number of messages transmitted over the expanse of the telegraph
system as a whole. Statistically speaking, wiretapping cases were rare. Yet
the public's familiarity with criminal wiretapping schemes still fueled
doubts about the telegraph industry's commitment to network security.
By the early 1900s, critics of Western Union began to express special concern about wiretapping incidents involving the company's current and
former employees, many of whom seemed all too willing to use their inside knowledge of the telegraph system for criminal gain. Blame for
Western Union's security woes eventually came to rest on the company's
policy of supplying service to bookmakers and bucket-shop proprietors,
which forced professional telegraph operators into routine contact with

the vices of gambling and speculation. According to a writer for the *Chicago Tribune*, the job of communicating with poolrooms had a "baneful effect" on the moral standing of the average telegraph clerk. "Of former operators of the [Western Union] race department," he observed, "at least two have been sent to Sing Sing, half a dozen are bookmakers, as many more are wire tappers, several have become confidence men, a few have been driven insane, and a hundred or more are wrecks from drink or are penniless because of their infatuation for betting. . . . Almost every wire tapper is a graduate of a poolroom."[68] Moral concerns like these eventually drove citizens in Chicago and New York to lobby against Western Union's ability to deliver service to poolrooms and bucket shops. The company resisted the pressure. The private market for real-time racing results and stock quotes was far too lucrative to abandon, regardless of how such information was used on the other end of the line.[69]

. . .

Perhaps the most telling measure of the nation's belief in the prevalence of telegraph tapping in the turn-of-the-century period, overblown or not, is how easily criminals and confidence artists seem to have manipulated the public. The crooked art of past-posting, in particular, spawned a unique financial scam that helped to bolster the wiretap's criminal profile even as it exaggerated its pervasiveness. Law enforcement officials called it the "wireless wiretap" game, or simply "the wire": a swindle concocted to defraud petty gamblers who were eager to bet large sums of money on racing results they believed to be tapped.[70]

Wireless wiretaps were at times baroque in their complexity. Fans of George Roy Hill's blockbuster film *The Sting* (1973), starring Paul Newman and Robert Redford, will be familiar with their inner workings. In the wireless wiretap game, a con man approaches a potential victim outside a poolroom and explains that he has an inside track on racing results through a telegraph tap. The con man then either introduces a second associate claiming to be a Western Union operator with a wiretap on a bookmaking establishment, or escorts the victim to a rented room filled with telegraph equipment alleged to connect to a poolroom line. (False wiretap setups like these were known as "wire stores." Reports from the period suggest that they were often outfitted with such an impressive array of signaling instruments—sounders, tickers, extension wires, and the like—that it was impossible for the uninitiated to realize their lines were all dead.)[71] The goal of the performance was to induce the victim to hand over a large sum of cash with the promise of substantial returns on upcoming races. Once the victim remits the money, the con men pick up shop and disappear. The act was so convincing that victims were often duped into making multiple bets on successive days, accumulating losses all while believing a lucrative payout was the next race away.

The origins of the wireless wiretap game are as murky as the origins of the wiretap itself. Several sources identify the notorious New York con artist Larry Summerfield as the scam's "Napoleon."[72] Summerfield is said to have first run a successful wireless wiretap sometime in the 1890s. His reputation soon grew, and the hoax spread. As a former New York district attorney wrote of the scheme's popularity, "One crook bred another every time he made a victim, and the disease of crime, the most infectious of all distempers, ate its way unchecked into the body politic."[73] By the early 1900s, wireless wiretap swindles began appearing in the crime logs of cities across America, with the largest incidents of fraud making front-page news.[74] Many victims appear to have been "rubes" unfamiliar with big-city life. But most anyone could become a sucker. According to the New York Police Department's 1914 manual of *Police Practice and Procedure,* which lists the wireless wiretap alongside several other confidence games then prevalent in the city, "the affair is staged with such care that the victim, even at the end, can hardly realize that he has been swindled."[75] Making matters more complicated, the targets of the scam were would-be swindlers themselves. Once defrauded, they could hardly turn to authorities to recover funds that were meant for illegal wagers. Reports of men and women bilked out of thousands of dollars in the "ancient game" of wireless wiretapping made headlines well into the 1920s.[76]

Throughout this period, the connections between wiretapping and crime drew widespread media attention and fed the popular imagination. The exploits of illegal wiretappers and wire store con men, already as much myth as reality, even gave rise to an entire subgenre of literary fiction: the "wire thriller."[77] Published widely in dime-store magazines and paperback editions around the turn of the twentieth century, wire thrillers were formulaic works of literature that used the nation's burgeoning communications networks as their central backdrop for narrative action. The plotline of the typical wire thriller would update the conventions of crime fiction for the age of the telegraph and the telephone: electronic experts, usually cast as outlaw heroes, employ high-tech methods to carry out elaborate criminal schemes, while resourceful rivals and lawmen work to foil their plans to exploit society's dependence on communications media. Almost all of these works feature wiretapping as a plot device. The fictional criminals at their center end up tapping lines to make fortunes, fool adversaries, and avoid punishment. Standout examples of the turn-of-the-century wire thriller subgenre include "Fighting Electric Fiends, or, Bob Ferret among the Wire Tappers" (1898), a short story published anonymously in the detective fiction rag *Nick Carter Weekly,* and the novelist Arthur Stringer's best-selling potboilers *The Wire Tappers* (1906) and *The Phantom Wires* (1907).[78] Later iterations include Frank L. Packard's *The Wire Devils* (1918).[79]

The plot of Arthur Stringer's *The Wire Tappers,* perhaps the most commercially successful work in the wire thriller canon, is instructive in its dogged insistence on the wiretap as a crook's tool. The novel follows the underworld escapades of Jim Durkin, a former post office telegraph clerk who has returned to New York City after a brief stint in prison. Desperate for money upon his release, Jim hires himself out to a criminal mastermind named MacNutt, agreeing to rig several wiretaps in the service of a multistate past-posting scheme. His partner in the conspiracy is an attractive female telegraph operator named Frances (see Figure 1.4). Jim and Frances take an instant liking to each other, and they eventually decide to make off with the winnings that the wiretaps for MacNutt have generated. Their first wiretapping scheme leads to a second: a plot to use counterfeit telegraph messages to burglarize the office of a jeweler. That scheme leads to a third, a plot to intercept the communications of the U.S. Department of Agriculture, which in turn leads to a fourth: a plot to manipulate the price of cotton at the New Orleans Stock Exchange, a thinly veiled allusion to the telegraph tapping scandal that made international headlines in 1899. That scheme begets a fifth, a sixth, and so on.

Jim and Frances amass a fortune throughout it all. They also fall in love, vowing to settle down together after planting one final tap. But MacNutt has followed their exploits through his own network of wiretaps, and the operators lose their money when he returns to exact retribution for the theft of his money. In the novel's climactic scene, Frances kills MacNutt in a gun battle, and she rushes to flee the country with Jim. The couple continues their wiretapping spree in *The Phantom Wires,* a sequel that Stringer published a year later, adhering to the same narrative formula.

"Quite motionless, waiting over the sounder, bent the woman." *Page* 51

Figure 1.4. "Quite motionless, over the sounder, bent the woman . . ."
A halftone illustration from Arthur Stringer's *The Wire Tappers* (1906)
depicts the novel's protagonist, Frances, listening to a telegraph tap.
KD 4383, Widener Library, Harvard College Library.

Today the most striking aspect of *The Wire Tappers* is how much of
its fictional universe seems to exist beyond the reach of the law and the
state, outside the ordered logic of civilized society. There are no authorities in pursuit of Jim and Frances; only the crime boss MacNutt has the
technical wherewithal to stop them. The federal government, whose feeble

presence is registered in the form of the U.S. Department of Agriculture, mostly functions as an entity to be spied on and defrauded. Every character in the book is a criminal, a victim, or both. In the novel's first two paragraphs, we learn that Jim has left prison and is ready to rejoin society. In Stringer's description he is reformed and "of the world again."[80] But by the end of the novel's third paragraph—less than a half-page later—his talents as a telegraph operator have already begun pulling him into another world, a darker world, back toward corruption, gambling, and vice. Frances follows the same downward trajectory. She begins as an innocent. She is a morally upstanding girl who has fallen on hard times. But as soon as she installs her first telegraph tap, an act of desperation rather than daring, she finds herself trapped in a cycle of criminal behavior from which she cannot escape.

For Stringer, the wiretap is the technological gateway to this dark other world, the only world that exists for Jim and Frances and everyone else: a world of greed, lawlessness, and violence. Tapping a line, in wire thrillers like *The Wire Tappers,* is all it seems to have taken to make an upstanding citizen break bad.

Chapter 2

Detective Burns Goes to Washington

THE CRIMINAL SCHEMES OF BUCKET-SHOP speculators and poolroom swindlers, wire store con men and wire thriller heroes, helped to brand wiretapping as a corrupt and dishonest activity throughout the late nineteenth and early twentieth centuries. The telegraph tap seemed a crooked tool of the lawless, rather than a respectable instrument of the law. Yet Americans of the period seldom addressed the evils of wiretapping in the language of privacy, or at least the language of privacy that we know and use today. Early debates about wiretapping, such as they were, revolved around a perceived need to protect the property of communications companies and ensure reliable service. The intangible rights of the customer were of secondary concern.

Historians typically regard Samuel Warren and Louis Brandeis's *Harvard Law Review* essay "The Right to Privacy" (1890) as the earliest formulation of privacy rights in modern America. The practice of wiretapping predates that canonical starting point by almost three decades. Warren and Brandeis would themselves evince only passing interest in its prevalence in their day.[1] Americans invoked privacy claims in response to a wide range of social phenomena in the interim: from the no-knock raids of state police officers to the federal government's attempts to broaden the amount of information collected in the national census; from the opening of mail to the censoring of postcards; and from the use of unlicensed images in media advertisements to the popularity and impertinence of the tabloid press.[2] But the ideas about privacy that gained currency in the era of Warren and Brandeis had much more to do with the ability of the individual to protect his or her property and reputation than with any urgent sense of the need to safeguard new forms of communication like the telegraph, still unfamiliar and legally untested, from the eyes and ears of imagined third parties. Telegraph messages weren't considered public during this period. But they didn't exactly seem private either.

It's worth remembering that privacy wasn't an operational feature of the telegraph system to begin with. The transmittal of every message, in practice, required the intervention of a real, live third party: the telegraph operator. This was one of many reasons early disputes about the government's power to subpoena messages centered on competing notions of the "confidentiality," "secrecy," and "inviolability" of the telegraph, abstract ideas that would exert only a glancing influence on Warren and Brandeis's account.[3] This is also why later attempts to enshrine communications privacy as a constitutional guarantee would look, for precedent, to the legal norms that developed around the postal system rather than the telegraph network.[4] In purely technical terms, conceiving of an exchange by telegraph as an exclusive conversation between two individuals, beyond the ken of meddlesome outsiders, entailed a minor leap of faith. Some nineteenth-century commentators declared the telegraph to be as sacred as the mail, but others believed that the medium's protocols rendered abstract claims of privacy meaningless. "There is no reason why telegraphic communications should not be made public when justice demands that it should be done," one federal lawmaker argued in 1877 in the wake of a dragnet government effort to access copies of telegraph messages. "I see nothing of privacy about them." Another reasoned that "there is nothing in the method of transmission that makes [telegrams] sacred or surrounds them with any special protection or privilege."[5]

Ideas like these applied to the new technology of the telephone too. Patented in 1876, the telephone spent its first few decades of social life as a specialty service for the well-to-do. Only around the turn of the twentieth century would corporate carriers begin to transform the technology into a medium of mass communication, supplanting the dominance of the telegraph and inaugurating what one contemporary observer dubbed the "telephonization" of American life.[6] The understanding that the telephone primarily enabled point-to-point communication, restricted to the callers on the line, was as slow to develop as it was for the telegraph. Eavesdropping was a feature of telephony from the beginning. Customer privacy was an invented ideal that came later.

Much like the telegraph system, the technical features of early telephone networks helped to produce this counterintuitive dynamic. The earliest telephone subscribers were connected on a single line that made it easy for them to hear each other's conversations whenever they placed a call. At times, an intelligible connection required yelling into the apparatus, which meant that anyone in close proximity was likely to hear at least one side of an ongoing exchange. Telephone providers would soon accommodate increased demand for service by linking groups of four or more subscribers to a central switchboard via "party lines." Such setups, which survived in some regions of the country well into the 1980s, were notorious for eavesdropping. In rural areas, especially, listening to neighbors gab on the party line was how many early subscribers kept up with community affairs. "Every country user did [it]," one observer of the telephone's early history recalled. "It was the way they got the news."[7]

Nor had the telephone system improved on its predecessor's reliance on the labor of human intermediaries. Throughout the late nineteenth century, and well into the early decades of the twentieth, telephone operators were instructed to listen to live calls in order to ensure a functional connection on the line. In the Bell System, which began expanding its reach into major U.S. cities in the 1880s, the operators who monitored connections were specially trained in what came to be called "civil listening": they were encouraged to concentrate on the vocal sounds of the callers themselves rather than on the content of their conversations.[8] Indulging in idle snooping or gossip was made subject to disciplinary action. But even when "civil" and well-regulated, operator eavesdropping remained intrinsic to the delivery of quality service. Telephone subscribers came to regard the arrangement, at best, as a benign inconvenience; at worst, as an irritating necessity. Either way, the type of privacy that we expect on our calls today wasn't a feature of early telephone networks. In 1916, the commissioner of the New York Police Department (NYPD), Arthur Woods, admitted as much in a written defense of the phone taps that officers in his force had been installing for more than two decades. "Telephone conversations from their very nature cannot be private . . . since the employees of the telephone company cannot help hearing parts of conversations and may, if they are inclined, easily hear all," Woods explained.[9] By that time, most carriers had eliminated the necessity of operator eavesdropping by introducing incandescent lamps that glowed on the switchboard whenever calls were connected.[10] Innovations like these were designed to save time and labor at the exchange, not protect the sanctity of the conversations over the wires.

• • •

Given the "semi-public" status of the new communications media of the late nineteenth and early twentieth centuries, it makes sense that the earliest wiretap laws in the United States would have little to say about individual privacy. The federal government didn't start regulating telegraph and telephone taps until World War I, when Congress enacted a temporary ban on wiretapping to ensure wartime communications security. (The measure was mostly ceremonial, and it expired as soon as the conflict ended.) That left the matter to the states, which were slow to react to the threat posed by wiretapping. The wiretap statutes that state legislatures ratified in the turn-of-the-century period were generally written with two goals in mind: protecting the proprietary equipment of telecommunications companies, and ensuring the delivery of uninterrupted service. Protecting the right to privacy wasn't yet part of the picture.

The early history of state-level wiretap law bears witness to this crucial fact, puzzling though it may seem in hindsight. The nation's oldest wiretap law, written in California, was enacted in 1862 to ensure what state policymakers termed the "fidelity" of the telegraph system: its proper and reliable functioning, which the installation of a wiretap impaired as a matter of course.[11] The same basic goal was explicit in California's subsequent efforts to revise its wiretap statute in accordance with the twentieth century's technological advancements.[12] In Connecticut, state officials banned wiretapping in 1889 after a private detective used a telephone tap to collect a "mass of damaging and spicy evidence" in a divorce case involving a prominent New Haven family.[13] As in California, the legislature conceived of the law as a way to protect telephone company property—so much so that the chapter of the Connecticut statutes containing the new wiretap ban was inserted between older provisions pertaining to "Injury to Property of and Trespass by Electric Companies" and "Willful Injury to Electric Railway Appliances," both of which were considered to be comparable offenses of malicious mischief.[14] The Connecticut Telephone Company had in fact lobbied for the law's passage.

Only in the years leading up to the watershed ruling in *Olmstead v. United States* (1928) did the language of privacy begin to seep into the logic of wiretap law at the state level.[15] But even Washington State's law against wiretapping—a critical sticking point in the Whispering Wires case—was enacted in 1909 to prohibit unlicensed third parties from "injuring" telephone lines and impairing service.[16] Such concerns about the sanctity of telecommunications infrastructure governed how the telephone industry itself regulated external technological innovations, whether or not they threatened customer privacy.

As a case in point, consider the magnetic telephone recorder—a technological forerunner to the answering machine. Magnetic recording devices were invented in Denmark in the late 1890s. Engineers in the United States perfected a version for commercial use in the mid-1930s. But the Bell System would spend decades thwarting attempts to bring magnetic telephone recorders to market. The company's fear wasn't simply that they could be used as makeshift wiretapping devices, recording calls while parties on the line were caught unaware. It was that attaching "foreign" equipment to the telephone network had the potential to damage proprietary corporate hardware—and, with it, the Bell System's monopoly over the use of telecommunications equipment.[17] The same stated interests were behind the industry's long-running effort to suppress the "Hush-a-Phone," a privacy-ensuring instrument invented in the 1920s to prevent bystanders from hearing calls conducted at home or in the office.[18]

Where law enforcement fit into the patchwork system of state-level wiretap regulations remained uncertain. Catching wiretappers in the act was time-consuming and difficult, and criminal cases against them seldom ended with convictions. Telecommunications companies thus typically had to investigate incidents of illegal wiretapping on their own. State and federal law enforcement agencies seemed content to sit on the sidelines throughout the 1900s and 1910s.

There was also the more philosophical issue of whether legal prohibitions against wiretapping applied to agents of the state, a basic matter of criminal procedure that would remain unresolved for almost half a century. Phone taps had long held a special allure for law enforcement. As soon as the telephone became an everyday convenience for crooks and law-abiding citizens alike, monitoring calls began to look like a practical way for the police to establish connections, follow leads, and gather evidence. But was it acceptable for an agent of the state to tap the wires in the service of a criminal investigation? Were police in jurisdictions with wiretap statutes breaking the law when they recorded phone conversations, and were the fruits of their eavesdropping admissible at trial? At the time the courts had little to say on such matters, so the police tapped anyway. It was easy to bend the rules when the rules left so much to interpretation.

In 1895, detectives in New York City began tapping telephones to collect evidence in criminal investigations. By 1915, the NYPD had established a six-man team of technicians dedicated to the job. Working out of an unmarked room on the third floor of an office building at 50 Church Street, in the heart of Manhattan's financial district, the NYPD "wiretap squad" monitored as many as 350 lines per year with the help of the New York Telephone Company.[19] The federal government's nascent Bureau of Investigation—rechristened as the Federal Bureau of Investigation (FBI) in 1935—also engaged in wiretapping in its initial years of operation, supplementing the tactics of political surveillance that many of its top officials had perfected during tours with the U.S. military in locations like the Philippines.[20] The practice became widespread at the Bureau during World War I and the "Red Scare" years, when fears of foreign plots and communist subversion ran rampant.[21] African American political leaders, labor organizations, and news outlets found themselves subject to Bureau of Investigation wiretaps more often than most in this period.[22] Thus began the U.S. government's sordid and lengthy campaign of telephonic harassment against Black America—one that would culminate in the tangled web of government wiretaps that ensnared Martin Luther King Jr., along with scores of other civil rights activists, four decades later.

Worries about the moral and legal status of wiretapping, however, kept government surveillance operations like these tightly under wraps. That left the wiretap free to take root in the domains of law enforcement and state security in the absence of discretion and control. Americans had little chance to debate or push back against official uses of the telephone tap, whether in the name of privacy or otherwise, when so few knew of their existence. In this way, the early decades of the twentieth century established a pattern in the politics of electronic surveillance that still holds true today: When the government taps a telephone, the American public is likely to hear about it only years later. By then the goalposts have moved, and the opportunities for meaningful accountability have dwindled.

．　．　．

It wasn't an official agent of the state but a private detective who brought the controversial prospect of law enforcement wiretapping into the public eye. His name was William J. Burns, a publicity-hungry private investigator who had emerged as "the most talked about detective in the United

States" by the early 1910s.[23] Burns's willingness to use wiretaps and other eavesdropping devices in criminal investigations was in large part what earned him that august distinction. The details of his story are worth recounting. His exploits helped to bridge the gap between the wiretap's seedy criminal past and its legally ambiguous institutionalization as a crime-fighting tool.

William John Burns was born in Baltimore, Maryland, in 1858 and raised in Zanesville, Ohio.[24] In 1873, Burns's father moved his family to nearby Columbus, where he abandoned his job as a tailor to join the local police force. Over time he rose through the ranks to become the city's police commissioner. Impressionable, ambitious, and savvy, the younger Burns would follow in his father's footsteps, first working in a tailoring shop and then venturing forth to start his own detective firm in 1888. On the surface, the two career paths couldn't have seemed more divergent. While tailors had been quietly serving customers for as long as men and women had worn clothes, the private detective trade was a relatively novel product of the territorial disputes and labor unrest of the post–Civil War years, which a decentralized state policing apparatus had proved ill-equipped to handle.[25] Yet the two professions were linked, at least notionally, in their devotion to miniscule details, observed and measured with precision. Burns found success in both. In 1891, he joined the U.S. Secret Service (then still a law enforcement agency, not yet delegated with its protective mission), working to combat counterfeiting syndicates and government corruption. After cracking a career case in Oregon in 1906 that returned more than a hundred indictments, Burns established a reputation as the "star of the secret service."[26] He became a household name at a time when most government agents worked in the shadows.

Encouraged by his sudden celebrity, Burns resigned from his government post to pursue more lucrative work in the private sector. In 1909, he joined the William Sheridan Detective Agency, a midsize investigatory outfit that he rebranded as the William J. Burns International Detective Agency the following year. The idea was to diversify the market for private detection services, which in the U.S. had long been dominated by a single firm, Pinkerton's—known for its nationwide network of undercover investigators, and for the brutal tactics they employed in the repression of organized labor. The two agencies were soon locked in a bitter corporate rivalry that spanned decades.

To establish his position at the vanguard of the detective profession, and to keep pace with the law enforcement community's growing embrace of Progressive reform, Burns worked to perfect and publicize methods of criminal investigation that appeared "scientific." He courted the press to lavish attention on the cases he solved. None of Burns's methods was more effective as a publicity generator than his use of the dictograph, an electronic instrument invented in 1905 to serve as a personal hearing aid and office intercom system.[27] Burns repurposed the dictograph as a surreptitious eavesdropping device. In his hands, the technology transformed into a cutting-edge tool of scientific detection—an instrument that could catch criminals in the act and capture auditory evidence in real time. The dictograph soon became Burns's calling card. As one laudatory press profile quipped in 1912, "Fictional detectives carry automatics and handcuffs. Burns carries a dictograph."[28]

Burns first employed the dictograph while investigating the labor activists accused of bombing the headquarters of the *Los Angeles Times* in 1910. He hid the device in the corner of a jail cell in order to overhear the suspects' conversations, several of which took place with their lawyers. In one of the earliest historical incidents of a cooperating witness "wearing a wire," Burns also persuaded an investigator employed by the defense to plant a dictograph in a meeting with the celebrated labor attorney Clarence Darrow, who was believed to have bribed a juror to secure an acquittal in the case.[29] Darrow was later indicted but found not guilty. The dictograph recordings were never introduced as evidence in either trial.

Most of Burns's electronic trickery in the *Los Angeles Times* investigation occurred behind the scenes. His first major pubic victory with the dictograph took place one year later, when the Ohio Manufacturers' Association retained the Burns Agency to investigate corruption in the Ohio legislature. To crack the case, Burns ordered two of his detectives to plant a dictograph beneath a couch in a room at the Chittenden Hotel in Cincinnati. A third detective, posing as a businessman, invited dozens of state officials there in an attempt to induce them to accept bribes. On the other end of the device was a stenographer from the office of the county prosecutor; he transcribed every word while sitting in the room next door. Several legislators caught in the eavesdropping sting were later convicted for bribery, with the secret dictograph recordings serving as a cornerstone of the prosecution's case.[30] In Burns's hometown of Columbus, the state capital of Ohio, bars served "dictograph cocktails" to celebrate the detective's victory.[31]

In the coming years, Burns would also publicize his success with the dictograph in *The Argyle Case* (1912), a stage play he used his own funds to produce, and *The Exposure of the Land Swindlers* (1913), a feature film in which he played himself in a starring role.[32] "The dictagraph [*sic*] is only one instance of the usefulness of modern science in detective work," he explained in a 1912 article on the significance of his experimental investigative methods. "The successful detective of crime has now largely passed out of the old system of haphazard guesswork, and is shaping itself along the lines of more strictly scientific study."[33]

Less ballyhooed in the coverage of Burns's electronic exploits, but no less crucial to his evolving approach to detection, was his occasional use of the wiretap. Burns had in fact recorded his own phone conversations with suspects in the 1911 Ohio legislature investigation. In a notable departure from the favorable coverage that the dictograph received in the case, the popular press reacted suspiciously to a Cincinnati judge's decision to permit the Burns wiretaps as evidence in the subsequent bribery trial.[34] In 1913, Burns tested a commercial wiretapping instrument aptly called a "Tel-Tap"—among other eavesdropping devices he began employing, such as the Detectifone (see Chapter 6)—after the inventor of the dictograph slapped the Burns Agency with a lawsuit over a rental fee dispute.[35] At the same time, lawmakers were starting to scrutinize Burns's methods, signaling a turn in the popular narrative surrounding his work.[36]

Burns's public reckoning came a few years later, in 1916, on the heels of a wiretapping scandal that made headlines in New York City—the first of its kind in the nation's history.[37] The origins of the controversy remain clouded, although the primary players involved suggest how far the wiretap had traveled since its days as a tool of speculators, gamblers, and con men. The scandal implicated three separate entities, all of which were revealed to have used wiretaps and other electronic surveillance tactics in the regular course of their "official" duties: the New York City Mayor's Office, which was accused of ordering law enforcement officers to tap the telephones of five Catholic priests suspected of charity fraud; the New York Police Department (NYPD), which was accused of tapping hundreds of telephones per year to track down criminals and snuff out labor disputes; and William Burns himself, who was accused of breaking into the offices of the prominent law firm of Seymour and Seymour, installing a listening device, and working with the NYPD to tap several of the firm's telephones in order to uncover evidence of criminal wrongdoing. The press focused much of its attention on Burns's role in the scandal. The electronic tactics that had won the detective acclaim suddenly seemed to have infiltrated the city's most trusted public offices. Was the publicity surrounding

Burns to blame? The reputation of the Seymour firm only added intrigue to the case. Its partners had recently brokered a $1 million sale of munitions to allied nations fighting in the First World War. Rumors spread (and later proved inaccurate) that Burns's eavesdropping operation encroached on sensitive "Government interests."[38]

The New York legislature convened public hearings on the wiretapping scandal in May 1916. George L. Thompson, a New York senator charged with regulating public utilities, chaired the proceedings. Whether testifying about the Mayor's Office spying on Catholic priests, about the NYPD spying on criminals and labor activists, or about the Burns Agency spying on the Seymour firm, the individuals called before the Thompson Committee would present arguments for and against the use of wiretaps in criminal investigations that remain recognizable today. On both sides, the reasoning departed from the concerns about corporate property and uninterrupted service that had first animated state-level wiretap regulations. The language of privacy was suddenly everywhere.

Many of the officials ensnared in the scandal would defend the wiretap as a necessary evil in the detection of crime and the prosecution of criminals. "We recognize the inviolable right of individual citizens to individual privacy in their telephone communications as in other relations of life," remarked the embattled mayor of New York, John Purroy Mitchel. "It is, however, absolutely essential to the protection of the community against the commission of crime that the police should use this powerful weapon. Conviction on conviction has been obtained [through wiretapping] which otherwise would have been impossible."[39] NYPD commissioner Arthur Woods similarly claimed that the prosecutorial ends justified the investigative means, following the broad outlines of the argument in favor of wiretapping that would motivate Justice William Howard Taft in the *Olmstead* decision twelve years later. In order to catch a crook, Woods asserted, the police needed to be allowed to "use the methods of a crook."[40] Burns adopted the same position when asked about the Thompson Committee hearings by the press. He was united with Mitchel and Woods in the belief that "rough methods" like wiretapping were necessary in the course of criminal investigations.[41]

At the same time, Progressive news outlets, labor organizations, and even the Pinkerton Agency itself came together to censure the parties involved in the scandal. Their rhetoric added an important new twist to the arguments against wiretapping. It wasn't just that tapping telephones was dirty and unethical, or that it had the potential to damage the integrity of the telephone system. It was that the practice seemed to trample on the rights of innocent telephone customers. In the wake of the Thompson Committee's opening round of hearings, for example, the editorial staff at the *Washington Post* proclaimed that wiretaps violated the U.S. Constitution's guarantees of due process: "Prison cells are occupied by men who for their own profit or convenience tapped wires and were caught at it. Tapping a man's telephone, stealing his personal and business secrets, without process of law or authority, is not so far removed from arresting a man and locking him up without a warrant."[42] Echoing the pleas of the labor groups that the NYPD was revealed to have tapped over the course of the 1910s, another article in the *Washington Post* demanded that lawmakers offer telephone networks the same privacy protections as the postal system: "The attempt to justify wire-tapping. . . . on the ground that it aids in the detection of crime is a fallacious and dangerous argument. . . . [W]e insist that wire-tapping is as vicious and indefensible as opening private mail."[43] The Pinkerton Agency began trumpeting a new corporate policy prohibiting the use of "any . . . device for 'listening in' on conversations" in response to the Thompson Committee hearings.[44] Such high-minded attempts to reinforce the integrity of the Pinkerton service, over and against Burns's invasive methods of detection, may not have reflected the company's actual practice.[45]

The Thompson Committee eventually recommended that the state of New York take measures to limit law enforcement wiretapping. It also recommended impaneling a grand jury to bring criminal charges against the Burns Agency detectives who worked with the NYPD in the Seymour investigation. There was one problem: New York's wiretap law, ratified in 1892, didn't appear to apply to police officials and licensed investigators working to detect crime. In the absence of controlling regulations, the grand jury was forced to limit its indictment of Burns to a simple case of breaking and entering. He was later found guilty and fined $100. The presiding judge refused to give the celebrity detective a prison sentence, reasoning that he had acted "from the very best of motives."[46]

Dissatisfied with the result of the Thompson Committee hearings, Burns's professional rivals soon began agitating for the revocation of his detective license. In the spring of 1917, Burns found himself before the New York State Comptroller's Office with the future of his business on the line. Prosecutors for the state claimed that Burns's use of wiretaps and eavesdropping devices both violated the rights of criminal suspects and disregarded basic norms of professional decency. "If there is operated in this country another Agency that combines equal arrogance, cynical contempt for rights and a willingness in achieving of any result for money, with disregard of methods, we have not heard of it, even among those who would naturally be expected to know of the existence of such a menace to society," charged the state's lead attorney, Meier Steinbrink. "The Burns Agency (which means Burns) is and has been all of these things; it has neither a competitor nor a peer in American social life. There seems to be no result obtainable by foul means or crime that they have not already obtained."[47] Steinbrink's accusations were actually the product of a secret dossier on Burns's electronic methods that the Pinkerton Agency had compiled to aid the state's case. A Pinkerton operative even sent Steinbrink a list of leading questions to ask the witnesses who were called to testify in the trial.[48]

The Burns Agency's licensing suit progressed in fits and starts over the course of the next several years. The protracted timeline played into Burns's hands. As the public's interest in the wiretapping scandal faded, the contradictions in his account of the Seymour investigation ended up falling on deaf ears. At the height of the Thompson Committee hearings, Burns had underscored to the press that his agency avoided wiretapping as a matter of principle. "We never have tapped a telephone wire," he told the *New York Times* in 1916. "We wouldn't dare. It is a felony in many states."[49] But by 1919, when Burns was finally compelled to speak on the Seymour affair under oath, his story had changed. In testimony before the New York State Comptroller's Office, Burns admitted that he had used a "Tel-Tap" to eavesdrop on phone conversations in a number of past cases. And while he hadn't personally tapped the Seymour firm's lines in the spring of 1916 (detectives at the NYPD took responsibility for that offense), he conceded that he had listened in on several calls that the police intercepted over the course of the investigation. Throughout the trial, Burns insisted that what transpired in the Seymour offices—from the picking of locks, to the installation of dictographs, to the tapping of telephones—was both necessary and legal.[50] Whatever the merits of his claims, the Burns Agency kept its license.

The Seymour case thus came and went: nothing had changed. Business at the Burns Agency returned to normal. Burns himself received little more than a slap on the wrist for his role in the wiretapping scandal. Despite the abiding concerns of labor unions and civil liberties watchdogs, and despite the unremitting protests of his rivals in the detective profession, he even found himself on the short list for a high-profile government job.

On August 18, 1921, William John Burns was appointed director of the U.S. Bureau of Investigation, the agency that would become the FBI. The private eye who carried a dictograph instead of a handgun took his talents to Washington.

. . .

On the surface, the New York City wiretapping scandal only reinforced the negative narratives about wiretapping that had circulated since the late nineteenth century. Tapping telephones still seemed dirty and unethical. What's more, many interested observers—from legal scholars to labor activists—had come around to the idea that the practice violated constitutional guarantees.

But there was a key difference after 1916. Now the state was openly wielding the powers of the wiretap against perceived criminal threats, rather than vice versa. Public officials were even claiming that such a reversal was a necessary step in the modernization of American police methods. Had the wiretap suddenly become an acceptable tool of the law?

The tepid fallout from the Thompson Committee hearings suggested that it might have, although the question remained untested in the courts.[51] With the notable exception of NYPD sergeant John Kennell—a troubled member of the wiretap squad who committed suicide following the revelations of his role in the city's tapping of the Catholic church—the key players involved in the New York City wiretapping scandal escaped the controversy unscathed.[52] John Mitchel avoided sanction. Arthur Woods kept his job as police commissioner. In 1917, a group of high-powered lobbyists submarined the New York legislature's effort to close the law enforcement loophole in the state's wiretap statute. Bolstered by the political victory, the NYPD wiretap squad continued monitoring phone lines from its headquarters at 50 Church Street. New York City would gain the humiliating title of "America's eavesdropping capital" in the years that followed.[53] Its reputation wasn't without merit.

Burns, for his part, hadn't merely escaped punishment in the wake of the Thompson Committee hearings. He had gained power and influence. During his three-year term at the Bureau of Investigation in Washington, Burns intensified his controversial methods of detection. He continued to approve of the use of telephone taps and eavesdropping devices, and he worked to establish a nationwide network of undercover spies. (That many of his spies already had jobs at the Burns Agency, sometimes retaining their private salaries while working for the government, was one of many unethical moves that raised the eyebrows of his critics.) The targets of Burns's most invasive investigative tactics, it turned out, weren't criminal but political: the labor movement, the Communist Party, even rival members of Congress. On at least one occasion he ordered a Bureau agent to tap the telephone of a senator who had criticized the administration of President Warren G. Harding.[54]

According to a widely circulated pamphlet published by the American Civil Liberties Union (ACLU) in 1924, the Bureau of Investigation had become "lawless in its methods" under Burns's direction. Burns's legion of eavesdroppers, informants, and spies was "analogous to the old secret police of the European autocracies whose main business was intimidation, espionage, and provocative acts."[55] Burns would resign from his post as a result of his role in the Harding administration's Teapot Dome bribery scandal. Notably, it was political corruption, not investigative excess, that forced his hand.

Nonetheless, the winds of administrative change appeared to be blowing. In 1924, the U.S. attorney general appointed to clean house after Teapot Dome, Harlan Fiske Stone, ordered an end to the Bureau of Investigation's political activities. Eliminating the use of wiretaps and other questionable methods of investigation became an easy way for Stone to improve the agency's tarnished reputation. By 1926, the Department of Justice had classified the installation of wiretaps as an offense analogous to the entrapment of suspects and the bribery of witnesses.[56] J. Edgar Hoover, then regarded as Burns's straight-laced successor, agreed with the policy—at least when he addressed the public. In an irony that belies the black record of surveillance abuse the FBI would accumulate in the coming decades, Hoover's earliest remarks on the subject of wiretapping were condemnatory. "We have a very definite rule in the Bureau that any employee engaging in wire tapping will be dismissed from the service," he told members of the House Appropriations Committee in December 1929. "While it may not be illegal, I think it is unethical."[57] Section 14 of the Bureau's new *Rules and Regulations Manual,* which explicitly listed wiretapping as an "unethical tactic," supported Hoover's statement.[58]

Still, Burns had opened the door for the institutionalization of government wiretapping. And while the imagined threat of communist infiltration gave Hoover the cover he needed to approve of the occasional wiretap in the wake of Burns's departure, it was the "great experiment" of Prohibition that caused the practice to break into the mainstream. The war on alcohol—not World War I, or the war against communism that followed—was what truly turned the telephone tap into an official tool of criminal investigation.

As we saw in the case of Roy Olmstead (see Chapter 1), the Prohibition Bureau's interest in wiretapping was a pragmatic accommodation to the mechanics of the bootlegging business. Most bootlegging ventures, even small-time outfits, relied on the telephone in their daily operations. By tapping them, federal agents could pinpoint the location of stills and stashes, identify the names and numbers of conspirators, and decipher the organizational hierarchies that governed them. Installing wiretaps was much more reliable and efficient than the old-fashioned methods of tailing and "buy-and-bust." That much was clear. But the Prohibition Bureau's adoption of the wiretap was also the product of a more immediate tactical reversal—a payback, of sorts, for the games that bootlegging syndicates often played to stay one step ahead of the law. The police relied on the telephone in their daily operations too. An oft-overlooked fact in the history of electronic surveillance is that some of the earliest cases of Prohibition wiretapping involved bootlegging gangs monitoring the telephones of law enforcement agencies, rather than vice versa. When the government started tapping the bootleggers, it was in part because the bootleggers did it first.

In 1922, for instance, the New York Prohibition Bureau reported a wiretap on its Manhattan headquarters. Responding to a series of leaks that had tipped off liquor syndicates in New York City and Washington, D.C., a federal agent told the *New York Times* that he "believed all along that the trunk telephone wires leading into this [New York] office were tapped by bootleggers. All they would have to do would be to make an indentation in the insulation and hook the instrument on. When the telephone was removed this opening could easily be closed so that the most expert eye would not detect anything."[59] The New York Bureau never found definitive evidence of a wiretap. But the U.S. Coast Guard did throughout the early 1920s. So frequently did bootlegging gangs tap Coast Guard telephones that government officials took to speaking in code over the agency's private lines.[60] Federal detectives claimed that members of Al Capone's syndicate in Chicago regularly eavesdropped on police calls well into the 1930s.[61] The Prohibition agents who first employed wiretaps were thus literalizing the basic rule of criminal investigation that Mitchel, Woods, and Burns had all espoused during the Thompson Committee hearings: to catch a crook, you act like one.

How often the Prohibition Bureau tapped telephones remains an open question. For obvious reasons, records of the federal government's wiretapping activities prior to the *Olmstead* decision are scarce. Critics of the practice intimated that telephone taps were common among run-of-the-mill Prohibition investigations.[62] But the reports that the Supreme Court's decision to hear *Olmstead* had left "hundreds" of pending federal trials hanging in the balance were most likely inflated.[63]

Once the Supreme Court gave the government the green light to wiretap in 1928, the Prohibition Bureau began keeping more consistent records of its wiretapping activities. By today's standards the annual number of federal wiretap cases wasn't large. But it was large enough to receive pushback, a result of deep-seated public dissatisfaction with the outcome of the *Olmstead* case. In 1928, the Prohibition Bureau reported just six investigations involving wiretaps, all of which were presumably put on record after the Supreme Court's decision in June of that year. The number jumped to forty in 1929, and then to sixty in 1930.[64]

In testimony before the House Committee on Expenditures in February 1931, the chief of the Prohibition Bureau's Special Agency Division, Dwight E. Avis, explained that the incremental increase in wiretaps was the product of a nationwide effort to "concentrate on larger cases." As evidence of the strategy's success, Avis cited the recent convictions of twenty-nine members of a bootlegging syndicate in Terre Haute, Indiana, and the indictments of four leaders of the notoriously violent Purple Gang in Detroit, Michigan.[65] Both cases rested on wiretap recordings. Enforcement victories like these led the Prohibition Bureau to begin supplying wiretapping equipment to its nine Special Agency Division offices around the country.[66] Chicago, New Orleans, New York, and Philadelphia all saw significant measures of wiretap activity after *Olmstead*. More than half of the Prohibition Bureau's wiretapping cases were, in fact, carried out in one city: Detroit.[67]

Yet official statistics can only tell us so much about the breadth and reach of government wiretapping during the nation's ill-begotten war on alcohol. In all likelihood the numbers reported to Congress by the Prohibition Bureau were a fraction of the sum total of wiretaps then in operation. There is a simple reason for this. More often than not, Prohibition agents tapped telephones solely to gather intelligence and generate leads. This meant that instances of their use didn't reliably show up in official records, whether in court or elsewhere. Only rarely—only in the largest federal cases—did the Prohibition Bureau monitor lines to gather evidence for trials. Avis himself admitted this to Congress when he testified that telephone taps were "useless if all the investigator does is listen to . . . conversations." Wiretapping was a dependable method for establishing leads, he explained. But successful Prohibition enforcement cases almost always involved following up on those leads with more traditional search methods, usually geared toward the seizure of tangible evidence.

"After listening in over a period of time on the syndicate's telephones," Avis continued, "the investigating agent is so thoroughly familiar with the syndicate's operations that open investigation subsequent to the breaking of the case becomes elementary. . . . Open investigation is then very simple, the agent being as well informed as to what the bootleg syndicate has been doing as the syndicate is itself. Consequently the investigator is not working in the dark, but in the light. He knows what has gone on and does not have to depend upon circumstances. He knows where and how to go about the securing of direct evidence."[68]

At trial, the "direct evidence" is what ended up mattering. Telephone tapping mostly functioned as a shortcut to acquiring it. Government affidavits and witnesses could thereafter attribute the leads in any given case to confidential informants, or bury the true source of the information altogether. The existence of the initial wiretap could remain undisclosed. Hoover's FBI would rely on this strategy to conceal its electronic surveillance activities for most of the twentieth century.

Considering the relative scarcity of Prohibition cases involving wiretap evidence, it seems crucial to look beyond *Olmstead v. United States*—the canonical case with which we began in Chapter 1—to understand how the practice of wiretapping found its way into the workings of American law enforcement. We need to examine the historical record more closely. It's likely that many more federal wiretap cases of the period resembled that of W. E. "Shorty" Peterson, a Seattle bootlegger whose business stretched into Portland, Sacramento, and San Francisco during the late 1920s and early 1930s. Peterson's venture was one of many smaller liquor syndicates serving the market that Olmstead had left behind. Peterson, of course, did most of his work on the telephone—and like the King of the Bootleggers before him, the Seattle Prohibition Bureau decided to bring him down by tapping his lines. Yet the resemblance between Olmstead and Peterson stops there. It's worth pausing to understand why.

Peterson ran most of his operation through a single telephone connection, PRospect-7119. The line was registered to his business partner, Oliver Crossen. On the morning of October 13, 1931, two Seattle Prohibition agents installed a wiretap on Crossen's phone. By 5:45 p.m. that day, they had already overheard several conversations about a distillery located at 3511 Columbia Avenue, in the city's Madrona neighborhood. At 7:06 p.m., the wiretappers struck gold: Peterson took a call and discussed the upcoming delivery of more than $5,000 worth of liquor by boat.[69] Rather than blow the wiretap immediately, however, the Prohibition Bureau chose to sit on the line, hoping to gather more information about Peterson's operations. On November 5, Seattle police raided the property at 3511 Columbia Avenue. There they seized "a large quantity of mash and other supplies and materials for use in the manufacture of whiskey," and they arrested two of Peterson's underlings, Harry A. Wilson and Carl F. Boston.[70] But the still wasn't active, and the $5,000 stash of liquor that Peterson had mentioned on the phone had already been sold.

Even by the modest enforcement standards of the period, this was a disappointing haul for a wiretapping case. The outcome of the November 5 raid put the Prohibition Bureau in a tricky position. Beyond the initial day of loose talk, which the two wiretappers were lucky enough to have overheard, Peterson and Crossen had been careful to avoid incriminating themselves over the telephone. After the raid, they began taking precautionary countermeasures. On November 7, Crossen instructed an associate to have the phone company check PRospect-7119 for a wiretap.[71] On November 13, he disconnected the line. (In the Prohibition Bureau's transcripts, the wiretappers describe this turn of events with evident disappointment: "Sound got weaker and weaker and line finally went dead. They found the tap.")[72] When Peterson and Crossen reconnected PRospect-7119 a week later, they made certain to instruct callers to avoid mentioning names, addresses, or other forms of identifying information. On the matter of wholesale shipments and deliveries, Peterson warned his associates not to "talk that stuff over the phone." On December 2, Crossen greeted a customer with the disclaimer that federal agents were "hot on my wire."[73] The Bureau had little meaningful information to work with as a result.

Scrambling to salvage the investigation, the Prohibition Bureau ordered a detective to tail Wilson, who posted bail shortly after his arrest. Within days, Wilson led the detective to a second distillery outside of the city—an outfit unrelated to Peterson's operation—as well as to the Consumers Compressed Yeast Company, a San Francisco-based business that turned out to have recently purchased a supply of materials typically used in the construction of whiskey stills. The discovery led to a second wiretap: this time at MAin-3330, the connection at the Compressed Yeast Company's Seattle office. The wiretappers monitored that line for two weeks, from February 25 to March 9, 1932.[74] On March 16, Seattle police executed a search warrant at the office building, seized the distillery construction materials, and discovered documents connecting Wilson to three other individuals who worked in a much smaller bootlegging syndicate. All four of the men were arrested. The evidence from the wiretaps and the raids didn't go any further, however. Peterson and Crossen went unindicted.

We can take a few lessons from the story of W. E. "Shorty" Peterson. The first is that wiretaps have never been a perfect medium of criminal investigation. They weren't in the era of Prohibition, and they aren't today. After *Olmstead v. United States,* and perhaps even before, bootleggers knew to watch what they said on the telephone. Peterson and Crossen were well aware that federal agents were monitoring their calls. The preventative measures they took after the November 5 raid kept them beyond the reach of the law, forcing the Seattle Prohibition Bureau to take its investigation in a much less ambitious direction. The wiretappers ended up in a very different place from where they started—a far cry from the Whispering Wires case, which yielded months' worth of actionable intelligence on a single investigative target, and a massive cache of recorded evidence to be used in court.

A second lesson from the Shorty Peterson story, more important for our present purpose, is that wiretaps are easy to hide from official records. Consider what happened when the Prohibition Bureau began preparing to bring the four arrested suspects to trial:

On April 12, 1932, one of the federal agents who monitored the phones at Prospect-7119 and MAin-3330, A. E. McFatridge, wrote a memorandum on the Peterson case to the director of the Seattle Prohibition Bureau.[75] In the document, McFatridge detailed the investigation from start to finish: from the wiretap at PRospect-7119 to the raid at 3511 Columbia Avenue; from the tailing of Harry Wilson to the discovery of the equipment stashed away at the Consumers Compressed Yeast Company office; and from the wiretap at MAin-3330 to the arrest of the four small-time bootleggers, Wilson included. McFatridge took pains to explain the successes and failures of the two wiretaps to the director of the Seattle office. After all, his initials were visible at the bottom of almost every page of the log of calls that he and his partner had transcribed.

That same day, McFatridge wrote a second memorandum on the Peterson investigation—this one addressed to the U.S. attorney in charge of prosecuting the case.[76] In this document, McFatridge offered an extended account of the raid at 3511 Columbia Avenue, the seizure of the equipment stashed away at the Consumers Compressed Yeast Company office, and the eventual arrest of the bootleggers. He went on to enumerate the materials submitted as evidence: copper pipes and copper tubes; orders, invoices, and receipts. There were fourteen exhibits in all, many of them documents signed by the four suspects who were to be charged with the crime of manufacturing liquor.

But the wiretaps? McFatridge never mentioned them to the prosecutor. The taps at PRospect-7119 and MAin-3330 were omitted from the memorandum to the U.S. Attorney's Office. The four suspects arrested in the investigation soon went to trial. It's likely that the prosecution never revealed what had brought them there in the first place.

. . .

It's impossible to know how many cases like Shorty Peterson's there were during the Prohibition years. Which is perhaps the point: official accounts can only tell us so much. At the very least, what we can say for certain is that Peterson's ordeal was more common than Olmstead's. In the size and scope of the surveillance operation at its center—eight separate wiretaps monitored over the course of five months, a 775-page typewritten transcript that helped to indict ninety-one criminal suspects—the case of the Whispering Wires was an exception, not the rule. It was just the sort of case that could test the boundaries of the law and make its way to the Supreme Court. It was just the sort of case that could shape public debates about privacy and surveillance for the next century. But it wasn't representative of the wiretap investigations typically launched by the U.S. government in the name of the Eighteenth Amendment. Olmstead's story is a good one. But so is Peterson's. The truth of history lies somewhere between them.

None of that stopped *Olmstead v. United States* from creating a firestorm. The national response to the Supreme Court's 1928 decision was divided, and those divisions mapped neatly onto larger divisions over the fate of America's war on alcohol.[77] Depending on where you stood on the political spectrum—whether you were a "dry" in favor of the Eighteenth Amendment, or you were a "wet" united in opposition against it—the practice of government wiretapping epitomized either the necessities or the perils of Prohibition enforcement. Drys praised the *Olmstead* decision. They defended wiretapping as a reasonable and just response to the mechanics of modern crime; the practice seemed like a necessary evil in the nation's effort to liberate itself from the scourge of drink. Wets, by contrast, denounced *Olmstead*. They rejected wiretapping as unethical and unconstitutional. To them, the practice seemed typical of the outrageous abuses of police power that the war on alcohol had begotten.

The public response to *Olmstead*—which rehearsed, in miniature, the public response to Prohibition—thus helped to set the general terms of the great American debate over government surveillance. If the arguments for and against Prohibition enforcement wiretapping sound vaguely familiar today, it's because we've adapted their terms to fit our times.

Notably, the *Olmstead* decision also instigated the earliest federal effort to pass a wiretap law during peacetime. In November 1929, John C. Schafer, a Republican congressman from Wisconsin, introduced the first of several legislative measures that congressional wets would propose in order to negate the effects of the Supreme Court's ruling. Schafer's bill was aimed at "prohibit[ing] the tapping of telephone and telegraph lines, and prohibiting the use of information obtained by such illegal tapping to be used as evidence in the courts of the United States in civil suits and criminal prosecutions, and for other purposes."[78] He wanted, in other words, an all-out ban. On the floor of the House of Representatives, Schafer introduced his resolution with rhetoric that drew equally on the legal and moral arguments against wiretapping that Brandeis and Holmes had provided in the *Olmstead* dissent a year earlier.

"So long as the Federal Government continues to permit the tapping of telephone and telegraph wires, it is guilty of tyranny equal to that of the most backward medieval despotisms," Schafer claimed, pausing as fellow congressional wets rose in applause. "If [the government is] permitted to continue this nefarious practice, the privacies of life and the homes of our people will be subject to public scrutiny at a time by disreputable as well as reputable Government agents and citizens. Any individual, be he a Government officer or not, who invades the privacies of the person and home of an American citizen by tapping telephone or telegraph wires, is one of the most despicable specimens of the human race."[79]

Overmatched by a dry majority in the House of Representatives, Schafer's resolution foundered. But his near-fanatical stance against wiretapping grew popular among congressional wets, particularly after the National Commission on Law Observance and Enforcement (the Wickersham Commission, as it was known) reported that Prohibition enforcement officers routinely relied on illegal police methods.[80] In 1931, Schafer joined forces with George H. Tinkham, a Republican wet from Massachusetts, to propose an amendment to the Justice Department's upcoming appropriations bill that would have prevented the Prohibition Bureau from using federal funds on wiretap investigations. The measure received the blessing of the ACLU, which began its long and storied campaign in opposition to government wiretapping in 1930.[81]

That proposal failed too. But it succeeded two years later, as soon as the wet cause gained a sizable majority in Congress. In February 1933, both the Senate and the House of Representatives denounced the "use of Government funds in committing crimes" and banned the employment of wiretaps in the enforcement of Prohibition laws.[82] It was the first time the federal government openly went on record against the practice.

The victory would prove short-lived. Just six days after the wiretapping ban was ratified, Congress sent the Twenty-First Amendment to the U.S. Constitution, repealing Prohibition, to the states. The war on alcohol ended ten months later—and with it ended the federal government's principled ban on wiretaps. The dirty business was just getting started.

Chapter 3

To Intercept and Divulge

FOR MORE THAN TWO AND A HALF DECADES, the history of wiretapping and electronic eavesdropping in the United States hinged on one word buried in an obscure federal statute. The word confounded government officials who were wary of wiretapping, and it infuriated civil liberties advocates who wanted to see the practice abolished. The word emboldened—and later embarrassed—the FBI in its clandestine effort to use electronic surveillance in criminal cases and espionage investigations. The word led to confusion, acrimony, and impassioned debate, eventually hastening a series of public scandals that captured the nation's attention in the early years of the Cold War.

The word that caused all of the trouble was *and*.

In June 1934, Congress passed the Federal Communications Act (FCA), a landmark law designed to regulate the technical advances in telegraphy, telephony, and radio that had reshaped the nation's communications industries in the years following World War I. The stated aims of the FCA were administrative. By all accounts, the most important outcome of the legislation, signed into law by President Franklin Delano Roosevelt, was the creation of the Federal Communications Commission (FCC) as an independent government agency.[1] But buried in Section 605 of the FCA was a series of seemingly innocuous provisions. Here is Section 605 in its entirety:

> No person receiving or assisting in receiving, or transmitting, or assisting in transmitting, any interstate or foreign communication by wire or radio shall divulge or publish the existence, contents, substance, purport, effect, or meaning thereof, except through authorized channels of transmission or reception, to any person other than the addressee, his agent, or attorney, or to a person employed or au-

thorized to forward such communication to its destination, or to proper accounting or distributing officers of the various communicating centers over which the communication may be passed, or to the master of a ship under whom he is serving, or in response to a subpoena issued by a court of competent jurisdiction, or on demand of other lawful authority; and no person not being authorized by the sender shall intercept any communication and divulge or publish the existence, contents, substance, purport, effect, or meaning of such intercepted communication to any person; and no person not being entitled thereto shall receive or assist in receiving any interstate or foreign communication by wire or radio and use the same or any information therein contained for his own benefit or for the benefit of another not entitled thereto; and no person having received such intercepted communication or having become acquainted with the contents, substance, purport, effect, or meaning of the same or any part thereof, knowing that such information was so obtained, shall divulge or publish the existence, contents, substance, purport, effect, or meaning of the same or any part thereof, or use the same or any information therein contained for his own benefit or for the benefit of another not entitled thereto: Provided, That this section shall not apply to the receiving, divulging, publishing, or utilizing the contents of any radio communication, broadcast, or transmitted by amateurs or others for the use of the general public or relating to ships in distress.[2]

Reading the statute carefully—a tall order, even for the initiated—underscores just how easy it would have been for lawmakers to overlook, or worse yet misconstrue, its crucial second clause: *no person not being authorized by the sender shall intercept any communication and divulge or publish the existence, contents, substance, purport, effect, or meaning of such intercepted communication to any person.* Herein lies a problem.

There are two main ways to interpret the second clause of Section 605, and your view of the story that follows will likely depend on the interpretation you favor. On one hand, it's possible to read the provision as a wiretapping ban. In effect, this is to read the "and" in the crucial "intercept . . . and divulge" construction as an *or,* which would imply that Congress intended to represent the interception and divulgence of private communications as independent acts, each equally prohibited in the eyes of the law. In this interpretation of Section 605, tapping a wire is a federal crime, and so is the public disclosure of wiretapped information.

On the other hand, it's possible to read the "and" in the "intercept . . . and divulge" clause more literally, as a conjunction stipulating that both acts need to occur for anything illegal to have taken place. In this alternative interpretation—an equally plausible, if somewhat finicky, interpretation—Section 605 functions as a prohibition against the "divulgence" of intercepted communications in a public venue such as a criminal trial or a newspaper column, not as a prohibition against the act of "interception" in and of itself. Here the statute serves more as a rule of decorum than as a strict legal prohibition: wiretapping is forbidden if and only if the contents of a wiretapped conversation are disclosed or made public. Quite obviously, this is an interpretation that calls into question the statute's force as a wiretap ban.

So was wiretapping illegal following the passage of the FCA? Or was it merely the divulgence of wiretapped information that was illegal? What did the "and" in "intercept . . . and divulge" really mean?

No one bothered to ask those questions when the FCA became law in 1934. There is little in the congressional record to suggest that government officials gave sustained attention to the language of Section 605 while the FCA was up for debate.[3] The statute received nothing in the way of public notice in the years following the bill's passage. But in December 1937, the "intercept . . . and divulge" clause emerged as an unexpected sticking point in a federal trial involving one of the largest bootlegging organizations in the United States to survive the fall of Prohibition. The case in question—*Nardone v. United States* (1937), or *Nardone I* in the shorthand of legal scholars—would come to dominate the national debate over wiretap reform during the 1940s and 1950s.

. . .

The events behind *Nardone I* read like a story ripped from the pages of a detective thriller.[4] Frank Carmine Nardone was the lieutenant of a New York–based bootlegging syndicate that had for years smuggled unlicensed alcohol into the United States from ports as far-flung as Belgium, Denmark, and Norway. In November 1935, federal authorities caught wind of a Nardone freighter carrying more than 24,000 gallons of liquor off the coast of Nova Scotia. Over the course of a four-month investigation, the U.S. Coast Guard tailed the ship as it traveled from Canada to the Caribbean and back again, anchoring to offload portions of its illicit cargo onto smaller land-bound vessels. On March 17, 1936, the game of cat-and-mouse ended in front-page headlines: more than 1,600 cases of alcohol were seized from a ship docked in Bridgeport, Connecticut, and three days later several members of the bootlegging syndicate—Nardone along with four others—were arrested at a fashionable restaurant in midtown Manhattan. A grand jury soon indicted Nardone and his accomplices.

As in many bootlegging investigations of the Prohibition era, the government's ability to build a case in *Nardone I* depended on a carefully planned wiretapping operation. (Federal investigators would hardly have been able to follow the movements of the ships in the bootlegging fleet without knowing their coordinates in advance, a problem that Nardone's haphazard reliance on the telephone easily solved.) Of the more than 500 telephone conversations police intercepted over the course of the investigation, seventy-two were introduced as evidence in the ensuing trial. Transcribed by federal agents and read aloud in court, the wiretaps provided the nail in the defense's coffin, and Nardone and his associates were easily convicted.

But Nardone's attorneys mounted an appeal based on a reading of Section 605. The contention was that the presiding judge had erred in admitting the prosecution's evidence. Because the case against Nardone rested on wiretapped conversations—and because those conversations were included in sworn testimony—the government had deigned to "intercept . . . and divulge" in the face of the FCA. According to the defense, the case should have been dismissed off the bat.

The argument was a long shot, but by December 1937 *Nardone v. United States* had made its way to the U.S. Supreme Court. In a surprise 7–2 decision, the majority ended up siding with Nardone, reasoning that the proof against him depended entirely on testimony offered in violation of the FCA. According to Justice Owen Roberts, who authored the Court's opinion, the "plain mandate" of Section 605 was to prohibit the divulgence of wiretapped conversations in a court of law. It was more tolerable to let criminals go "unwhipped of justice," Roberts argued, than to condone investigative practices and courtroom procedures that were "inconsistent with ethical standards and destructive of personal liberty."[5] At face value the decision seemed to eliminate much of the ambiguity surrounding the FCA. Under the "intercept . . . and divulge" provision, it was illegal to disclose the contents of wiretapped conversations in federal court proceedings. Wiretaps were inadmissible as evidence.

Considering the confusion that followed, it's worth noting that the two justices who dissented in *Nardone I* immediately recognized the practical implications of the Court's decision. Roberts, for his part, had stopped short of specifying whether Section 605 prohibited the interception of private communications in the absence of public divulgence. But according to Justice George Sutherland, who authored an exasperated dissent, the Court's decision was still tantamount to a wiretap ban. By forbidding one side of the "and" in Section 605 ("divulgence"), the majority was implicitly casting aspersion on the other ("interception"). What's more, Roberts had singled out law enforcement wiretapping as a "grave wrong" in his argument.[6]

Sutherland went on to predict that *Nardone I* would function, in practice, as a procedural deterrent against government wiretapping, a move that seemed to spell disaster for the nation's war on crime. "The decision just made," Sutherland wrote, "will necessarily have the effect of enabling the most depraved criminals to further their criminal plans over the telephone in the secure knowledge that even if those plans involve kidnapping and murder, their telephone conversations can never be intercepted by officers of the law and revealed in court."[7] Sutherland saved his most scathing critique of the ruling for the final lines of his dissent, which questioned the common sense of the seven justices who voted to overturn Nardone's conviction: "My abhorrence of the odious practices of the town gossip, the peeping Tom, and the private eavesdropper is quite as strong as that of any of my brethren. But to put the sworn officers of the law, engaged in the detection and apprehension of organized gangs of criminals, in the same category is to lose all sense of proportion. . . . [W]e well may pause to consider whether the application of the rule which forbids an invasion of the privacy of telephone communications is not being carried in the present case to a point where the necessity of public protection against crime is being submerged by an overflow of sentimentality."[8]

Even from the perspective of a skeptic, then, the Supreme Court's decision in *Nardone I* seemed to tip the judicial scales toward the first interpretation of Section 605—the interpretation that construes the statute as a wiretapping ban. Read any other way, the ruling is virtually meaningless. What's more, any doubt about the scope of the FCA that lingered after *Nardone I* appeared to vanish two years later, when the Supreme Court handed down two more wiretap rulings. The first, *Weiss v. United States* (1939), prohibited the divulgence of intrastate communications in federal court proceedings.[9] This closed an important loophole in the

Court's initial construction of Section 605. The second ruling was a reconsideration of Nardone's bootlegging case, which federal prosecutors attempted to retry on different grounds earlier that year.

Perhaps even more decisive in its outcome, the 1939 reprise of *Nardone v. United States*—often referred to as *Nardone II*—ended in yet another reversal, this time because key witnesses were permitted to consult wiretap transcripts to refresh their memories before testifying in court.[10] In effect the Supreme Court was doubling down on its interpretation of Section 605, widening the scope of a controversial decision—*Nardone I*— that was "not the product of a merely meticulous reading of technical language," but a "translation into practicality of broad considerations of morality and public wellbeing."[11] Crucially, *Nardone II* didn't simply shore up the Court's position on the exclusion of wiretaps in criminal trials. The ruling also condemned the bare act of "interception" itself, because it deemed inadmissible even collateral information derived from wiretapped conversations. In one of the earliest known uses of a designation that would become notorious in the history of criminal procedure, the majority opinion in *Nardone II* characterized evidence gleaned from wiretaps as "fruit of the poisonous tree."[12] Few could now question which part of the "intercept . . . and divulge" clause carried the poison.

And yet this was only the beginning of the story. Despite the force of the Supreme Court's interpretation of the FCA, Section 605 proved nothing short of catastrophic as a deterrent against wiretapping. Debate about the true meaning of the law would end up raging for decades. In 1958, a team of legal experts captured a sentiment that had become commonplace when they described Section 605 as a "statutory Frankenstein," created and applied with monstrous disregard for practical consequence.[13] Other observers who took stock of the nation's wiretapping problem in the period criticized the FCA as a "flop," an "illusory safeguard," and the root cause of a "black record of excess."[14] In 1966, U.S. Attorney General Nicholas Katzenbach would tell *Life* that it was impossible to imagine a law "more totally unsatisfactory."[15] What went wrong?

As we'll see, the Supreme Court's interpretation of the ambiguous "and" in Section 605 ended up raising more questions than it answered. Behind the scenes, it also opened the door for rampant wiretap abuse. It took the shock of two salacious wiretapping scandals for the nation to realize the consequences of its muddled electronic surveillance policies: a case involving an alleged Soviet spy, and a case involving a crooked private investigator. But by the mid-1950s the monster had already done its damage. Lawmakers at both the state and federal levels found it difficult to lay the groundwork for reform. Taking stock of the disastrous reign of Section 605 ultimately helps to reveal the embattled political origins of the "surveillance state" in America—a state that, in many respects, was at war with itself from the very beginning.

• • •

At first glance, the Supreme Court's rulings on wiretapping during the late 1930s seem to mark a turning point in the history of U.S. wiretap policy. As we saw in Chapter 1, the Court's controversial 5–4 decision in *Olmstead v. United States* (1928) held that the practice of wiretapping falls outside of the protections provided in the Fourth and Fifth Amendments to the U.S. Constitution.[16] Despite widespread political opposition, *Olmstead* defined government wiretap authority for the balance of the 1930s: tapping was legally sanctioned as a method for obtaining criminal evidence, and federal agencies made use of it in a variety of investigative capacities, both with and without administrative oversight. The *Nardone* and *Weiss* decisions signaled a change in course. Using Section 605 as a lever, a more progressive judicial majority had realigned the Court with the dissenting opinions in *Olmstead,* which famously disparaged wiretapping as a violation of civil liberties. By the end of the decade, the deck was stacked against the government.

Legal experts took note of this shift almost immediately. According to New York attorney Frederick F. Greenman, who published a book-length study of federal wiretap law in 1938, the majority opinion in *Nardone I* reflected a "deep-seated view . . . that the *Olmstead* case was not correctly decided."[17] After *Nardone II* and *Weiss,* commentators of all political stripes followed Greenman's lead in characterizing the Supreme Court's new position on wiretapping as a rejection of the *Olmstead* precedent.[18] (President Roosevelt must have vaguely seen this shift in the works when he granted Olmstead a full presidential pardon on Christmas Day, 1935.)[19] For the *Harvard Law Review,* the basic message of the *Nardone* and *Weiss* decisions was that "uncontrolled and indiscriminate wire tapping" was to be condemned as a violation of civil liberties.[20] Many observed that this message was in keeping with a groundswell of popular opposition to the practice that emerged in the wake of congressional hearings on illegal wiretapping in early 1940.[21]

Federal courts likewise applied *Nardone* and *Weiss* in accordance with the view that the Supreme Court had turned against the *Olmstead* precedent. In a development that proved disconcerting for government officials, three federal judges dismissed high-profile criminal cases involving wiretapped evidence in the two years between *Nardone I* and *Nardone II*.[22] In 1938, government agencies like the Treasury Department, which had long relied on wiretaps in the investigation of racketeering and tax evasion cases, began working to stem the tide by lobbying members of Congress to pass a law reauthorizing government tapping.[23] *Nardone II*'s ban on wiretap leads seemed strict enough to make some observers wonder whether Justice Sutherland's predictions about the nation's fight against crime were already coming true. According to an article published in the *Michigan Law Review* in May 1940, the shift from *Olmstead* to *Nardone* was giving "undue protection to criminals" in the federal justice system. After weighing the desire to protect civil liberties against the enforcement of criminal law, the Court appeared to have wedded itself to the view that "the unethical aspect of wiretapping by officers of the law is serious enough to warrant such activity being curbed as far as is possible."[24]

In short, by 1940 there wasn't much uncertainty about the Supreme Court's stance on electronic surveillance. Whatever the doctrinal significance of the *Nardone* rulings, it was clear that the Court had moved to deter the practice of wiretapping through the red tape of criminal procedure. Yet even as government agencies scrambled to conform to a more restrictive legal regime, the ground was shifting. Behind closed doors, officials at the Department of Justice and the FBI were working to limit *Nardone*'s influence. Deliberately concealed and chillingly cynical, their efforts to reframe and undermine the Supreme Court's decisions blunted the force of Section 605 until the FCA was rendered a dead letter.

The historian Athan Theoharis's research on classified FBI documents helps to illuminate the origins of the government's effort to undermine the *Nardone* rulings during the late 1930s and early 1940s.[25] According to Theoharis, officials at the Department of Justice and the FBI began toying with an alternative reading of Section 605 in direct response to *Nardone I*. On December 22, 1937, just two days after the Court handed down its decision, FBI assistant director Edward Tamm sent a confidential memorandum to J. Edgar Hoover summarizing a policy meeting with Assistant Attorney General Alexander Holtzoff. Tamm opened the missive by reassuring an apprehensive Hoover that the public had "misinterpreted" the Supreme Court's construction of Section 605. He went on to relay a tortured exchange with Holtzoff about the FBI's ability to maintain its wiretap authority in the face of the decision.

In Holtzoff's view, *Nardone I* wasn't actually intended to curb wiretapping. The ruling prohibited "intercepting *and* divulging or publishing," he explained, not "interception or divulgence." As a result, the "*interception* of telephone or telegraph messages by telephone tap or otherwise is not in itself a violation." In theory this meant that the FBI could tap with impunity as long as the contents of intercepted messages went undisclosed. Even if Section 605 functioned differently in practice, Holtzoff assured Tamm that prosecuting a government violation of the FCA "could only be done with the approval of the Department [of Justice] in Washington and . . . *certainly the Department would not authorize any prosecution against its own employees.*"[26]

Holtzoff's logic was shaky. Clearly he was well aware that the Justice Department would be treading a fine line. But Hoover was sold. He appended a note to Tamm's memorandum confirming that the Bureau would stand by its existing wiretap policy. The only concession to *Nardone I* was his request that investigating agents avoid relying on wiretaps as criminal evidence: "No phone taps without my approval, & as previously we will not authorize any except in extraordinary cases & then not to obtain evidence but only for collateral leads."[27] A week later Hoover sent a letter containing the same instructions to the agents in charge of the FBI's local field offices.[28] The idea was that it was lawful to tap as long as federal prosecutors didn't attempt to use wiretap transcripts or recordings in court.

Nardone II foreclosed even this interpretation of Section 605. After the 1939 decision, "collateral leads" from wiretapping were just as much a part of the FCA prohibition as wiretaps themselves. In accordance with the new precedent, U.S. Attorney General Robert Jackson issued an internal order banning all FBI wiretaps shortly after he took office in 1940.[29] The practice might have waned at the Bureau if President Roosevelt hadn't then taken an extraordinary step to intervene. In a confidential memorandum to Jackson dated May 21, 1940—a document that would become infamous when it was declassified a decade later—Roosevelt encouraged Jackson to reverse his internal ban, granting the Justice Department executive authority to use wiretaps in national security cases.

"I have agreed with the broad purposes of the Supreme Court decision relating to wire-tapping in investigations," Roosevelt wrote to Jackson, referring to the shift in policy that *Nardone II* signaled. "The

Court is undoubtedly sound both in regard to the use of evidence secured over tapped wires in the prosecution of citizens in criminal cases; and is also right in its opinion that under ordinary and normal circumstances wire-tapping should not be carried on for the excellent reason that it is almost bound to lead to abuse of civil rights." But in hindering the government's ability to tap in the interest of "national defense," Roosevelt continued, the Court had overstepped its bounds. He closed the memorandum by claiming that threats of espionage, sabotage, and "'fifth column' activities" were prevalent enough to make an executive exception to the ruling:

> I am convinced that the Supreme Court never intended any dictum in the particular case which it decided to apply to grave matters involving the defense of the nation. . . . You are, therefore, authorized and directed in such cases as you may approve, after investigation of the need in each case, to authorize the necessary investigating agents that they are at liberty to secure information by listening devices direct to the conversation or other communications of persons suspected of subversive activities against the Government of the United States, including suspected spies. You are requested furthermore to limit these investigations so conducted to a minimum and to limit them insofar as possible to aliens.[30]

Roosevelt's secret directive to permit FBI wiretapping was a swift and decisive blow to the Supreme Court's rulings in the *Nardone* cases. Many would later rationalize the move as a strategic attempt to protect government counterintelligence operations, which were likely to increase in importance as the nation inched toward direct involvement in World War II.[31] But the fact is that Roosevelt's "national defense" exception for government wiretapping—a policy later known simply as the "Roosevelt doctrine"—was also a political response to an ongoing public debate about the ethics and extent of wiretapping itself.

Roosevelt sent the confidential memorandum to Jackson on the very same day that the Senate Committee on Interstate Commerce held the first in a series of high-profile hearings on the tapping of state politicians in New York, Pennsylvania, and Rhode Island. Perhaps more strikingly, when Congress began assessing emergency proposals for wiretap reform in the months following the Pearl Harbor attack, Holtzoff, Hoover, and Roosevelt all offered public statements in support of legislation designed to authorize government tapping in national security investigations.[32]

None of them so much as hinted that such a policy was already in place. Looking back, the duplicity is almost breathtaking: while members of Congress debated the implications of annulling the *Nardone* decisions, representatives of the Justice Department and the FBI—and even the president himself—contributed to the hearings knowing full well that the *Nardone* decisions were already annulled.

Thus a secret federal wiretapping policy was born: a policy derived from a strained reading of legal language, a policy adopted in flagrant disregard for judicial precedent, and a policy hidden in plain sight from the American public. Following the president's confidential memorandum of May 1940, national security wiretaps were permitted at the FBI when approved by the attorney general. For federal agents on the ground, circumventing *Nardone* was thereafter as easy as corroborating wiretap evidence through other means, or attributing wiretap leads to "confidential informants" in court. The program flourished during World War II. To cover its tracks, the Justice Department chose not to maintain records of the FBI's electronic surveillance activities in the years that followed.[33]

Roosevelt's secret wiretapping directive remained in place for the remainder of his presidency. After the war, as Theoharis has also shown, Hoover made another extraordinary move to expand the FBI's wiretap authority. In a July 1946 letter to President Harry Truman, Attorney General Thomas Clark informed the new administration of the Roosevelt doctrine and asked for its renewal in cases "vitally affecting the domestic security, or where human life is in jeopardy."[34] To add weight to his appeal, Clark reproduced the entirety of Roosevelt's 1940 memorandum in the letter. But at Hoover's encouragement he excised the crucial final sentence of the original document ordering the FBI to limit its wiretapping activities to investigations involving "alien" security threats. Truman signed off on the letter unaware that he was giving the Bureau a longer leash.

．　●　．

Ironically, one of the most valuable FBI surveillance targets to emerge in the years following the renewal of Roosevelt's secret wiretapping directive was a Justice Department employee. Her name was Judith Coplon, a twenty-seven-year-old analyst in the Department's Foreign Agents Registration Section who in 1948 came to be suspected of spying for the Soviet Union.

Along with Alger Hiss, Klaus Fuchs, and Julius and Ethel Rosenberg, all of whom were brought to trial in the same period, Coplon remains synonymous with the U.S. government's effort to combat espionage in the early years of the Cold War. She is also the first Soviet spy known to have been identified on the basis of decrypted KGB cable messages, a fact that wasn't revealed until 1986, when one of the principal players in the Venona Project, a top-secret Army code-breaking program, published a tell-all memoir.[35]

Despite Coplon's renown, historians have long overlooked the significance of her case to the evolution of U.S. wiretap policy.[36] Coplon's convictions for espionage and conspiracy in 1949–1950, both of which were overturned on procedural technicalities, offered the nation its first glimpse of the government surveillance activities that had become routine under the Roosevelt doctrine. But Coplon's case did more than simply reveal what was otherwise hidden. As the star player in a drama that paradoxically cast her as both perpetrator and victim—an illegal spy, illegally spied on—Coplon served as a touchstone for the volatile debate over FBI wiretapping in the early years of the Cold War. By the mid-1950s she had emerged as a powerful political symbol. Civil liberties activists who protested the FBI's surveillance programs often did so in Coplon's name. Likewise, government officials who advocated for wiretap authorization openly attempted to exploit the public outcry over her botched conviction. In short, the struggle over Judith Coplon was nothing less than a contest over the future of communications privacy in the United States. Her case seemed to bring Section 605 to its breaking point.

Judith Coplon was born in Brooklyn, New York, in 1921, the daughter of a toy maker and a milliner.[37] A promising student from an early age, Coplon went on to attend Barnard College, where she majored in history and served as an editor for the student newspaper, the Barnard *Bulletin*. Despite her fleeting involvement with a chapter of the Young Communist League on campus, and despite penning several left-leaning columns for the *Bulletin* throughout her tenure as an undergraduate, Coplon secured a job in the New York office of the Justice Department's Economic Warfare section shortly after she graduated in 1943. Sometime in the fall of 1944, a KGB agent recruited her to the Soviet cause.[38] Within months her value as an intelligence asset escalated when she was promoted to a position with security clearance at the Justice Department's headquarters in Washington, D.C. For the better part of the next four years, Coplon's job allowed her to supply detailed information to the KGB about the FBI's counterintelligence activities. KGB reports, decoded by the Venona Project and declassified decades later, praised Coplon as a "very serious, modest, thoughtful young woman who is ideologically close to us." Another Soviet cable noted that she "treats our assignments very seriously and conscientiously and considers our work the main job in her life."[39]

It was through Venona that Coplon first came to the attention of the FBI.[40] The discovery of her KGB connections prompted an elaborate counterintelligence investigation. But the Bureau's near-rapacious interest in ensuring a criminal conviction would prove the undoing of an otherwise ironclad case.

After informing the Justice Department that one of its employees was potentially working for the Soviets, FBI officials put Coplon under round-the-clock surveillance in January 1949, tailing her on two separate trips to visit her parents in New York. When both of the trips turned out to include suspicious late-night meetings with a Russian foreign national later identified as Valentin Gubitchev, investigators appeared to have stumbled onto Coplon's foreign contact. Eager to catch the suspected spies in the act, the FBI encouraged Coplon's supervisor at the Justice Department to present her with a "hot and interesting" document containing fabricated information about a classified counterintelligence briefing.[41] On March 4, 1949, Coplon took the bait, again traveling to New York to visit her parents, and again arranging a late-night meeting with Gubitchev. This time the suspects abruptly broke off their rendezvous. After an extended pursuit through the city's subway system, the FBI decided to close in without a court-issued warrant. Coplon and Gubitchev were arrested, questioned, and searched—again, all without a warrant—and the investigating officers discovered a cache of government documents folded neatly in Coplon's purse. One was the bogus memorandum her boss had floated to her earlier that same day. The purpose of the aborted meeting was obvious.

Coplon was booked on a litany of espionage and conspiracy charges in both Washington and New York, a confusing legal arrangement that forced the government to put her on trial in two separate jurisdictions. In the Washington trial, a circus-like affair from start to finish, Coplon's erratic defense attorney, Archibald Palmer, attempted to establish his client's innocence by mounting what reporters covering the case called a "love defense": Coplon was in an illicit relationship with Gubitchev, the story went, and their evasive activities on the evening of March 4 were the result of their shared fear of discovery.[42] In a shamelessly desperate twist to the tale, Palmer also attempted to convince the jury that Coplon was planning to use the confidential documents in her purse as fodder for a semi-autobiographical novel she was writing about the life of a female government employee. Federal prosecutors easily undercut both stories. No drafts of Coplon's alleged novel-in-progress existed. She had also spent several nights in a hotel with a Justice Department attorney, Harold Shapiro, in the weeks between her clandestine meetings with Gubitchev. Scandalized and outraged, the jury found the defense's argument untenable. Coplon was convicted and sentenced to ten years in prison.

Throughout the Washington trial, Palmer went out of his way to insinuate that the FBI had established its case against Coplon on the basis of an illegal wiretapping operation. During direct examination, for instance, Palmer asked Coplon a series of leading questions about a "regular crackling" on her home telephone in the days leading up to her arrest, a sly allusion to an old wives' tale about the presence of a wiretap on the line.[43] The prosecution's objections to this line of inquiry were quickly sustained. Later in the proceedings, Palmer directly confronted several FBI agents involved in the investigation about wiretapping and electronic eavesdropping. All of them professed ignorance of any such activity by the Bureau.

Palmer pushed harder on the wiretap angle when the scene shifted to New York in December 1949. In the preliminary hearings of Coplon's second trial, the dam of secrecy broke. Over the course of six momentous weeks, the presiding judge in New York compelled the prosecution to submit a damaging series of affidavits that disclosed the existence of an extensive FBI wiretapping operation on Coplon. From January 6, 1949, to September 27, 1949—dates that spanned the opening of the Coplon investigation, the period following her arrest, and even the early phases of her first trial in Washington—the FBI maintained four separate wiretaps pertaining to the case: a tap on Coplon's home telephone, a tap on Coplon's office telephone, a tap on the home telephone of Coplon's parents, and a tap on Gubitchev's home telephone.[44] Official documents also revealed that investigators routinely monitored a microphone hidden in Coplon's desk at the Justice Department.[45] A total of ninety-six federal agents in Washington and New York were involved in the wiretap operation, all with the attorney general's approval.[46]

The FBI's surveillance activities in the Coplon case were unusual, in terms of their scope and their apparent disregard for bureaucratic protocol. The revelations of the wiretap operation quickly cast a pall of perjury over the proceedings in Washington. As the pretrial hearings wore on, the details of Bureau misconduct only worsened.

Testimony in the early days of January 1950 revealed that the FBI agents had listened to telephone conversations between Coplon and her father, who fell ill and died in the weeks following her arrest; to conversations between Coplon and Shapiro, who maintained a working relationship with the prosecution team despite his direct involvement in the events surrounding the case; and, perhaps worst of all, to conversations between Coplon and Palmer, a flagrant violation of attorney-client privilege that occurred no fewer than fourteen times in the days following her arrest.[47] Ironically, investigators were also party to a discussion between Palmer and Coplon's brother about a possible tap on the Coplon family telephone.[48] The web of "technical surveillance," as it was then called, was even revealed to extend to Coplon's friends and colleagues, giving the investigation the appearance of a dragnet fishing expedition.[49]

The final bombshell came on January 12.[50] A last-minute round of affidavits disclosed the existence of an internal FBI memorandum, issued on November 7 and initialed by Hoover himself, ordering the destruction of all records pertaining to the Coplon wiretaps. The "TIGER Memo," as the document came to be known—a reference to the Bureau's code name for the Coplon electronic surveillance operation—appeared to corroborate the defense's allegation that the FBI was working to conceal the source of its proof to avoid complications in the New York trial.[51] The prosecution countered that the destruction of wiretap records was part of an "automatic procedure" at the Bureau: unless documents and recordings had "pertinency and value," they were said to have been "destroyed . . . 60 days after they were made."[52] But testimony offered at the close of the pretrial hearings directly contradicted this claim. Arthur Avignone, the FBI agent who oversaw the TIGER operation, admitted that the request to destroy the records was "unusual . . . a deviation from the normal routine."[53] Another agent testified that no such routine existed.[54]

All told, the New York hearings exposed a disquieting pattern of electronic surveillance activity at the Bureau—"dirty business," in Oliver Wendell Holmes's famous phrase, conducted and concealed to bolster the prosecution's case, ensure a high-profile conviction, and possibly thwart judicial inquiry.[55] But Coplon's second trial pressed on. In accordance with the rules of evidence established by the *Nardone* decisions, the government was successful in persuading the court that the FBI had obtained the vital part of its proof against Coplon independent of the TIGER operation.[56] The wiretapped conversations, the violations of attorney-client privilege, the destruction of administrative records: all were deemed irrelevant to the proceedings, and Coplon was once again convicted and sent to prison. The difference, this time, was that it was almost impossible to ignore how she had arrived there.

The most prominent figure to step forward and cry foul was James Lawrence Fly, a high-profile New York attorney then serving as the director of the American Civil Liberties Union (ACLU). Tall, lanky, and notoriously quick-tempered, Fly had spent a tumultuous five-year stint (1939–1944) in Washington as chairman of the Federal Communications Commission. Much of his early work for the FCC involved forcing recalcitrant federal agencies to conform to the *Nardone* decisions. Over time, the frustrations of the job drove him to wage a public crusade against government wiretapping that occasionally bordered on zealotry.

Convinced that Section 605 was key to protecting the "integrity of our entire scheme of communications," as he put it in a March 1941 letter to President Roosevelt, Fly first drew the ire of the intelligence establishment by blocking a wartime law that would have granted the federal government emergency wiretapping authority.[57] He left Washington at the end of World War II, a decision in part necessitated by a FBI smear campaign against him.[58] But during the late 1940s, Fly so often pestered federal officials about Section 605 that many would joke that he took pains to live up to his name.

Above all else, the Coplon ordeal confirmed Fly's belief that lax enforcement of the Federal Communications Act had encouraged widespread wiretap abuse in Washington. On the eve of Coplon's second trial, Fly wrote an urgent letter to U.S. Attorney General J. Howard McGrath to express skepticism about the government's testimony and implore the Justice Department to probe the FBI's "extensive record of . . . illegal [wiretap] activity" in other cases.[59] When McGrath refused to dignify Fly's charges in writing, Fly unleashed a furious torrent of letters, petitions, and editorials intended to pressure the government to acknowledge its misdeeds. On January 5, 1950, Fly sent a short statement to the editors of the *World Telegram* and the *Baltimore Sun* denouncing the FBI's defiance of federal wiretap law in the Coplon case. "Where has there ever been swept up such a collection of . . . misleading conduct, and of repeated violations of the law and of Constitutional guaranties?," he wrote.[60] Two days later, the *Washington Post* ran an expanded version of Fly's editorial comment.[61] On the heels of the release of the TIGER memorandum, the article reached national syndication.

Fly's charges gained enough traction to force the Justice Department to respond. On January 9, 1950, McGrath released a press statement confirming what many in Washington had long suspected: that the FBI regularly engaged in wiretapping in the course of national security investigations.[62] In an unanticipated twist, however, McGrath's statement also alluded to an unspecified set of "policies condoning limited wiretapping," crafted by Roosevelt and followed by all three of McGrath's predecessors as attorney general.[63] Hoover offered a similarly fuzzy justification for the Bureau's wiretap operations in a letter to the ACLU two days later.[64] To diffuse a volatile situation, McGrath agreed to declassify Roosevelt's May 1940 memorandum. But when asked about the future of the federal government's secret wiretapping policy, he remained defiant: "In view of the emergency which still prevails and the necessity of protecting the national security, I can see no reason at the present time for any change."[65]

The government's half-hearted attempt to justify the excesses of the Coplon investigation only emboldened Fly's campaign. In late-January 1950, Fly worked with the ACLU to file an *amicus curiae* brief in Coplon's New York trial, arguing that the FBI's violations of federal wiretap law were grounds for the case's dismissal.[66] At the same time, he began quietly circulating a petition that called for an independent review of the FBI's "lawless conduct" in other cases.[67]

In early February 1950, Fly went so far as to file a lawsuit against three giants of the telecommunications industry—AT&T, the New York Telephone Company, and the Chesapeake & Potomac Telephone Company—in order to recoup excess charges billed to him during his years as a subscriber in Washington and New York. Because the FBI was in the regular business of listening to the conversations of ordinary Americans, he argued, telephone users were actually entitled to pay party-line rates.[68] (Perhaps unsurprisingly, Fly's lawsuit never went to trial.) Days later he published another screed against government wiretapping for *The New Republic*.[69] In draft form he titled the essay "The Police State."[70]

· · ·

If Fly's efforts to raise awareness of government wiretap abuse seem reckless, even quixotic, in the context of a McCarthy-era climate hostile to political dissent, it's worth noting that his charges against the FBI found a welcome audience in the early months of 1950. Hundreds of prominent figures included their names on Fly's petition to investigate the FBI's surveillance activities: attorneys in New York, lawmakers in Washington, and even former first lady Eleanor Roosevelt, the widow of the president whose secret wiretap policy was at the center of the scandal.[71] Even though many of Fly's petitioners recognized the need to continue prosecuting Coplon for espionage ("If the Coplon gal is guilty I want her punished," Fly underscored in a February 1950 letter), the Bureau's crimes seemed grave enough to take a stand in her name.[72]

Perhaps more importantly, Fly's public crusade against the FBI also paved the way for Coplon's bid to overturn her New York conviction, an effort that ended up having massive implications for the debate over government wiretapping in the early years of the Cold War. Filed in October 1950 by Albert Socolov, Samuel Neuberger, and Leonard Boudin, a new defense team hired to replace the unpredictable Archie Palmer, Coplon's appeal reiterated most of the complaints that Fly had lodged in the ACLU's *amicus curiae* brief eight months earlier. The main difference was that Coplon's attorneys chose to underscore not just the illegality of the FBI's wiretapping operation, as Fly had done so forcefully in the wake of the TIGER revelations, but the illegality of the initial arrest itself. After all, compounding the long list of enforcement transgressions in the case, the investigating agents appeared to have committed a cardinal Fourth Amendment violation: detaining and searching Coplon without a warrant.

The slight shift in emphasis proved decisive. In December 1950, New York Second Circuit judge Learned Hand made the controversial decision to invalidate the results of Coplon's second trial. Hand's 7,000-word opinion for the court—an extraordinary document, seldom consulted in the history of U.S. wiretap law—corrects a number of long-standing misconceptions about the nature of Coplon's judicial victory. As such, it's worth a closer look.

For Hand, the Coplon appeal boiled down to three separate allegations. First, that the FBI had arrested Coplon illegally. Second, that the government had unlawfully introduced wiretap evidence against Coplon obtained in violation of Section 605. And third, that the presiding judge had unfairly barred Coplon's defense team from consulting all of the wiretap records pertaining to the case. Contrary to the sensationalized accounts that would come to dominate coverage of the appeal in the months that followed, it was primarily on the merits of the first allegation that Hand decided to overturn the conviction. Despite conceding up front that Coplon's "guilt is plain," he used the opening paragraphs of the opinion to argue that the FBI's failure to secure a warrant for Coplon's arrest provided ample justification for vacating the results of the New York trial.[73]

The second and third allegations were more complicated. Hand acknowledged that the judge in the New York trial had erred in using claims of "national security" privilege to prevent the defense from consulting a portion of the available wiretap records. But he also went out of his way to refute the charge that the TIGER operation had, in and of itself, tainted the prosecution's case. As Hand pointed out, Coplon and her attorneys were "allowed to examine all the recordings . . . of the 'taps' taken at her home, and many of those taken at her office; and nothing in any of these could have constituted 'leads' to any of the evidence introduced at the trial."[74] And insofar as the contents of the FBI's phone taps were summarized in many of the pretrial affidavits, it stood to reason that Hoover hadn't ordered the destruction of records with the "sinister purpose" of burying the existence of the entire TIGER operation.[75]

In the end, Hand reasoned that the presiding judge was justified in his initial ruling that the issue of illegal wiretapping was irrelevant to the proceedings. In an attempt to drive home this crucial point, he concluded the opinion by distinguishing his decision from a referendum on federal wiretap policy: "Perhaps . . . the powers of the Bureau to arrest without warrant should be broadened; and perhaps it would be desirable to set limits . . . to the immunity from 'wiretapping' of those who are shown by independent evidence to be probably engaged in crime. All these are matters with which we have no power to deal, and on which we express no opinion; we take the law as we find it; under it the conviction cannot stand."[76]

It's hard to overstate the significance of Hand's reasoning, especially because it has become commonplace for historians to claim that Coplon's judicial victory was the result of an illegal wiretapping operation.[77] The TIGER revelations fanned the flames of an ongoing public debate over the FBI's use and abuse of electronic surveillance. But the messy history of federal wiretap policy—from Section 605 to the *Nardone* rulings, and from the *Nardone* rulings to the Roosevelt doctrine—was merely an ancillary consideration in the final decision on Coplon's appeal. To this point, Hand never once questioned the legality of the FBI's TIGER operation in his 7,000-word opinion on the case. The "intercept . . . and divulge" clause isn't mentioned; the *Nardone* decisions appear only in passing. Coplon walked free because of an illegal arrest, and because of a minor procedural oversight in the handling of wiretap records. She didn't walk free because of illegal wiretapping.

Notwithstanding these crucial facts, misinformation about the resolution of the Coplon trials spread like wildfire. Within weeks of Judge Hand's decision, lawmakers who favored government wiretapping began working to capitalize on the outrage surrounding the appeal. Their efforts amounted to little in terms of concrete policy change. But they solidified the myth—still widely regarded as historical fact—that Coplon's victory was the result of a loophole in federal wiretap policy.

New York congressman Kenneth Keating, for example, responded to the Coplon decision by calling for the legalization of all government wiretapping. According to Keating, Hand's decision was an "absurd and dangerous" precedent, one that both protected the rights of known spies and waylaid the government's efforts to combat espionage.[78] J. Edgar Hoover sent a misleading memorandum to Attorney General McGrath in 1951 similarly claiming that the outcome of the Coplon case had the potential to stymie the FBI's fight against communism. "Without legislation modifying the effect of Section 605 of the Communications Act of 1934," Hoover complained, "allegations will be made that this Bureau is engaging in illegal practices. . . . and prosecutions may be dismissed if such information is not produced."[79] McGrath shared Hoover's fears, and he soon turned to Keating to help propose a new federal law authorizing wiretapping in national security investigations.[80]

McGrath and Keating's efforts began to yield dividends in the spring of 1953, when four new wiretapping bills reached the floor of the House of Representatives for debate. Three of the proposals gave the FBI and other government agencies wiretap authority in espionage and national security cases, essentially codifying Roosevelt's executive policy of the 1940s.[81] The fourth, known as H.R. 5149, went in a more radical direction, recommending an outright repeal of Section 605 and offering a blank check to the FBI in its wiretapping operations.[82] Controversially, the proposal also contained a "retroactive" clause, designed to enable the government to reopen federal cases that had been foiled or shelved as a result of the *Nardone* rules.[83]

That Coplon was the target of the House's legislative push in 1953 was an open secret. In his opening statement in the hearings on the four proposals, Keating spun Coplon's appellate victory as the direct result of the indeterminacy of Section 605:

> The laws and court decisions affecting wiretapping have left the whole situation in a hopeless muddle. . . . This problem was brought out dramatically in the trial of Judith Coplon. Her attorneys turned the trial into a fiasco, and won out for her on appeal, to a large extent because this law [Section 605] is so vague and unsatisfactory. . . .
>
> A lot of people have raised a fuss about the dangers of a police state, and invasions of rights, which tend to cloud our thinking and obscure the issues. . . . It is both foolhardy and inexcusable to give [traitors and spies] the protective privileges afforded by our present laws. If we are to cope successfully with the menace they present, we must untie the hands of those charged with the responsibility of apprehending these vicious characters who infest our precious land.[84]

Despite the flimsiness of Keating's claims, almost every official who testified in the House's exploratory hearings on wiretapping that summer cited the outcome of the Coplon case as pretext for giving federal agencies wiretap authority. This was true even for representatives of political organizations whose opposition to electronic surveillance was otherwise entrenched. "The labor movement has long resisted any abridgement of the constitutional guarantees against self-incrimination," testified Andrew Biemiller, a member of the American Federation of Labor's National Legislative Committee. "However, we recognize compelling reasons of national security in the present world conflict with communism, which makes it desirable to permit wiretapping by authorized Government agents in cases involving espionage and permit the utilization of such evidence in court actions."[85] For Biemiller, as for so many others who supported the House's wiretapping proposals in the early 1950s, Coplon's hard-won freedom was one of those "compelling reasons."

A compromise between the House's four wiretapping bills passed by a vote of 377 to 10 in April 1953.[86] But it faced an uphill battle in a less hawkish Senate the following year. Coplon again loomed large over the public hearings on the proposed legislation, and the fate of Congress' "Wiretapping for National Security" resolution ultimately came to rest on conflicting accounts of the factors that had led to the success of her appeal. Early in the proceedings, McGrath's successor as U.S. attorney general, Herbert Brownell, took to the Senate floor to argue for the abolition of Section 605 and the *Nardone* rules. Like Keating and Hoover before him, Brownell invoked Coplon as a symbol of a dangerously imbalanced system, one in which known criminals could "escape punishment merely because they resorted to the telephone to carry out their treachery."[87] Yet this was a selective interpretation of history—and, predictably enough, it was James Lawrence Fly who put his neck on the line to set the record straight. Called to contribute to the Senate hearings three weeks later, Fly began his testimony by offering his usual condemnation of wiretapping's "dragnet character."[88] Yet the closing minutes of his testimony took an unexpected turn when he began assailing Brownell's "misleading statements" about Section 605's influence on the Coplon case. "The [Circuit] Court's decision . . . specifically denies that [the FBI] had the evidence to convict Judith Coplon through wiretapping, or any evidence to convict through wiretapping," Fly implored, waving a bound copy of Learned Hand's opinion before the committee holding the hearings. "The Attorney General [Brownell] found that; and it ill behooves him . . . that he made an improper contention."[89]

As Fly pointed out, Coplon had been "wired for sound, tapped and bugged in every way," but the fruits of the TIGER operation had neither established her guilt nor proved her innocence.[90] This was a troublesome fact, and it raised more troublesome questions: What's the use of wiretapping when wiretap evidence fails to matter in even the most clear-cut of cases? Why let the government tap at all?

No one could offer a persuasive answer, and the new "Wiretapping for National Security" bill ended up stalling in the Senate in the weeks that followed. As in the early months of World War II, Fly had played a role in derailing another key piece of federal legislation designed to authorize government wiretapping. The Justice Department didn't officially drop the charges against Coplon until 1967.[91]

Chapter 4

The Wiretapper's Nest

JUDITH COPLON'S CASE REVEALS a number of important truths about the history of wiretapping in the United States at midcentury. Americans have long remembered the late 1940s, 1950s, and 1960s as a period of intensified government surveillance—an era in which the imperatives of the Cold War led the nation's intelligence bureaucracy down dark and dangerous paths. This was, we're told, a period of political witch-hunts and civil liberties rollbacks. A period that set the stage for the worst excesses of what later came to be known as the American "surveillance state": bugs and wiretaps, informants and detentions, loyalty tests and letter openings, black-bag jobs and blackmail.

No one can deny the accuracy of that narrative—or its renewed relevance, given the realities we now confront under the watchful eye of the NSA. But above all else, the aftermath of the Coplon affair demonstrates that the surveillance state was by no means monolithic or even coherent at the moment of its apparent origin. In the protracted fight over Coplon's appellate victory, anti- and pro-wiretapping factions stepped forward to reveal what Washington insiders like James Lawrence Fly seem to have known all along: that when it came to the issue of electronic surveillance, every branch of the American federal government—executive, legislative, judicial—was operating in a state of turbulent contradiction. While some lawmakers campaigned in support of government wiretap authority during the late 1940s and 1950s, others worked tirelessly to stem the tide of reform. And even though the FBI capitalized on the law's ambiguities and continued the dirty business of tapping telephones in its hunt for spies and subversives, the justification for its investigative methods remained feeble as long as Section 605 and the *Nardone* decisions remained on the books.

What's more, resistance to government wiretapping was both vocal and widespread throughout the postwar period. As we have already seen, the efforts of Fly and the ACLU powerfully shaped the course of public deliberation on federal wiretap policy during the early Cold War years. The strength of their case against electronic surveillance is one reason Section 605 held sway as long as it did. By 1954, few missed the irony that the Justice Department's opportunistic push to authorize national security wiretapping was a push for the nation to embrace surveillance tactics that were known to be common among its communist enemies. As one U.S. congressman asked in an editorial in *Newsweek* on the "Hot Wire-Tapping Debate," "If we adopt totalitarian methods to combat the activities of totalitarian agents, we thereby compromise our democratic principles. . . . What has a nation profited if it gains security and loses its liberty?"[1] That a "Hot Wire-Tapping Debate" endured at all, at the height of the anti-communist craze, itself suggests that the state's power to snoop and spy was far from total in this period.

The incoherence of the federal government's position on electronic surveillance intensified when New York congressman Kenneth Keating's "Wiretapping for National Security" bill bounced back to the U.S. House of Representatives in the spring of 1955. Despite the Senate's decisive rejection of the resolution, many officials in Washington still believed that the need for government wiretapping was more pressing than ever. "Let us not delude ourselves any longer," U.S. Attorney General Herbert Brownell pleaded in the pages of the *Cornell Law Quarterly*. "We might just as well face up to the fact that the communists are subversives and conspirators working fanatically in the interests of a hostile foreign power. . . . If we are to be safe, the wires of America must cease being a protected communications system for the enemies of America."[2]

As dire as Brownell's vision must have seemed, a wiretap authorization bill never passed. Section 605 remained intact. When the House of Representatives reopened the floor to debate wiretap reform in March 1955, ranking members put forward two new legislative proposals, neither of which called for the repeal of the "intercept . . . and divulge" clause or for an exception to the *Nardone* rules in national security investigations. Instead, the House called—somewhat improbably—for the elimination of wiretapping in all of its forms, both sanctioned and unsanctioned. "The present situation is intolerable," New York congressman Emanuel Celler remarked at the start of the House's floor debate.

"Wiretapping is a vicious cancer. It requires heroic treatment, not palliation or plasters. It must be cut out."[3]

The swing in the House's deliberations was the result of a single event, a new scandal that temporarily sidelined the effort to legalize wiretapping in Washington. A month earlier, on February 11, 1955, an anonymous tip had led two New York Police Department (NYPD) detectives and two New York Telephone Company investigators to an apartment on the fourth floor of a residential building at 360 East 55th Street in midtown Manhattan. In the back bedroom of the unit, the group discovered a cache of stolen wiretapping equipment that turned out to have direct lines into six of New York City's largest telephone exchanges: PLaza 1, 3, and 5; MUrray Hill 8; ELdorado 5; and TEmpleton 8. The connections blanketed an area of Manhattan running from East 38th Street to East 96th Street, a swath of the city's most expensive real estate.

"There wasn't a single tap-free telephone on the east side of New York," professional wiretapper Bernard Spindel remarked of the arrangement.[4] (Spindel was in all likelihood the source of the anonymous tip.) News of the discovery made the front page of the *New York Times* a week later.[5]

The midtown Manhattan "wiretap nest," as the 55th Street listening post came to be known, remains one of the largest and most elaborate private eavesdropping operations ever uncovered in the United States. Subscribers whose phones were tapped at the time of the raid included a range of New York commercial interests, with assets both large and small: a modeling agency and an insurance company; an art gallery and a lead mining company; and perhaps most sensationally, two publicly traded pharmaceutical corporations with competing patent interests. (The two firms, Bristol-Myers, and E. R. Squibb, were at the time locked in a nasty legal battle over the commercial rights to the antibiotic tetracycline. Evidence later revealed that representatives from a third firm, Pfizer, had employed the wiretap nest to spy on both entities, paying more than $60,000 in cash for the service.)[6]

Yet contrary to the popular image of the phone tap as either a technology of state surveillance or a tool of corporate espionage, the vast majority of the lines ensnared in the 55th Street operation turned out to be owned by private individuals.[7] Some—like the burlesque artist Ann Corio, whose phone conversations were recorded in a dragnet search for incriminating information on prominent midtown residents—were the targets of blackmail. Others—like the New York socialite John Jacob Astor VI, who wanted someone to keep tabs on his wife—were involved in messy civil suits and divorce cases. By all accounts, the setup had the technical capacity to monitor as many as a hundred telephone lines at the same time. Between 50,000 and 100,000 individual subscribers were alleged to have been tapped over the course of fifteen months.[8]

Four men were eventually indicted in conjunction with the raid on the 55th Street wiretap nest: John G. Broady, an attorney and private investigator; Warren B. Shannon, a freelance electrical technician; and Walter Asmann and Carl R. Ruh, two rogue employees of the New York Telephone Company.[9] In the course of the ensuing criminal trial, Shannon, Asmann, and Ruh were all granted immunity in exchange for testifying against Broady, who emerged as the brainchild of the operation. Broady ended up receiving an unusually harsh prison sentence—four years, twice as long as the penalty suggested by New York's penal code—and at the close of the proceedings the presiding judge broke custom by publicly chastising the principals in the case: "In my many years as a judge, I have made it a rule not to excoriate defendants when imposing a prison sentence. However, the public interest requires some comment concerning this case. . . . Illegal wiretapping is a slimy activity, which directly and adversely affects our social and economic life. It cannot be condemned too strongly." The prosecuting attorney made a similar statement to the press: "If we are to safeguard the civil rights of our citizens, including the sacred right to privacy, we must take a stern view of the crimes for which this defendant [Broady] stands convicted."[10]

The gravity of the response to Broady's conviction only heightened the suspicion that there was more to the story than met the eye. A number of strange details from the early newspaper reports on the case remained unexplained at the end of the trial. The freelance electrician initially indicted for the crime, Warren Shannon, turned out to have been living in the apartment at East 55th Street for more than a year. Although he was at home with his wife when investigators arrived on February 11, no arrests were made, and no wiretapping devices were confiscated. When the NYPD returned to the scene a week later, much of the equipment used in the operation had disappeared.

Fishier still was the lag between the initial NYPD raid on February 11 and the *New York Times'* report on the story, published a week later, on February 18. News of the wiretap nest never appeared in the police blotter. And although the two telephone workers involved in the scandal were placed on "administrative leave" the day after the raid, officials at New

York Telephone waited until the *Times* story appeared to make an internal statement to employees on the matter.[11] As many observers noted in the wake of the scandal, the wiretap nest might well have escaped public notice had the New York City Anti-Crime Committee, an independent citizens' advocacy group, not received a last-minute tip that interested parties were in the process of initiating a cover-up. When the Anti-Crime Committee contacted the city's district attorney, Frank Hogan, on February 17, six days after the raid, Hogan found himself in the embarrassing position of having to admit that NYPD officials hadn't informed him of the situation. Indictments came only after the story made the front page of the *Times* the following morning.

"There was some reason, either best known to the phone company, or to the police department, why it was agreed immediately that no charges be preferred—despite the fact that right under their noses there were . . . clear-cut violations of the law," John O'Mara, director of the New York City Anti-Crime Committee testified, darkly, to a House Judiciary Subcommittee later that year.[12] What was it that the players involved—the NYPD, the New York Telephone Company, or both—didn't want the world to know?

For Charles Grutzner, the lead crime reporter at the *New York Times,* the 55th Street case was symptomatic of a broader "mushrooming of the wiretap practice" in the state of New York, a development that telephone company officials would have had obvious qualms about publicizing.[13] Other outlets floated a more elaborate theory, at once probable and paranoid. Considering the size and longevity of the 55th street operation (established, sources said, in December 1953), it seemed possible that NYPD officials were aware of its existence prior to the February 11 raid. Had dishonest cops agreed to look the other way in exchange for the ability to shake down local criminals via wiretap? Such an arrangement would certainly have been consistent with earlier grand jury inquiries into police corruption in NYPD gambling and vice investigations. The fact that the case involved New York Telephone employees only reinforced this conjecture. Bell system linemen were long rumored to have had a hand in the city's illegal wiretap trade.[14]

According to one observer, the attempted cover-up of the 55th Street scandal was the American public's first glimpse of "the 'Big A,' or The Alliance—a group made up of corrupt cops, telephone men, and expert illegal wiretappers in the private eye racket . . . [that] deals in outright blackmail, selling information, and . . . does much of its work for big businessmen who want to get the jump on a competitor." The midtown Manhattan tap nest was one of many private listening posts around the country ("Los Angeles, Chicago, Philadelphia, Detroit, Boston, Miami, and Washington all have wiretap centers comparable to the cozy set-up recently exposed in New York"), and the shadowy "Alliance" had a vested interest in keeping their workings under wraps.[15] The rumors of conspiracy and corruption now seem far-fetched. But at the time the story was plausible enough to occasion internal handwringing among Bell system providers. In a company bulletin dated March 9, 1955, New York Telephone assured nervous stakeholders that there was "no foundation" to national reports that there was a "corrupt alliance between telephone employees, the police, and illegal wire-tappers."[16] In the months that followed, several regional subsidiaries quietly instituted a policy to limit employee access to company equipment that could be used for unlawful purposes.[17]

. . .

Conspiracy or not, the 55th Street "wiretap nest" was itself an unsettling image. That four men could set up shop in a midtown apartment, commandeer an array of stolen electronic devices, and tap into thousands of lines servicing some of the most high-profile addresses in New York City—the story seemed to confirm creeping anxieties about the invasive reach of modern communications systems, their susceptibility to manipulation and control.

To quell further public uproar, the New York state legislature in Albany appointed Anthony P. Savarese, an assemblyman with connections to the New York City Anti-Crime Committee, to convene an emergency joint commission on the illegal interception of electronic communications. Charged with cutting through the "miasma of hearsay" surrounding the tap-nest scandal and recommending corrective legislation, Savarese began his work in late February 1955.[18] He filed a hotly anticipated preliminary report the following year. But the commission's official findings only served to bolster the sense that wiretapping was more entrenched and pervasive than the national debates over Section 605 and the Judith Coplon case had made it seem. While policymakers in Washington were busy squabbling over the meaning of an obscure federal law, a wide range of illegal electronic surveillance practices had invaded state jurisdictions like New York.

For the members of the Savarese Commission, the 55th Street tap nest was emblematic of that imbalance. Their aggressive approach to solving the state's wiretapping crisis seemed to provide, at least for a time, new models for national wiretap reform.

For the most part, the Savarese Commission's fact-finding mission revealed little that the American public didn't already know. After months of exhaustive investigative research and closed-session interviews, the Commission's findings merely corroborated prominent studies of the pervasiveness of illegal wiretapping that federal authorization advocates had previously dismissed as partisan or overblown. Where Savarese made his mark was in understanding why illegal wiretapping was pervasive in the first place—and in offering stopgap policy solutions to help prevent further growth of the wiretap trade.

According to the Commission's March 1956 report, the 55th Street scandal was the product of a host of developments that had made the New York telephone system "vulnerable to tapping": technological advances that made phone taps both easier to plant and harder to detect; corruption among state police officers and low-level employees in the telecommunications industry; and the unfettered expansion of the private investigation field in the years following World War II.[19] Yet the Commission's most enduring conclusion—echoed in later studies like Samuel Dash's influential 1959 report *The Eavesdroppers* (see Chapter 5)—was that any honest effort to curb illegal wiretapping in America would have to start at the state and municipal levels. Section 605, whatever its true meaning, wasn't the problem.

To be sure, the failings of New York state wiretap law were legion. A comprehensive court-order system had governed the phone tap protocols for New York law enforcement agencies since 1938. Although many policy experts considered the system a model for federal wiretap reform, the Savarese Commission discovered that judicial oversight was easy to circumvent, and existing criminal laws offered the state little room to prosecute police officers who chose to tap wires illegally.[20] The foundations of New York's laws against private wiretapping (i.e., wiretapping conducted by individuals acting outside of the state's "sovereign authority") were even shakier. The New York penal code prohibited any attempt to "cut, break . . . or make connection with any telegraph or telephone line, wire, cable, or instrument," a clear sign that wiretapping without the written permission of a state judge was a criminal offense.[21] The problem was that the statute was written in 1892. Six decades' worth of techno-

Figure 4.1. Advertisement for Argonne AR-20 Induction Coil Telephone Pickup (ca. 1950).
NMAH Trade Literature Collection, Smithsonian Institution Libraries.

logical advancements had all but rendered it obsolete—so much so, the
Commission noted, that almost every attempt to prosecute illegal wiretapping in the state of New York since 1892 had failed on technical
grounds.[22]

One major challenge to New York's 1892 wiretap law, frightful for
midcentury observers to behold, was the rise of what was known as *induction* wiretapping, a newfangled eavesdropping technique that didn't
require a physical connection to a telephone line. With the help of simple
magnetic circuits called "induction coils"—essentially spare radio parts,
available at most any hardware store—the induction method amounted,
somewhat paradoxically, to a wireless wiretap. In the words of one electronics manufacturer, "Simply slip [an induction coil] under the base of a
desk phone or lay on top of a ringer box of wall phones" and achieve
"optimum results" (see Figure 4.1).[23]

Tiny, cheap, and almost impossible to detect in action, induction coils
were in wide use in wiretapping operations of all sorts by the late 1930s,
and nowhere more so than in New York.[24] In part this was because the

state's penal code had explicitly defined illegal wiretapping as an unwarranted *physical* connection to a telephone line. As the Savarese Commission pointed out, echoing many commentators who nitpicked the 1892 law in the period, it was impossible to bring criminal charges against wiretappers caught using induction coils when they never so much as touched the phone company's equipment.

The 55th Street operation had relied on wiretapping techniques that were more primitive than induction. But the Savarese Commission went to great lengths to show that even simple wiretap installations were impossible to prevent and prosecute according to the letter of the law. For most of the twentieth century, both private surveillance experts and law enforcement officials mostly relied on what was known as the *direct* wiretap method. As its name suggests, this technique involved connecting directly to the circuitry of the telephone system, either by scraping away the insulation along the route of a phone line and appending an extension wire, or by attaching an amplifier and headphones to a telephone junction box, where multiple residential lines met and joined the system's main frame.

Direct wiretapping was tedious work that became both more and less difficult to carry out in the postwar years. More difficult, because installing a direct wiretap required the ability to pinpoint a subscriber's cable "appearance" location, identifying information that became increasingly hard to determine as the telephone system expanded its labyrinthine reach. By World War II, telecommunications providers had also wised up to security concerns, adding locks to the most obvious direct tap locations, such as basement junction boxes.[25] But direct wiretapping proved less difficult to carry out in this period for almost the exact same set of reasons. The sprawl of the telephone system also meant that communications hardware and infrastructure—and, more importantly, the employees who managed them on a daily basis—were impossible to oversee in their entirety. For the right price, the Savarese Commission discovered, anyone who wanted to find a line to tap could bribe a phone company employee for the relevant cable appearances, or even for direct access to the main frame, just as John Broady had when setting up the tap nest. What's more, most wiretappers knew how to locate lines from experience.

"90 per cent of all tappers today are old telephone company men," reported William J. Mellin, a retired government investigator who claimed to have tapped more than 15,000 lines during his forty years of work for the Internal Revenue Service.[26] Mellin's estimate would have the ring of hyperbole if the Savarese Commission hadn't come to the same conclusion.[27]

As we'll see in Chapter 5, the wiretap enforcement challenges that the Savarese Commission discovered in New York were generally consistent with nationwide trends. Most urban jurisdictions around the country dealt with similar problems, albeit on a smaller scale. What truly distinguished the Empire State in the 1950s—what made it America's "eavesdropping capital," in the words of the privacy law expert Alan Westin—was yet another loophole in state wiretap law, one that raised doubts as to whether the sort of wiretapping that the NYPD discovered at East 55th Street was even illegal at all.[28]

The loophole was the result of a curious court decision involving a Brooklyn businessman named Louis Appelbaum, who sued his wife for divorce in 1949. The evidence in the suit was partly based on telephone conversations that Appelbaum had permitted Robert La Borde, a notoriously prolific New York private investigator, to record on his home line. The presiding judge dismissed the divorce suit and went on to charge both Appelbaum and La Borde for violating the state's wiretapping law. Both men were convicted. But an appellate court reversed the ruling in 1950, reasoning that telephone subscribers maintained a "paramount right" to tap their own lines.[29]

The language of the appellate court's opinion in *People v. Appelbaum* (1950) was unambiguous in its support for what would become known as "one-party consent" eavesdropping: "When a subscriber consents to the use of his line by his employee or by a member of his household, or by his wife, there is a condition implied that the telephone will not be used to the detriment of the subscriber's business, household, or marital status. . . . In such situations, the subscriber . . . may have his own line tapped or otherwise checked so that his business may not be damaged, his household relations impaired, or his marital status disrupted."[30] For a resident of New York in the early 1950s—a man, most likely, because the gendered language of the ruling perversely implied that men had more claim on subscriber's rights than women—it was entirely legal, under *Appelbaum*, to record any conversation made on your home telephone. It was also entirely legal to hire someone else to do it for you.

The Savarese Commission spent most of its investigative energy working to understand the effects of the *Appelbaum* decision, eventually coming to the conclusion that it had encouraged a "lively, active, and lucrative" private eavesdropping industry throughout New York State.[31]

According to the Commission's March 1956 report, the case had thrown into confusion what was left of New York's 1892 wiretap law. It had also created a growing market for an urban professional whose doings had long preoccupied studies of electronic surveillance nationwide: the wiretapper-for-hire—or, more colloquially, "private ear." These were men (again: almost all were men) with a uniquely modern expertise. They knew how to tap any telephone, and they knew how to locate any telephone that was tapped. The tools of their trade were cheap, easy to use, and virtually impossible to detect in action. *Appelbaum* gave them license to bring their work, long maligned as dirty and disreputable, out into the open.

After 1950, in the words of the Savarese Commission, New York private ears were "immune practitioners in a nonhazardous occupation."[32] They went about their business as freely as plumbers, housepainters, and insurance salesmen.

Reliable facts and figures about the private eavesdropping industry that prospered under *Appelbaum* are difficult to find. The Savarese Commission conducted months of closed-session interviews to create a thumbnail sketch of the men who were offering freelance wiretapping services around the state of New York, and from its rough outlines emerges a picture of the sort of work that fed the growth of their profession in the postwar period. Most were either proficient in electronics early on, tapping their first lines by the age of twelve or thirteen, or had received special technical training while serving in the military. Most had gone on to find paying jobs in telecommunications, law enforcement, or freelance private investigation, three professional fields that expanded dramatically after World War II. And in the course of their regular duties, most had the opportunity to discover that telephone lines were easy and lucrative to tap—easy and lucrative enough, in any event, to turn wiretapping into a dedicated career, despite the risks that occasionally came with it. In 1955, the year of the 55th Street scandal, private wiretapping contractors were reported to net as much as $250 per day in Brooklyn and Manhattan.[33] The jobs with the most legal exposure commanded the highest rates.

The New York City Anti-Crime Committee, which conducted its own study of illegal wiretapping in 1955, estimated that there were no more than a dozen private ears openly offering wiretap services in New York City.[34] But the Savarese Commission found that the numbers were in fact much larger statewide, particularly given that uninitiated private investigators could dabble in phone tapping with lucrative returns.[35] The biggest names in the profession—Robert La Borde, John Broady, Bernard Spindel—tended to make their money monitoring telephone lines for New York businesses. Many more found work in the domestic sphere, helping to litigate civil and marital disputes.

The Savarese Commission discovered that divorce wiretapping was far and away the most common job for private eavesdropping specialists in the 1950s. Because New York divorce laws were "adversarial," requiring one party to show fault in the other before the state could terminate a union, wiretap recordings that captured evidence of infidelity could have a dramatic effect on the outcome of individual cases.[36] This was why John Jacob Astor VI had turned to John Broady—Astor believed that a wiretap would prove that his wife was having an affair with another man.[37] The Savarese Commission found the arrangement to be surprisingly common. New York's private ears tapped more lines to monitor cheating spouses than their counterparts in law enforcement did to gather criminal evidence.

Seen in this context, the 55th Street tap nest wasn't a one-off. Instead it was the natural extension of a professional activity that followed the laws of the market like any other. In the wiretap nest, Broady had realized the dream of most any entrepreneur: he had centralized the means of production, and had hired a dedicated staff to carry out the bulk of the labor for him. Private wiretapping operations never again achieved the size and scope of the 55th Street tap nest. But, as we'll see in Chapter 6, the phone tap trade only strengthened its economic foothold as electronic eavesdropping technologies advanced in the late 1950s and early 1960s. Divorce cases remained the primary driver of the market.

The Savarese Commission's report would inaugurate a new day for wiretapping in the Empire State—or so it seemed on the surface. In July 1957, after more than two years of legislative wrangling, policymakers in Albany added an amendment to the New York penal code that expanded the state's definition of illegal eavesdropping to include both direct and induction wiretapping, and levied hefty fines on phone companies that failed to report violations of the new law.[38] The amendment also closed the *Appelbaum* loophole, prohibiting one-party consent eavesdropping and barring the use of wiretap recordings or transcripts in civil court proceedings. But when the Savarese Commission recommended tightening oversight of law enforcement wiretapping, police officials pushed back, and lobbyists in Albany eventually pressured the legislature to keep the state's court-order system intact. The resulting compromise seemed to place New York law enforcement beyond the reach of reform.

"This new legislation makes great progress toward the control of private eavesdropping; it makes little progress in the control of eavesdropping by law enforcement officers," Savarese warned in his foreword to the Commission's final report.[39] A decade later the U.S. Supreme Court's landmark ruling in *Berger v. New York* (1967) would bring further scrutiny to New York's court-order system of police wiretap oversight. The consequences were explosive.

The legacy of the 55th Street scandal in New York was thus as mixed as that of the Coplon scandal in Washington. By the end of the decade, it seemed as though both everything and nothing had changed. When Congress held exploratory hearings on "Wiretapping, Eavesdropping, and the Bill of Rights" in the winter of 1959, ranking members of the Senate Subcommittee on Constitutional Rights wrote to Wellington Powell, New York Telephone's vice president of operations, to testify about the outcome of the wiretap nest case. In an official letter later introduced into the congressional record, Powell expressed optimism about the success of the Savarese Commission's effort to curb illegal wiretapping in New York.

"The new laws have strengthened privacy of communications by providing new sanctions and by eliminating loopholes and administrative difficulties under old laws," he reported. To bolster the new legal regime, New York Telephone had also "added more specially trained personnel to [its] special agents' forces" and intensified "indoctrination and supervision concerning security practices."[40]

But between the lines, Powell's letter offered an ominous set of statistics that underscored just how unworkable the twin ideals of privacy and security were in the field of telecommunications. In Manhattan alone, the New York Telephone Company managed 75,000 terminal boxes. Those 75,000 boxes connected to more than 4,000 miles of cable, and those 4,000 miles of cable contained more than 3 million miles of telephone wire. The entire New York Telephone System serviced an estimated 7,900,000 handsets. In a communications network so unmanageably vast, preventing an isolated illegal act was nothing less than a Sisyphean task. Neither the state of New York nor the telephone company that served it could guarantee a future without wiretaps, much less a future without wiretap nests. "Exposure is considerable," Powell warned.[41]

． ． ．

Section 605 of the Federal Communications Act remained on the books, unrevised, until 1968, when Congress finally moved to bring order to the nation's wiretapping laws. What, in the end, was the statute intended to accomplish? What did the bothersome "and" in "intercept . . . and divulge" really mean?

The answers remain elusive. Scholars who study electronic surveillance today disagree about the meaning of Section 605. Some claim that the law was intended as a wiretapping ban, arguing that the language of the "intercept . . . and divulge" clause implied that both police officials and private citizens were prohibited from tapping electronic communications.[42] (If Section 605 had actually permitted government wiretapping, why would officials as powerful as Roosevelt and Hoover have gone to such extraordinary lengths to get the practice legalized?) Others continue to claim that the law was merely intended as a rule of evidence, arguing that the Supreme Court deliberately sidestepped the issue of "interception" in order to avoid interfering with the work of federal law enforcement.[43] (If Section 605 had in fact prohibited wiretapping on some level, wouldn't the Court have simply come out and said so in one of the two *Nardone* opinions?) The astonishing thing is how frequently both sides of the argument retread the terrain of the old debates. At the very least, the lack of consensus in the scholarly literature on the subject should help us sympathize with the nation's predicament at midcentury. If the meaning of the law ultimately depends on the partisan ambitions of the people who wield it, finding a common language is well-nigh impossible. Even—or perhaps especially—with words as simple as *and*.

Perhaps the final lesson of Section 605 is that the meaning of the law doesn't matter at all?

Charles Einstein was one observer who thought so, albeit for slightly different reasons. A modestly successful writer of pulp fiction in the early 1950s who went on to make his name as a sports journalist, Einstein, like many Americans of the day, followed the twists and turns of the Coplon and Broady scandals as they unfolded in the pages of the nation's leading newspapers. In the 55th Street tap nest affair Einstein saw the beginnings of a novel, a "behind-the-scenes, headlines-fresh story of the 20th Century's latest and lowest crime."[44] He went to work turning reality into fiction almost as soon as Broady went to trial. The 25-cent paperback that resulted—*Wiretap!* (1955)—remains one of the best-selling works of fiction about eavesdropping ever published in the United States. The book's plot captures much of the national mood surrounding the prospect of electronic surveillance reform at midcentury.

Wiretap! opens with the assassination of a crooked judge in a fictional northeastern town called Aimerly, a city with a rising rate of crime and a thriving phone tap trade. Citizens of Aimerly believe the murder to be the work of "The Syndicate," a racketeering organization run by a ruthless Irish crime boss named Andy Fennell. But when the trail leading to the judge's killer goes cold, police in Aimerly are confronted with a series of strange occurrences that distract the public's attention from the ongoing murder investigation.

The most important of these comes in the opening paragraphs of the novel's second chapter, when an Aimerly detective stumbles onto a wiretap nest clearly meant to resemble the one found by real-life authorities months earlier on the fourth floor of 360 East 55th Street:

> Alf Hazlitt stood there, young and new in his policeman's uniform; but after a time, assuring himself all the while it was being done in the course of duty, he opened the door to the apartment and looked in.
>
> At first, what he saw did not register. To the right, and just under the casement window of the large room in which he stood, was a large telephone switchboard, complete with jacks, toggles, and headsets. Against the far wall, five tape recorders sat side by side on the floor; four of the five were closed, but the fifth still had the cover up, the tape wide and thick on the ready spool. While closed, the other recorders in their suitcase-type equipment were not locked; two of the headsets hung by their wire alongside the switchboard panel, and the jacks lay in wiry, rubbery disarray at the base of the switchboard as if they had been pulled all too hurriedly from their connections.[45]

Einstein's readers wouldn't have missed the allusion to the 55th Street affair. But for good measure he has the novel's protagonist, a hardboiled crime investigator named Sam Murray, throw in a direct reference a few chapters later. "Same thing they had in New York," says Sam after learning of the raid on the apartment in Aimerly. "Tap nest. That's what the papers called it."[46] If the book's basis in the wiretapping scandals of the early 1950s wasn't already apparent, Einstein later reveals that one of the primary targets of the fictional tap nest is "Judith Chasen," a local lounge singer whose name comes suspiciously close to that of the era's most well-known victim of an illegal wiretap operation.

A few days later, Sam arrives in town to investigate wiretapping on behalf of the state's Anti-Crime Committee, whose members—as in New York at the time—want to pressure the state legislature to enact new wiretap policies. Sam's first point of contact is Harry Millburn, a private eye known on both sides of the law for his electronic expertise. Harry proceeds to give Sam the inside dope on wiretapping in the city. Most of the early chapters in the book consist of Harry explaining the nuts and bolts of electronic surveillance operations, at times in torturous technical detail. "Don't believe what you read in the papers all the time, crime-buster," Harry tells Sam at one point. "You can't tell whether or not somebody's listening in on your phone. Period."[47]

Following Harry's advice, Sam traces the wiretap nest back to crime boss Andy Fennell. Along the way he discovers that the entire city of Aimerly is "wired for sound," as the promotional copy on the book's lurid front cover describes it (see Figure 4.2). The murdered judge has been signing illegal wiretap orders for the city's district attorney. The district attorney has been working with Harry to eavesdrop on The Syndicate. And Harry has been working for Fennell all along, tapping the phones of local politicians and using the recordings as blackmail. Everything comes to a head when Sam discovers that he's also the target of an electronic surveillance sting. Sam is ultimately forced to choose whether to expose Fennell's stranglehold on the city or protect his own professional reputation.

The final chapters of *Wiretap!* wrap up the novel's murder mystery, but Einstein elects to leave Sam's inquiry into the city's "wiretapping situation" unresolved. He reserves the most memorable lines of the book for crime boss Andy Fennell, who in a climactic scene warns Sam about the futility of his fictional home state's efforts to curb illegal wiretapping:

> You and I know how the story goes. . . . We do our work, each of us in his own way, and we always can tap a phone if we feel like it. The ability to do it is always there. When your state legislature investigation gets through with wiretapping, it'll still be there. You know why? Because legislation can't enforce in the business of wire-tapping. Science is always five giant steps ahead of the law. When they invented the automobile, they also invented the getaway car. When they invented the phone, they invented the tap.[48]

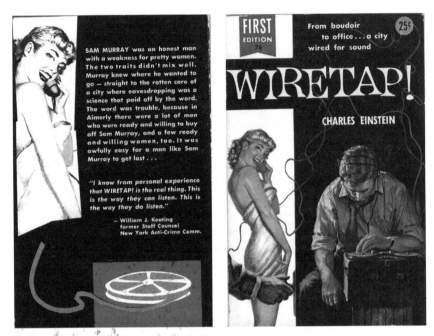

Figure 4.2. Jacket cover for Charles Einstein's pulp novel *Wiretap!* (1955). Note that the promotional copy on the back cover is offered by William J. Keating, the staff counsel of the New York City Anti-Crime Committee. Reproduced from the author's collection.

When they invented the phone, they invented the tap. For Einstein, electronic surveillance seems to have represented an inevitable byproduct of modern communications advances. Perhaps this is why the novel ends as unsatisfyingly as it does. In the immediate aftermath of the Coplon and 55th Street scandals, a solution to the growing American wiretapping epidemic didn't seem possible. Wiretaps were simply part of the wires.

The ambiguous ending of *Wiretap!* seems to capture many of the ironies of the age of Section 605: the age of *Nardone v. United States*, the Judith Coplon affair, and the 55th Street wiretap nest scandal. The "intercept . . . and divulge" provision might have been intended to prohibit wiretapping, or it might have been intended to prohibit the disclosure of wiretapped information. But the reality of the postwar period—a reality that Einstein no doubt understood—is that the law did neither in practice. In the wake of the Coplon case, policymakers in Washington argued fiercely over the fate of the *Nardone* rules and the Roosevelt doctrine. But federal agencies wouldn't begin to face political consequences for the abuse of wiretaps in national security investigations until the 1970s. In the wake of the 55th Street controversy, state and municipal governments around the country likewise passed a flurry of wiretap reforms, many of which sought to prohibit the private use of electronic surveillance equipment. But at least in New York, the sense among those who knew best was that aggressive policy measures amounted to little more than sound and fury.

"You can't legislate . . . against illegal wiretapping," warned New York District Attorney Edward Silver, echoing the sentiments that close Einstein's novel. "They did it before there were statutes and they will do it regardless of what you do."[49] On the other side of the law, private ears like Bernard Spindel offered equally worrisome predictions about the spread of the wiretap trade in the face of new policies: "Never before have so many people been willing to pay so much to find out what others are thinking and doing. Never before have we been so capable of accomplishing these desires. Whatever legislation may be enacted . . . is already many years too late."[50] Futility was the order of the day. "Most experts believe that no matter what legislation is enacted, the unhappy outlook as of now is that wiretapping is here to stay and will increase," *Newsweek* reported in an article on "The Busy Wiretappers" in the spring of 1955.[51]

The tumultuous decade that followed proved all of the predictions right.

Part II

THE BUG IN THE MARTINI OLIVE

Chapter 5

Eavesdroppers

IN JULY 1956, SIX MONTHS AFTER a New York judge sent John G. Broady to prison for his role in the 55th Street wiretap nest conspiracy, the Pennsylvania Bar Association Endowment (PBAE) commissioned a nationwide study of illegal investigative techniques among American law enforcement agencies. The main point of concern was wiretapping.[1]

At the time, Pennsylvania was one of several states around the country that lacked coherent electronic surveillance regulations. Members of the PBAE's board believed that a national fact-finding mission had the potential to help state lawmakers establish an effective policy for police agencies and private citizens alike. The man appointed to direct the study was Samuel Dash, a prominent Philadelphia prosecutor whose stint as the city's district attorney had given him a firsthand look at electronic surveillance abuses on both sides of the law. Two decades later, while serving as chief counsel of the Senate Watergate Committee, Dash would see many of those abuses come full circle.

After receiving a $50,000 grant from the Fund for the Republic, Dash hired a team of part-time researchers to help complete the study.[2] Their ranks included Richard Schwartz, a communications engineer at the University of Pennsylvania, and Robert Knowlton, a legal historian at Rutgers University. The group worked together for sixteen months in twelve different cities, consulting electronic surveillance experts with direct experience in the field: police officers, law enforcement officials, district attorneys, judges, phone company employees, electrical engineers, private investigators, even convicted criminals. "Every person we interviewed was an eavesdropper himself," Dash recalled, "a person who either employed it, authorized it, or actually engaged in installing the tap. . . . [W]e got our information right from the people who knew most about it."[3] Dash himself interviewed more than 300 individuals over the course of the investigation. Only a handful agreed to go on the record.

The result of Dash's efforts was *The Eavesdroppers,* a colossal 483-page report co-authored with Knowlton and Schwartz. Rutgers University Press published it as a stand-alone volume in 1959. *The Eavesdroppers* uncovered a wide range of privacy infringements by state authorities and private citizens—a much bigger story than the PBAE had anticipated. While law enforcement agencies were tapping lines in violation of state and federal statutes, phone companies were underreporting wiretap statistics to maintain public confidence in their services. While American businesses were stockpiling electronic equipment to spy on employees and gather competitive intelligence, private investigators were using frightening new tools to eavesdrop on wayward lovers and loose-lipped politicians.

On the eve of the book's release, Dash told members of Congress that the American public's refusal to confront the long history of eavesdropping had exacerbated the looming privacy crisis. "The context of this whole subject clearly shows that we are not dealing with a new problem," he testified to a Senate Subcommittee on Constitutional Rights in July 1959. "We are dealing with a problem that is at least 100 years old. At least from the very beginnings of electronic communication there were interceptions of electronic communications. Each generation seems to forget the problems of the past and considers this their own unique problem."[4]

The Eavesdroppers ended up having an outsized influence on national debate. According to the legal scholar Alan Westin, whose own watershed study *Privacy and Freedom* (1967) made extensive use of Dash's research, the American public's attitude toward wiretapping during the 1940s and 1950s had drifted ambivalently between naive "fascination" and "nervous awareness."[5] But the PBAE report brought things to a fever pitch. In the 1960s, journalists and academics began citing the book as irrefutable evidence of mounting wiretap abuse. The U.S. Supreme Court consulted it in two of the period's most important Fourth Amendment decisions.[6] And federal lawmakers entered portions of the finished text of *The Eavesdroppers* into the record of every major congressional hearing on communications privacy held between 1959 and 1968, after which new legislative reforms appeared to render the book obsolete.[7] By the end of the decade, commentators of all political stripes were frantically sounding the alarm of a full-blown "electronic listening invasion," largely on the basis of Dash's findings.[8] The godless march of technological progress finally appeared to have trampled the nation's bedrock values. No one seemed safe.

This chapter uses the publication of *The Eavesdroppers* to chart this momentous swing in popular perception during the late 1950s and early 1960s: from benign interest in wiretapping to anxiety over the collapse of the American right to privacy, from fascination with electronic eavesdropping technology to panic over its seemingly limitless reach. In the process, I argue for a more historically grounded understanding of wiretapping and electronic eavesdropping than scholarly orthodoxy has seemed to allow. The 1950s and 1960s have long been remembered as a period of intensified government surveillance in the United States. As a result, scholars in a variety of fields have exerted a great deal of energy uncovering the clandestine eavesdropping programs that the federal intelligence bureaucracy brought to bear against political dissidents at the height of the Cold War.[9] Yet a closer look at the history of wiretapping and electronic eavesdropping in the age of *The Eavesdroppers* reveals, as we saw in Chapter 4, a slightly different story—one in which ordinary Americans located privacy's death in a far more diverse range of phenomena than we'd probably recognize today: in the work of state and local law enforcement agencies, which wiretapped extensively in low-level criminal investigations; in the exploits of private investigators and eavesdropping specialists, who capitalized on new technological innovations to expand their industry's reach; and, perhaps most importantly, in the contradictions of state and federal lawmakers, who sent conflicting messages about the legitimacy of electronic surveillance activities that had dogged the nation's communications systems for more than a century. The cultural career of *The Eavesdroppers* charts an "unofficial" history of postwar eavesdropping that has long been obscured by academic and popular fascination with the clandestine doings of the FBI, the CIA, and other government agencies.

. . .

Like the Savarese Commission report that preceded it, *The Eavesdroppers* was a residual product of the period of confusion that followed the passage of the 1934 Federal Communications Act (FCA), which provided that "no person . . . shall intercept any communication and divulge or

publish the existence, contents, purport, effect, or meaning of such intercepted communication to any person."[10] As we saw in Chapters 3 and 4,
the ratification of the FCA sparked wide-ranging debate over the legality
and extent of wiretapping in the United States. The discussion played out
in newspaper editorials and legal journals, television programs and pulp
novels, courtroom testimonies and congressional hearings. More than
seventy-five articles about wiretapping appeared in mainstream American
magazines between 1934 and 1955 alone.[11]

Yet in 1956, when the PBAE commissioned its nationwide eavesdropping investigation, fact was still difficult to separate from fiction. "There
is in America considerable public awareness of wiretapping," Dash wrote
in the opening paragraphs of *The Eavesdroppers,*

> but most people are completely confused about it. . . . They have
> read, for instance, that wiretapping is rampant and wiretappers
> swarm all over the city, wildly tapping every phone within their
> reach. At the same time, they have read that the telephone com-
> pany says wiretapping does not exist, and people are only imag-
> ining that phones are tapped.
>
> We are given the impression at times that the police and district
> attorneys use wiretapping in all their investigations and have vir-
> tually placed a dragnet of wiretaps on every suspect's phone in the
> city. On the other hand, in states where police wiretapping is pro-
> hibited by law, police say they never use wiretapping. In states
> where police wiretapping is permitted by law, we read statements
> by the police that they hardly ever use it, and then only in the in-
> vestigation of serious and major crimes, affecting life or the secu-
> rity of the country.
>
> People are, naturally, disturbed by reports on the development
> of electronic devices that can spy on them no matter where they
> run to hide. They have read about a little electronic gadget which
> can be pointed at telephone booths to pick up both sides of tele-
> phone conversations, permitting the user of the device to listen
> in or record the conversations at his pleasure. There have been
> accounts of a super-sonic ray which can be beamed at a wall or
> window to retrieve voice-sound vibrations resulting from conver-
> sations taking place within a building. They have also read that
> this is poppycock.[12]

In short, although by the mid 1950s most Americans shared a vague sense that eavesdropping was a "national problem," few seemed to know what the problem actually entailed. Dash used the lingering uncertainty as his starting point for the PBAE investigation. *The Eavesdroppers* was the first concerted attempt to debunk the myths—the "poppycock," in Dash's words—and uncover the facts.

The published report that resulted from Dash's efforts was divided into three parts, each written by a different member of the PBAE research team. In part 1, Dash outlined the history of wiretapping in the United States and presented the results of his fact-finding mission. He limited his report to nine metropolitan jurisdictions that best represented the range of the study's findings: Baton Rouge, Boston, Chicago, Las Vegas, Los Angeles, New Orleans, New York City, Philadelphia, and San Francisco. In part 2, Richard Schwartz, the engineering expert in the group, explored the modern technologies that made wiretapping and electronic eavesdropping possible, from the tape recorder to the parabolic microphone. In part 3, Robert Knowlton surveyed relevant judicial decisions and legal statutes dating back to the English common-law prohibitions against eavesdropping first described in Blackstone's *Commentaries.*

The mix of voices between the three sections of *The Eavesdroppers* gave the study the appearance of offering divergent opinions on the wiretapping issue. But Dash stressed that he and his co-authors were united in offering an objective account, free of political biases and policy recommendations. Even so, the published product suggested something of the group's ideological leanings. Along with a sensational blurb about the book's contents ("the unknown story of wiretapping today—its victims, its practitioners, the technologies, and what the law says about it"), the front jacket of *The Eavesdroppers* featured the image of a keyhole-shaped silhouette, out of which peered the glowering red eye of a peeping Tom. The metaphor was mixed, but it didn't obscure the underlying message: the mechanisms that guaranteed the American right to privacy, from locks to laws, no longer offered protection.

For Dash's part, asserting that wiretapping was a violation of constitutional guarantees was something of an extraordinary political reversal. In the early 1950s, Dash had climbed the ranks of the Philadelphia District Attorney's Office, in part due to his willingness to use electronic surveillance in criminal investigations. In 1955, he penned an article for the *Dickinson Law Review* claiming that wiretapping wasn't "an unfair or coercive procedure," but instead was an "'ear witness' of crime—almost, if not as good as, an eye witness."[13] The argument against this line of reasoning was well established. But Dash took pains to point out that critics of law enforcement wiretapping usually made their case either by enumerating lurid local incidents of wiretap abuse, or by limiting their accounts of the problem to cities that were popularly regarded as wiretapping "hotbeds," such as Washington, D.C., and New York City.[14] In less publicized jurisdictions, according to Dash, the benefits of wiretapping far outweighed the costs. When he was called before the House Judiciary Committee in June 1955 to testify about his experience using electronic surveillance in Philadelphia, he expressed his opposition to the passage of restrictive federal wiretap laws in no uncertain terms: "We would be powerless in Philadelphia today to combat organized crime and rackets if we could not wiretap. . . . Without wiretapping, Philadelphia's crime problem would be indeed a serious one."[15]

Writing *The Eavesdroppers* changed Dash's opinion. Instead of focusing on sensational incidents and wiretapping hotbeds, the PBAE investigation took what amounted to a bird's-eye view, aggregating data from multiple jurisdictions to reveal the scope of wiretap use on a national scale. The approach gave the authors an opportunity to distinguish trends that had eluded both sides of the debate over electronic surveillance in the 1950s.

According to Dash, the source of the American "wiretapping crisis"—now identified as such—wasn't the vulnerability of the telephone system or the proliferation of unlicensed private surveillance experts (two of the most common explanations during the 1940s and 1950s, per New York's Savarese Commission report), but the contradictions inherent in American wiretap policy. As Dash explained in the opening section of *The Eavesdroppers,* by far the most widely cited of the book's three parts, wiretapping was in fact illegal on the federal level. Despite the fuzziness of the "intercept . . . and divulge" rule, the Supreme Court had plainly affirmed an anti-wiretapping interpretation of the Federal Communications Act in the *Nardone* decisions of the late 1930s. True, many government agencies had managed to circumvent the federal wiretap ban—that much was well-known. But Dash's research drew attention to the fact that wiretapping policies at the state level also remained chaotic and unenforceable. "Permissive" states like Louisiana and Massachusetts upheld relatively lax wiretapping laws; "prohibition" states like Illinois and California barred wiretapping outright. (California's wiretap restrictions were the nation's oldest and most stringent, dating back to the 1860s.) Some states had well-established systems for regulating the use of wiretaps by law enforcement agencies and private citizens, but a disconcerting number still remained "virgin jurisdictions," with no wiretapping laws on the books.

As Dash pointed out, the "confused and mixed up situation" between states and the federal government created some confounding legal paradoxes.[16] At a time when a federal official could be sentenced to prison for wiretapping, a state official could wiretap with impunity, even when openly committing a federal crime in doing so. Criminal prosecutors could similarly present recorded telephone conversations as evidence in any state court in the country, regardless of what federal statutes said—or didn't say—about the divulgence of wiretapped information. Licensed private detectives, many of whom specialized in electronic eavesdropping and tapped for hire in insurance investigations and divorce cases, occupied the same legal no-man's-land.

The contradictions between the laws of individual state jurisdictions—and between the laws of the states and the laws of the federal government—made it virtually impossible to prevent wiretap abuse. To borrow the language of one of the PBAE investigation's earliest collaborators, when it came to wiretapping there was a vast gulf between the "law-on-the-books" and the "law-in-action."[17] And as *The Eavesdroppers* demonstrated, the law-on-the-books barely mattered at all. Because the crime of wiretapping was exceedingly hard to detect and prosecute, the federal government rarely enforced its wiretapping ban: only three individuals had been brought to federal trial for violating the "intercept . . . and divulge" rule since the passage of the Federal Communications Act in 1934.[18] The enforcement of state wiretapping laws was just as rare. The upshot was that police officials and private investigators resorted to electronic surveillance without fear of reprisal, regardless of whether they did the dirty business of wiretapping in "permissive," "prohibition," or "virgin" jurisdictions.

"There is one point on which all states seem at least tacitly to agree," U.S. Senator Edward V. Long later wrote in *The Intruders* (1966), an electronic surveillance exposé directly inspired by Dash's research. "Regardless of the legal situation, little is done to prevent tapping. . . . [N]ot only is there legal confusion on both state and federal levels, but there is intentional disregard of the law by government agents, and failure to meet enforcement responsibilities under the law."[19]

In Boston and New Orleans, two of the most laissez-faire districts in the country, Dash discovered that police tapped extensively in criminal investigations but seldom reported doing so in official documents. They also avoided using wiretap evidence in criminal trials. The code of silence was part of a gentleman's agreement with local telephone companies, who had a vital interest in preserving the public fiction of secure electronic communications. In New York City, where wiretapping was regulated by a well-developed court-order system, plainclothes police officers were planting as many as a hundred illegal taps a day, a number well in excess of official estimates. New York district attorneys were also hiding microphones in police interrogation rooms to gather information about criminal suspects.[20]

Similar problems of oversight plagued law enforcement agencies in Chicago and Los Angeles, two jurisdictions where wiretapping was prohibited by law. Dash revealed that the Chicago Police Department maintained an intelligence unit of more than forty officers whose sole duty was to tap wires, despite the best efforts of the Illinois State's Attorney's Office to discourage illegal eavesdropping. In Los Angeles, police agencies were routinely skirting California's stringent electronic surveillance laws by hiring freelance wiretapping experts—"private ears"—to assist with criminal investigations. The collusion between the two sides was an open secret. Dash was amused to discover that more than sixty private detectives in the Los Angeles area openly advertised illegal wiretapping services in the city's Yellow Pages (see Figure 5.1). Accompanying some of these entries were racy promotional illustrations that featured men with headphones listening to the telephone conversations of unwitting female callers.[21]

The incoherence of the nation's eavesdropping laws opened other avenues for wiretap abuse, as well. The fact that electronic surveillance experts operated in "a carefree manner . . . unafraid of law-enforcement interference" was well-known by the time the PBAE commissioned its study, particularly since the recorded telephone conversation had emerged as an effective weapon in divorce cases and civil disputes during the 1940s and 1950s.[22] Dash, by contrast, discovered that one of the newest, and most lucrative, fields for the professional eavesdropper was corporate espionage. A Harvard Business School study reached the same conclusion in 1959: a growing number of American companies were hiring wiretapping specialists both to spy on competitors—and to spy on themselves.[23] Under the condition of anonymity, business executives around the country told Dash that they routinely tapped their office telephones if they sus-

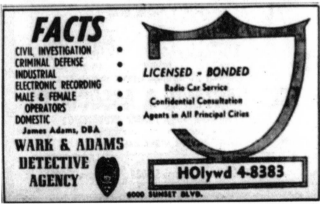

Figure 5.1a,b. Samuel Dash discovered that many licensed private detective agencies openly peddled "sound recording" or "electronic recording" services throughout the late 1950s. These two advertisements ran in the August 1958 Central Los Angeles Yellow Pages, despite the fact that wiretapping and electronic eavesdropping were illegal in the state of California. City Directories of the United States, Library of Congress.

pected employees of stealing, or if they feared the influence of unions among their workers. Some admitted to planting recording devices in office restrooms. In an ironic twist, Dash discovered that telephone companies weren't averse to tapping their own lines for this same purpose.[24] Some of the nation's most skilled wiretappers turned out to have been telephone company linemen, who were illegally hiring out their services to law enforcement agencies and private investigators, thereby completing the sinister circle.[25]

The litany of excesses was shocking—taken together, they seemed to tell a much bigger story about the erosion of time-honored American values. One early reviewer of *The Eavesdroppers* wrote that Dash, Schwartz, and Knowlton's research depicted a "thoroughly unpleasant society."[26] Another lamented that "only America . . . could produce a book of this sort."[27]

Yet the details about carefree private ears and prying business executives weren't what led readers of *The Eavesdroppers* to make such sweeping pronouncements. Above all else, it was the book's portrayal of the police. "There is something disturbing . . . about the position that law enforcement officers should be permitted to violate the very law they are sworn to uphold. That is hypocrisy," wrote a reviewer in the *American Bar Association Journal*.[28] The idea of the police tapping wires in open defiance of state and federal laws gave many the impression that the nation was careening inexorably toward totalitarianism. According to *Commentary*, "When police install taps or bugs, they are in effect laying traps, and taking over some of the functions of a secret, or political, police. When they go so far as to . . . bug a conversation between a suspect and his lawyer in a prison cell or when they have to rely on eavesdropping informers, as is generally necessary in tapping cases, the health of the society is obviously jeopardized. It is the old question of power and corruption."[29] Missouri senator Thomas C. Hennings Jr., chair of the Senate Subcommittee on Constitutional Rights during the 1950s, likewise called *The Eavesdroppers* an "appalling" read, based on the book's allegations of secret police wiretapping.[30]

And yet, as damning as Dash's findings seem in hindsight, the public response to *The Eavesdroppers* was far from uniform. The book inspired as much support for wiretapping as outraged opposition, and the argument in defense of the practice similarly hinged on the charges that Dash had leveled against American law enforcement.

To understand why the electronic surveillance abuses detailed in *The Eavesdroppers* would have found some justification in the late 1950s and early 1960s—indeed, to understand why the PBAE investigation provoked debate at all—it's necessary to take a detour through another curious wiretapping case in New York City. It involved two brothers who were convicted of a crime they weren't suspected of having committed. Their case played an important role in stoking the fires of the controversy that ensued.

.   .   .

During the early 1950s, Salvatore and Angelo Benanti ran a small but lucrative drug organization out of the Reno Bar, a notorious mafia haunt on Manhattan's lower east side.[31] In March 1956, New York Police Department (NYPD) detectives filed an official request to tap the Reno's telephones in an effort to mount a viable state narcotics case against the Benantis. A judge granted the wiretap application in accordance with New York law, and the NYPD went to work. For a long month and a half that spring, the investigating officers picked up little in the way of useful information. But on the afternoon of May 10, 1956, a detective on the wire overheard Salvatore make an urgent call to an unknown recipient about the delivery of "11 pieces" later that evening. A drug deal seemed imminent. Around 7:00 p.m., officers saw Angelo leave the Reno and drive off in a light-green Chevrolet coupe. They followed the car for several blocks and pulled it over.

When the investigating officers searched Angelo Benanti's vehicle, they quickly located the "11 pieces" his brother had mentioned on the phone. The problem was that "pieces" in question were eleven five-gallon containers of alcohol—contraband that had nothing to do with the trafficking of narcotics.

With the prospect of a blown wiretap at hand, the officers arrested Angelo on the grounds that the containers in the car lacked the alcohol tax stamps required by federal law. This was a relatively minor charge, an awkward throwback to the bootlegging wiretaps of the 1920s. But the untaxed alcohol at least gave the NYPD the power to turn both Benantis over to the Federal Alcohol and Tobacco Tax Division of the U.S. Treasury Department and proceed with an indictment.

What started as a state narcotics investigation had accidentally gone federal. The turn of events put the prosecution in an uncomfortable position. The state of New York had an extensive court-order system in place for authorizing law enforcement wiretaps. Established in 1938, it was the nation's first. This meant that the incriminating information about Salvatore—the "11 pieces" conversation, tapped and recorded—was legal to use as evidence in a New York state court of law. But the case against the Benantis was no longer under New York's jurisdiction. On the federal level, per Section 605 of the FCA and the *Nardone* rulings, the interception and divulgence of telephonic communications was illegal. If federal prosecutors revealed the source of the information that had led to Angelo's arrest and tied the crime to Salvatore, the entire case against the Benantis would be jeopardized.

When Salvatore Benanti went to trial in August 1956, the prosecution's initial strategy was to keep the existence of the Reno Bar wiretap under wraps. (Angelo pled guilty in exchange for reduced jail time soon after his arrest.) But Salvatore's attorney, George Todaro, sensed that something was amiss as soon as the investigating officers were called to testify. What else but a wiretap could explain why the NYPD had arrested the Benanti brothers, known drug dealers, for an offense as inconsequential as the transportation of untaxed alcohol? Shouldn't the prosecution's evidence be dismissed as fruit of the poisonous tree?

The presiding judge didn't think so. The federal officers who handled the case had nothing to do with the planting of the Reno Bar wiretap, which was itself sanctioned by a court order. Accordingly, on October 9, 1956, Salvatore Benanti was convicted and sentenced to eighteen months in prison. Todaro appealed the decision. As he summarized the argument against Benanti's conviction, "Despite the warrant issued by the New York State court pursuant to a New York law, we have no alternative other than to hold that in tapping the wires, intercepting the communication made by the appellant [Salvatore Benanti] and divulging at the trial what they had overheard, the New York police officers violated the federal statute."[32] A Second Circuit judge affirmed the conviction in May 1957, spurring another round of appeals. The U.S. Supreme Court agreed to hear the case late that same year.

What happened next surprised many. On December 9, 1957, the Supreme Court unanimously ruled that the NYPD officers who originally pursued the Benanti investigation had violated Section 605 of the Federal Communications Act, despite the presence of a wiretap order signed by a New York State judge. Salvatore Benanti's conviction was overturned. The language of the Court's opinion in *Benanti v. United States* (1957) answered any questions as to whether the ruling was meant to clarify the long-standing uncertainty surrounding Section 605's relationship to state wiretap law. Writing for the Court, Chief Justice Earl Warren explained that the "plain words of the statute created a prohibition against any persons violating the integrity of a system of telephonic communication and . . . evidence obtained in violation of this prohibition may not be used to secure a federal conviction." Section 605 was thus an "express, absolute prohibition against the divulgence of wiretapping," and the court-ordered tap on the Reno Bar was merely "another example of the use of wiretapping that was so clearly condemned under other circumstances."[33]

The implication of the Court's reasoning was momentous: it was a federal crime to tap a wire and divulge its contents, regardless of what the states had to say about the matter. After two decades of turning a blind eye to the conflict between state and federal wiretap laws, the U.S. Supreme Court had finally signaled an end to law enforcement wiretap abuse in all of its forms.

As legal experts were quick to point out, a second ruling handed down later that same week—*Rathburn v. United States* (1957), a case involving police listening to phone conversations with the consent of one of the parties involved—partially mitigated the Court's anti-wiretapping stance. What's more, the language of the *Benanti* opinion itself did little to encourage federal authorities to begin prosecuting state police officials for violating Section 605.[34] Nevertheless, the decision had dramatic short-term consequences. Within weeks, New York State courts were forced to postpone several cases involving warranted wiretaps. Three other states with court-order systems—Maryland, Oregon, and Louisiana—began to reevaluate the legality of their policies.[35] The disruption was so pronounced that noted trial lawyer Edward Bennett Williams could draw a crass comparison between *Benanti*'s influence on state affairs and that of *Brown v. Board of Education* (1954).

"As you know, there is a statute here [in New York] which permits the interception of telephonic communications when authorized by warrants issued out of your courts," Williams wrote to a New York Civil Liberties Union staffer in advance of a public appearance. "I do not believe your state statute is valid. . . . I don't think New York can ignore the law of the land on wiretapping any more than Arkansas can ignore it on integration of the school system."[36]

New York State Supreme Court Justice Samuel H. Hofstadter went so far as to stop signing wiretap requests in light of the *Benanti* ruling. In an extended memorandum on the subject, Hofstadter declared that "when state officers indulge in wiretapping they are violating Federal law and subject themselves to Federal prosecution." According to Hofstadter, *Benanti* implied that the stain of illegitimacy went all the way up the chain, from the police officers who tapped the wires to the judges who signed the wiretap requests: "Clearly a judge may not lawfully set the wheels in motion toward the illegality by signing [a wiretap] order. . . . [T]he warrant itself partakes of the breach, willful or inadvertent, of the Federal law."[37] A *New York Times* columnist underscored the same point in the weeks following the decision: "Any state official who stands up and offers wiretap evidence at a trial will now be proclaiming the commission of a Federal crime, with whatever moral sanctions that implies."[38]

Historians of electronic surveillance law tend to overlook *Benanti v. United States,* leaping straight from the *Nardone* decisions of the late 1930s to the Supreme Court's landmark rulings in *Berger v. New York* (1967) and *Katz v. United States* (1967), the two cases that provide much of the precedent for the system of police wiretapping under which we live today.[39] When the *Benanti* ruling is remembered, if at all, it's typically as a small but necessary step toward the abolition of what was then known as the "silver platter" doctrine, a counterintuitive rule of law that gave federal authorities the freedom to use evidence seized illegally by state police.[40] But viewing *Benanti* solely from the perspective of doctrinal legal history obscures its political import. In its day the ruling was a bombshell, provoking backlash from liberals and conservatives alike.

On January 17, 1958, a month and a half after the Court's decision, six members of the U.S. Senate Rackets Committee proposed a bill exempting state law enforcement officials from the scope of the new wiretap ruling.[41] (Coincidentally, one of the two members of the committee who refused to sign off on the bill was a charismatic young senator from Massachusetts whose presidential administration would later become the target of a congressional eavesdropping inquiry.)[42] The exemption had little hope of passing in Congress. Instead, the proposal was part of a broader political campaign designed to convince the American public that *Benanti* was a dangerous blow to law enforcement. The idea was timeworn. As we have seen, attempts to portray restrictions on eavesdropping as harmful to public safety are almost as old as eavesdropping itself. But the wiretapping-as-necessary-evil argument became something of a noisy refrain in the months following the *Benanti* decision, and nowhere more so than in the state of New York, where the ruling appeared to have the most impact.

In December 1957 the Executive Committee of the New York State District Attorney's Association drafted an official resolution asserting that *Benanti* "deprives State prosecutors and other law enforcement officers of one of the most effective weapons in combating serious crimes and organized criminal activity."[43] One month later, the more liberal-minded staff at the *New York Times* ran an official editorial statement condemning the Supreme Court's move to curb state law enforcement wiretapping. According to the *Times,* although "no one who believes in the individual's right of privacy likes the idea of law-enforcement officers . . . secretly listening in on telephone conversations"—and although wiretapping usually elicits "strong moral revulsion" among most Americans—the *Benanti* decision had robbed police agencies of a "necessary weapon" in the war on crime.[44]

Edward S. Silver, an outspoken district attorney for King's County, New York, cast the ruling in a much darker light: "If, for any reason, the Senate or the Congress doesn't do something to correct the *Benanti* decision, we just will not be able to use wiretapping in our law enforcement, and it simply means that a lot of people are going to have carte blanche in their criminal operations. You will not be depriving me but the people who elected me to fight crime in our county. If I am compelled to hunt lions with a peashooter, so be it."[45] By 1960 Silver was already claiming that *Benanti* had "seriously crippled law enforcement."[46]

• • •

Benanti v. United States received little more than passing mention in the text of *The Eavesdroppers,* a curious fact to consider given that Dash, Schwartz, and Knowlton were still in the initial phases of their research when the Supreme Court handed down the ruling. In the book's final section, Knowlton briefly remarked that the decision had the potential to bring about a "long overdue appraisal of the relationship between the federal and state laws."[47] But the book's authors otherwise seemed reluctant to wade their way into a controversy still unfolding. Dash, for his part, ended up taking a much more measured stance on the ruling than his colleagues in New York. In 1959 he went on record supporting the decision as an important step in reconciling the contradictions between state and federal wiretapping laws. But he also predicted that the new judicial precedent was unlikely to have any lasting effects on the ground because the Court had stopped short of formulating its opinion as a rule of evidence.

"Even though the *Benanti* decision may make wiretapping illegal in a State," Dash reasoned, "if that State has a rule of evidence admitting illegally seized evidence in court, then wiretapping, though illegal, may still be the basis for a prosecution. . . . A police officer does commit a crime, a Federal crime, by wiretapping, but that evidence in certain states throughout the country may still be used."[48]

Notwithstanding the temporary stays on court order wiretaps in New York, state law enforcement agencies could, in this interpretation of the ruling, proceed with business as usual. *Benanti* thus seemed likely to go the way of the *Nardone* decisions, which were neutralized by loopholes soon after they were handed down in the late 1930s. The doomsday scenario of organized crime running amok was overblown.

As it turned out, *Benanti v. United States* influenced the practice of police eavesdropping in an altogether different way—an issue that we'll consider in Chapter 6. For the time being, what's crucial to highlight here is how rapidly the Supreme Court's 1957 decision fanned the flames of an ongoing debate over the place of wiretapping in American law enforcement. Dash found his research caught in the middle. Despite his attempt to minimize *Benanti*'s relevance to the abuses that *The Eavesdroppers* uncovered, the new precedent ended up casting a long shadow on the public reception of the PBAE study. In the wake of the Supreme Court's most controversial wiretap ruling in decades—a ruling that to many seemed to hamper the police's ability to combat sophisticated criminal organizations and national security threats—*The Eavesdroppers* came to look like something other than a neutral fact-finding investigation. A handful of reviewers lauded the book as a crucial first step in bringing order to the chaotic field of American wiretap policy (according to the *American Bar Association,* Dash deserved "special credit for having brought to this thorny subject intelligent thinking"), but a far more vocal contingent yoked the book to the *Benanti* decision, denouncing *The Eavesdroppers* as the latest in a series of attacks on American law enforcement.[49]

The attempt to discredit *The Eavesdroppers* in light of the *Benanti* ruling began before the study even appeared in print. Consider, for instance, Edward S. Silver's comparison of fighting organized crime after

Benanti to hunting lions "with a peashooter." The analogy was part of Silver's July 1959 testimony before the U.S. Senate Subcommittee on Constitutional Rights, and it came as a sudden digression in a much longer screed against Dash, who had testified earlier that same day. Silver wanted to cast "very serious doubt" on Dash's portrayal of law enforcement wiretapping. His criticisms insinuated that *The Eavesdroppers,* which was slated for publication two months later, ran the risk of stoking "fanciful and imaginary fears . . . of what is going to happen in our country if we permit a lawful wiretap."[50] In Silver's opinion, both Dash and the U.S. Supreme Court had wrongly collapsed the "distinction between law enforcement agencies who are fighting crime and private groups or persons who use wire tapping for nefarious purposes," thereby criminalizing a form of police work that had become increasingly necessary in the electronic age.[51] To solve this problem, Silver somewhat improbably proposed a moratorium on the use of the terms "wiretapping" and "eavesdropping" in public discourse about police surveillance. Both words seemed to have unfairly sinister connotations. "If we had a word or term that meant the 'scientific devices to combat crime' the very use of that term would make most people understand a lot more clearly what law enforcing people have in mind," Silver wrote a year later.[52] In titling *The Eavesdroppers* as he did, Dash appeared to be playing to popular misconceptions about the police's ability to listen in on the conversations of American citizens. In Silver's view, he was kicking the nation's law enforcement agencies when they were already down.

Frank S. Hogan, an outspoken district attorney in New York City, shared many of Silver's sentiments. Hogan was particularly stung by *The Eavesdroppers.* He had offered to cooperate with Dash in the early stages of the PBAE investigation, only to find the policies of his jurisdiction slammed in the book. Smarting from the personal slight, he defended himself by dismissing the final product as a work of fiction:

> It is unfortunate . . . that much of this discussion [about wiretapping] has been dominated by emotional considerations and fed by a wealth of irresponsible and inflammatory 'data.' Such recklessness, from purportedly objective sources, has served to harm rather than secure the interests of the public. . . . The authors of *The Eavesdroppers* have firmly aligned themselves with this damaging contingent of critics by painting, particularly concerning New York City, a picture bearing no factual resemblance of the true situation,

as known by reputable officials. They grind their axe upon a wheel of untruth and far-fetched speculation.[53]

For Hogan, *The Eavesdroppers* and the *Benanti* ruling were allied forces: both defended civil liberties at the expense of civic safety. The language he used to discredit Dash's study came directly out of the mounting criticism of the Supreme Court—to be sure, the "damaging contingent of critics" who were hurting the cause of law enforcement in New York could have just as easily referred to the nine justices who unanimously voted to overturn Salvatore Benanti's conviction.

Squabbles over the content of *The Eavesdroppers* continued into the early 1960s, and assessments of the book's findings eventually began to reflect the growing schism between anti- and pro-wiretapping camps in the ongoing national debate over electronic surveillance reform. When Yale Kamisar, a criminal procedure expert at the University of Minnesota, convened a symposium on the nation's "Wiretapping-Eavesdropping Problem" in the winter of 1959–1960, for example, he could only imagine Dash, Schwartz, and Knowlton's research as the starting point for the discussion.[54] Kamisar sensed that *The Eavesdroppers* had caused "not inconsiderable controversy" in the months following its publication. For the foreseeable future, the issue of wiretap abuse among law enforcement officials and private citizens would have to be "approached in defense or attack of the work of these men."[55]

Kamisar's symposium appeared in the April 1960 issue of the *Minnesota Law Review*. In the interest of political balance, it featured "pro-wiretapping" contributions from Edward S. Silver and the private investigator Harold Lipset, both of whom defended the right to eavesdrop using the latest electronic technologies. The representatives of the "anti-wiretapping" camp were Thomas C. Hennings, chair of the Senate Subcommittee on Constitutional Rights; Edward Bennett Williams, chair of the American Bar Association's Committee on Criminal Defense Procedures; and Kamisar himself. Clear rhetorical patterns emerged on both sides of the debate. Criticism of the book gave way to condemnation of *Benanti v. United States* and support for law enforcement tapping. Silver, for instance, began his contribution to the symposium by blasting Dash's study ("an innuendo-splattered thriller") and concluded with a gloss on the necessity of violating privacy in special cases to protect the public interest ("If you don't give the police every reasonable chance to do the job, you're only cutting off your nose to spite your face").[56] By contrast, defenses of *The Eavesdroppers* were tantamount to arguing for a ban on eavesdropping in all of its forms. The three articles that unequivocally praised the book all closed with strident calls for a national legislative crackdown.

By the early 1960s, then, *The Eavesdroppers* had become a bellwether for the course of wiretap reform in the United States. Whatever else Dash, Schwartz, and Knowlton's research revealed about the scope of the "wiretapping-eavesdropping problem," the book laid bare an unbridgeable divide in the public debate on the issue, projecting two different courses for the nation to take in the coming decade. There were those who believed that wiretapping and electronic eavesdropping should be permitted in special cases, particularly cases involving organized criminal activities and national security threats. And there were those who believed that wiretapping and electronic eavesdropping should be banned altogether. While few made the argument that the field should remain unregulated, the two camps shared little common ground. A legislative stalemate seemed likely. When federal lawmakers brought four different wiretap reform bills to Congress early in the spring of 1961, none went forward.[57]

State legislatures around the country had better luck at reaching consensus, but the new policies they crafted merely reinforced the same basic schism. On one end of the spectrum were the states that elected to capitalize on apparent loopholes in the *Benanti* ruling. Between 1957 and 1959, Maryland, Massachusetts, Nevada, and Oregon passed laws permitting the use of wiretap evidence in state courts, indirectly sanctioning state police wiretapping despite its newly firm status as a federal crime. On the other end of the spectrum were the states that followed the Supreme Court's cue in *Benanti*. California, Florida, Indiana, Illinois, and New Jersey all either passed new laws banning wiretapping at the end of the 1950s, or made moves to shore up old statutes that had the same effect. Pennsylvania—Dash's home state, the source of the PBAE investigation—ended up taking the latter approach.[58]

At the dawn of the new decade, the running tally of state wiretapping laws reflected the difficulty the nation would face in bringing its contradictory electronic surveillance policies into alignment. Six states had statutes authorizing law enforcement officers to tap telephones under various circumstances, while thirty-three states prohibited wiretapping outright. Eleven states still had nothing on the books.

Despite the apparent impasse, privacy law expert Alan Westin believed there was new cause for optimism. *The Eavesdroppers* had exposed too much to let the issue recede once again into the background of public consciousness. "Whichever approach emerges from the current debate as the dominant public reaction, the most satisfying aspect of the wire tapping revolt . . . is the proof that Americans value their constitutional privacy too highly to let it ebb away before an advancing technology or the forays of official and unofficial intruders," Westin remarked in "Wiretapping: The Quiet Revolution" (1960), a tellingly titled essay for *Commentary*. "This is a promising start to what must be the next step—a full-scale federal clean-up in the 1960s and extension of state control laws to fifty jurisdictions. When this is accomplished, the tools for protecting privacy in the electronic era will be at hand."[59]

Westin had a few more years to wait for that next step, perhaps longer than he might have expected. A more unsettling problem was on the horizon, a problem at once both big and small. True to form, the authors of *The Eavesdroppers* were among the first to see it looming in the distance.

Chapter 6

Tapping God's Telephone

EAVESDROPPING TECHNOLOGIES OF VARIOUS SORTS have been around for centuries. Prior to the invention of recorded sound, the vast majority of listening devices were extensions of the built environment. Perhaps nodding to the origins of the practice (listening under the *eaves* of someone else's home, where rain *drops* from the roof to the ground), early modern architects designed buildings with structural features that amplified private speech. The Jesuit polymath Athanasius Kircher devised cone-shaped ventilation ducts for palaces and courts that allowed the curious to overhear conversations. Catherine de' Medici is said to have installed similar structures in the Louvre to keep tabs on individuals who might have plotted against her. Architectural listening systems weren't always a product of intentional design. Domes in St. Paul's Cathedral in London and the U.S. Capitol building are inadvertent "whispering galleries" that enable people to hear conversations held on the other side of the room. Archaeologists have discovered acoustical arrangements like these dating back to 3000 BCE. Many were used for eavesdropping.[1]

The earliest electronic eavesdropping technologies functioned much like architectural listening systems. When installed in fixed locations—under floorboards and rugs, on walls and windows, inside desks and bookcases—devices like the Detectifone, a technological cousin to the more common Dictaphone, proved predictably effective (see Figure 6.1). According to a promotional pamphlet published in 1917, the Detectifone was "a super-sensitive device for collecting sound in any given place and transmitting it by a wire thru any given distance to the receiving end, at which point the person or persons listening are able to hear all that is said at the other end. . . . It hears everything, the slightest sound or whisper. . . . The result is the same as though you were present in the room where the conversation was being carried on."[2] Such devices were typically marketed

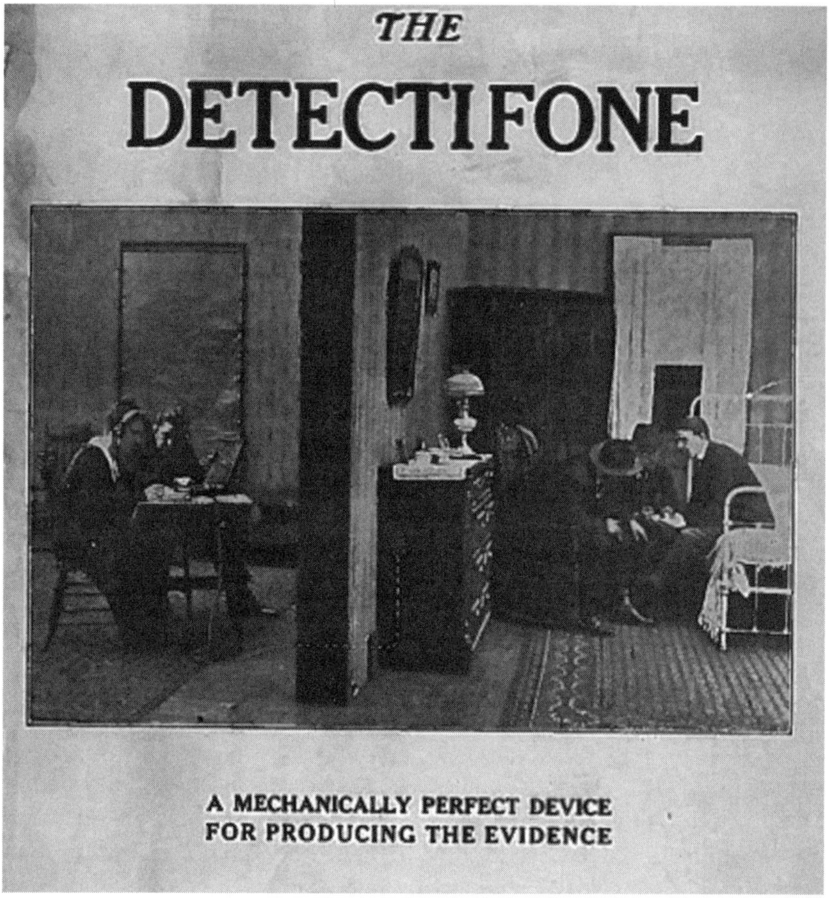

Figure 6.1a (above) and b (opposite). "The Detectifone: A Mechanically Perfect Device for Producing the Evidence" (1917). NMAH Trade Literature Collection, Smithsonian Institution Libraries.

At the left — Anderson System No. 24 — Leather case; size, eight inches by eleven inches by two inches; weight, three and one-half pounds.

Below — Anderson System No. 31 — Made in units of six, twelve, eighteen, twenty-four keys, etc., as high as ninety-six keys for use in *permanent* installations where a large number of transmitters are located in various parts of factories, stores, banks, etc. By raising the key it enables the observer to listen in at any part of the building. Equipped with socket to receive connections for either one or two double receiver head sets.

as investigative tools for private detectives and law enforcement agencies. But manufacturers also envisioned more pedestrian uses for the technology: verifying the loyalty of business associates, corroborating statements made under oath, even monitoring patients in hospitals and insane asylums.[3]

The devices that we now think of as "bugs" emerged much later. (In fact, the word *bug* didn't gain traction as a nickname for a concealed eavesdropping device until after World War II.)[4] During the late 1940s, electronic innovations made it possible for eavesdroppers to miniaturize listening technologies like the Detectifone. This made them easier to hide. It also freed them from the strictures of the built environment, dramatically expanding their reach. Reports of an American bugging epidemic began circulating in the early 1950s—first, as glimpses of the man-made miracle of electronics miniaturization began to appear in newspaper exposés, trade magazines, and Hollywood films, and later as congressional subcommittees revealed scandalous eavesdropping tools on the floor of the United States Senate. The numbers were impossible to substantiate, but by 1960 all accounts suggested that the bug had outstripped the wiretap as the professional eavesdropper's weapon of choice.[5] The electronic listening invasion had begun.

. . .

The middle section of *The Eavesdroppers,* written by the University of Pennsylvania engineer Richard Schwartz, was intended to account for this new development in the world of electronic surveillance. Brusquely titled "Eavesdropping: The Tools," Schwartz's chapter took stock of the miniaturized listening devices that professionals were using in the field. In the process, he told a more disconcerting story about ordinary technologies turned against the society that had created them. There were induction coils that allowed eavesdroppers to listen to telephone conversations without making physical contact with telephone wires. A special brand of conductive paint, invisible to the unaided eye, could redirect phone signals to outside lines. There was a new class of microphones engineered to be smaller than sugar cubes and thinner than postage stamps. These could be secreted away in surprising locations: wall sockets, picture frames, packs of cigarettes. They transformed everyday items into covert listening machines.[6]

Then there were the technologies of remote listening, futuristic gadgets that seemed to defy the laws of physics. Tiny radio transmitters embedded in briefcases or wristwatches could broadcast conversations to eavesdroppers lying in wait elsewhere. Directional microphones shaped like satellite dishes and shotguns could intercept conversations from thousands of yards away. Schwartz even reported on the development of an eavesdropping laser beam, long rumored to be on the open market. Unfortunately—or fortunately, depending on how you looked at the situation—this was the only device he discovered to be apocryphal.[7]

In public interviews Schwartz admitted that he knew next to nothing about bugging prior to joining the Pennsylvania Bar Association Endowment (PBAE) investigation.[8] By reputation, the field was prohibitively complex, an art known only to a specialized few. But his research quickly proved such narratives false. "Although there is a tendency to regard electrical and electronic eavesdropping as a kind of black magic," he explained, "the devices employed, no matter how dramatically described by private detectives to lawyers and the public, use no principles unknown to communications and electronic engineers."[9] The technologies at the center of the bugging revolution weren't the stuff of spy thrillers and science fiction novels. They were "available to anyone who wishes to get them," and innovations in the electronics industry were only making them easier and more effective to use.[10]

There were two main reasons electronic listening devices were on the rise during the late 1950s and early 1960s. The first was that bugging was still something of an uncharted legal frontier in postwar America. Although devices like the Detectifone had been on the market since the early 1900s, available to police officials and private citizens alike, it wasn't until the early 1940s that the U.S. Supreme Court had the chance to consider their legal implications. In *Goldman v. United States* (1942), FBI agents had attached a warrantless "detectaphone" (the spelling referred to a generalized eavesdropping device, not necessarily one made by the Detectifone brand) to a wall adjoining the office of an attorney suspected of fraud, using what they overheard to secure his conviction.[11] The Court upheld the verdict on the grounds that the government's use of a listening device wasn't a violation of the Fourth Amendment. The detectaphone hadn't physically trespassed onto Goldman's property, and the criminal evidence that the FBI "seized" through its use—i.e., oral speech—wasn't materially tangible. (The government hadn't intercepted and divulged any private information communicated by wire, so Goldman's predicament didn't appear to fall under the jurisdiction of the Federal Communications Act, either.)

The *Goldman* ruling appeared to signal that the use of concealed listening devices could continue unabated, at least in the hands of American law enforcement. The Court's subsequent decisions in *On Lee v. United States* (1952) and *Irvine v. California* (1954), similar cases involving concealed microphones, didn't help matters.[12] In 1959 Samuel Dash could still assert that electronic eavesdropping remained an "almost untouched area" in state and federal jurisprudence.[13] Edward Bennett Williams agreed at the start of the 1960s that bugging laws were "even more chaotic and outdated" than wiretapping laws.[14]

The years following the publication of *The Eavesdroppers,* which Dash later remembered as a "period of heightened awareness" of electronic listening devices, only seemed to muddy the waters.[15] In *Silverman v. United States* (1961), the Supreme Court was presented with a case not unlike *Goldman:* without a warrant, FBI investigators had inserted a microphone attached to a long metal rod—a so-called spike mike—into the baseboards of a suspect's home, using its acoustical properties to transform the residence's ductwork into a massive amplifying system. (Readers of Patricia Highsmith's classic 1952 novel *The Price of Salt,* which the filmmaker Todd Haynes adapted as *Carol* in 2015, will remember the "spike mike" as the listening device that Carol's estranged husband uses to confirm her romantic affair with the protagonist Therese.) What the FBI heard through the ducts enabled federal prosecutors to convict Silverman. This time, however, the Supreme Court unanimously ruled to overturn the verdict. The spike mike, as a physical extension of the investigating officers, had trespassed onto Silverman's private property, making the oral evidence against him—now distinguished as a "tangible thing"—illegally obtained.

Silverman was a victory for the Fourth Amendment—and a victory for *The Eavesdroppers,* which was given pride of place in the footnotes of the Court's opinion.[16] But the ruling notably balked on the more pressing question of the lawfulness of eavesdropping devices themselves. According to Justice Potter Stewart, who wrote for the Court, the details of the *Silverman* case didn't require consideration of the "frightening paraphernalia which the vaunted marvels of an electronic age may visit upon human society," much less the FBI's use of the spike mike.[17] All that mattered was that the investigating officers had physically intruded on Silverman's property. Bugging thus remained an open field. For police and private investigators whose eavesdropping work had been hindered by the *Benanti v. United States* decision four years earlier, the road was clearer than ever. The safest course of action—that is, the least illegal—was to replace wiretaps with bugs.[18] As the years wore on, supply followed demand.

And yet, as with this history of telephone tapping in this period, the law only tells us part of the story. The Supreme Court's earliest rulings on electronic eavesdropping—*Goldman, Irvine,* and *Silverman*—also happened to coincide with a flurry of technical innovations in the electronics industry. The ambiguity of the law made state and federal officials much less equipped to keep pace with the developments that ensued. This was the second reason for the bugging epidemic of the 1950s and 1960s: electronic eavesdropping technology had suddenly raced ahead, eluding the law's grasp and leaving eavesdroppers to operate without fear of reprisal.

According to the professional eavesdropper Bernard Spindel, after World War II the "modern world of gadgetry" had created bugging techniques that "def[ied] the imagination and also detection." The rate of innovation in the field could already, in 1955, be compared to "the rapidity in changes of ladies' fashions," which all but ensured that regulatory efforts would lag behind the times.[19] State and federal authorities agreed that there was little the government could do to stem the tide. "We've had some complaints about bugging, but we've never been able to catch anyone at it," Curtis B. Plummer, chief of the FCC's Bureau of Safety, admitted to the *Washington Post* in 1962. "We have reason to believe that most of the people who do this kind of work are thumbing their noses at us. . . . They just go ahead, knowing there's a small chance of getting caught."[20] (In hindsight, Plummer's remarks have a ring of prophecy to them: the FCC passed a resolution prohibiting the use of bugging equipment four years later, only to discover that the ban had no discernible effect.)[21] Reports in venues as diverse as *Time, Life, Business Week,* the *Chicago Tribune,* and *Popular Mechanics* came to the same conclusion.[22] Bugging had become too effective to regulate through ordinary channels.

Behind the rapid advances in electronic eavesdropping was a single technological innovation: the transistor. Pioneered by researchers at Bell Laboratories in the late 1940s, transistors provided the means to make electronic components smaller, enabling the development of a host of technological devices that helped to reshape postwar American society: the calculator, the portable radio, the hearing aid, and—perhaps most importantly—the integrated circuit and the personal computer. Scholars typically identify the transistor as the breakthrough that made the "information age" possible.[23] But there was an ominous side to the technology, often overlooked in historical accounts of its social applications. Transistors were easy to construct, and by the late 1950s they were cheap and easy to acquire. When electrical engineers and surveillance experts realized their potential, they ushered in what Harold Lipset later remembered as a period of "extreme miniaturization" in the field (see Figure 6.2).[24]

In *The Eavesdroppers*, Schwartz reported that transistorizing a bug halved its size without changing its overall manufacturing cost.[25] The resulting eavesdropping devices, some no bigger than the head of a matchstick, seemed nothing short of miraculous: bugged television sets, staplers, doorbells, and flower arrangements; bugged shirt buttons, tie clips, hat bands, and lighters; even bugged lipstick tubes and cavity fillings. As Alan Westin explained, these weren't "'Buck Rogers' developments, technically possible but still on the drawing boards." They were "already in use, and spreading across the nation with cancerous speed."[26]

All told, the combination of rapid technological innovation and crawling legal control yielded a situation that one federal official described as "total anarchy."[27] Following Dash and Schwartz's lead, lawmakers in Washington soon turned their attention to the bugging epidemic. The congressional hearings that ensued, led by Edward V. Long and the Senate Subcommittee on Administrative Practice and Procedure, mostly served to expand on the territory that *The Eavesdroppers* had already covered. But disturbing new details about the pervasiveness of electronic eavesdropping came to light—at first in pieces, and then seemingly all at once. In 1960, the U.S. ambassador to the United Nations disclosed that a listening device had been lodged inside the state seal of the American Embassy in Moscow for the better part of a decade.[28] News reports suggested that as many as one out of three divorce cases in major American cities involved a conversation intercepted by a hidden microphone, and as many as one out of five businesses had purchased top-of-the-line audio surveillance equipment to spy on competitors.[29] A torrent of books and articles on the electronic eavesdropping crisis appeared, some written by former professionals in the field. The titles suggested that the nation had at long last reached a point of no return: "Bug Thy Neighbor" (1964), *The Privacy Invaders* (1964), *The Naked Society* (1964), "The Big Snoop" (1966), *The Intruders* (1966), *The Electronic Invasion* (1967), *The Ominous Ear* (1968), *The Third Listener* (1969).[30] And in the midst of the mounting

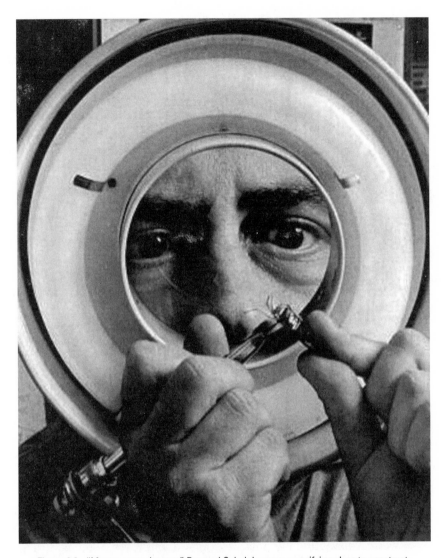

Figure 6.2. "Master eavesdropper" Bernard Spindel uses a magnifying glass to construct a miniature listening device. Arthur Schatz / The LIFE Picture Collection / Getty Images.

anxiety, a private detective with a flair for the dramatic appeared before Congress and pretended to sip a dry martini throughout his testimony. The olive in his glass contained a listening device, designed to record conversations at a range of up to forty feet. At the end of the proceedings, he played back his opening statement for rhetorical effect.[31]

. . .

Much of the information about bugging detailed in *The Eavesdroppers* came from two anonymous sources who perhaps knew more than anyone else about the subject. One was J. Arthur Vaus, a reformed mob wiretapper whose improbable conversion to Billy Graham's ministry captured the nation's attention in the early 1950s.[32] The other was Harold Lipset, a private investigator whose inventive use of electronic listening equipment made him a household name in the early 1960s.[33] (It was Lipset who performed the "bug in the martini olive" trick on the floor of the U.S. Senate in February 1965.) The two men followed similar paths to national notoriety. Both perfected the tools of their trade in the Army during World War II. Both began their careers doing freelance investigative work. And both ended up in the public eye during periods of heightened concern over threats to privacy.

Yet despite their similarities, Vaus and Lipset also seem to embody two different registers of the electronic surveillance debate at midcentury. During the 1950s, Vaus's bugging exploits were generally regarded as miraculous feats of human ingenuity, wonders of the electronic age. By contrast, when Lipset took center stage in the mid 1960s, Americans tended to view electronic eavesdropping in a more dystopian light, and his technological achievements accordingly seemed to portend the demise of the right to privacy. Taken together, their careers help illustrate the shift from wonder to alarm in popular discourse about eavesdropping during the postwar period.

J. Arthur Vaus, also known as "Big Jim" Vaus, grew up in Los Angeles, the son of a Baptist minister.[34] He exhibited an interest in electronics early on, engineering his first listening device in an attempt to play an elaborate prank on his older sister. When Vaus enlisted to serve in the military during World War II, his skills made him an ideal candidate for the U.S. Army Signal Corps. He was assigned to work on developing radar defense technologies. Despite an auspicious start, Vaus was eventually caught stealing military equipment and sentenced to ten years in an Army prison. An early release at the end of the war allowed him to return home and open a small electronics repair shop in Hollywood.

Business at the shop was slow. Vaus found more remunerative work when the Los Angeles Police Department (LAPD) approached him to assist on criminal cases on a part-time basis. As Vaus explained the arrangement, "to see and not be seen was a constant problem of the police," and his homespun line of bugging equipment, manufactured in the back of his Hollywood shop, provided a convenient—if illegal—investigative solution (see Figure 6.3).[35] "We had one gadget that made it possible for us to listen in to telephone conversations so that we didn't need to go anywhere near the telephone in question," Vaus recalled, looking back on the devices he built for the LAPD.

> We could sit back, on the other side of town, and hear all that was being said on that telephone—not only hear what was being said, but determine what telephone number was being dialed, without being near the phone and without any physical connection. We had another gadget that made it possible to take sound out of a room, through solid walls. From a listening post several miles away we could hear all that was being said. . . . We had what we called a "talking cane," where we couldn't get inside by any other means or method, for instance, in a hotel or office building where it was necessary to get conversation out. . . . By pressing the cane right up against the wall, the sound would come right through the cane, through a little wire up into our sleeve into a little pocket amplifier, and then into an earplug that permitted us to hear all that was being said in the room. Not only hear what was going on, but with another device that we carried we could transmit the same conversation to an automobile outside where recorders would make a permanent record of the conversation. Of course, if we couldn't gain interior access to the building, under the cover of darkness, we frequently operated the same device, taking this cane, which was also a telescoped affair, we could extend it some four or five stories in height, press it up to a windowpane, take the sound through the windowpane into the cane, down to the amplifier, and then again, out to the car to be transmitted there for recording. . . . It's always interesting what goes on when people don't know they're being listened to.[36]

Figure 6.3. "Automatic Impulse Indicator" (pen register) constructed by Jim Vaus, date unknown. Papers of J. Arthur Vaus, Archives of the Billy Graham Center, Wheaton College.

Vaus was the first to admit that his miniature microphones and "talking canes"—variations on the detectaphones and spike mikes used in the *Goldman* and *Silverman* cases—seemed "a little on the fantastic side" to the uninitiated.[37] But they were real, they were effective, and they enabled him to corner a growing market for private surveillance work in and around Hollywood during the late 1940s.

On top of his freelance electronic efforts, Vaus was hired in this period to tap the phones of businesses and record the conversations of state political candidates. He also accepted a $500 offer to bug Mickey Rooney's wife, whom the comedian suspected of infidelity. (Rooney was right, it turned out.) Vaus's skills eventually came to the attention of the legendary Los Angeles mob boss Mickey Cohen. After Cohen made Vaus a lucrative retaining offer, Vaus abandoned part-time police work and jumped headfirst into the world of organized crime. Over the course of eight eventful months, he made a small fortune tapping lines, planting bugs, and operating state-of-the-art electronic security equipment for Mickey Cohen's syndicate.

All of that would change on the afternoon of November 7, 1949. In between sessions of a grand jury investigation of the LAPD Vice Squad—a high-profile inquiry that hinged on secret telephone recordings that Vaus was alleged to have made on Cohen's behalf—Vaus went to see *Pinky* (1949), Elia Kazan's popular melodrama of racial passing, starring the actress Jeanne Crain. Perhaps it was the pressure of the grand jury investigation. Or perhaps, as Vaus's wife, Alice, later suggested, it was Crain's portrayal of "a girl who pretended to be something she was not, just as Jim was bluffing about being . . . as good as the next fellow."[38] Whatever it was, Vaus abruptly left *Pinky* and drove straight to Billy Graham's "Canvas Cathedral" on the corner of Washington and Hill Streets, just outside of what was then L.A.'s business district. After listening to one of Graham's fiery sermons, Vaus knelt in the sawdust, prayed for forgiveness, and was reborn.

"I made a decision to accept Christ and his teachings," Vaus told the *Los Angeles Times* the following week. "This has brought an entire change in my life. I have given up all my former work in crime and politics. . . . I propose . . . to live fearlessly for Christ and make amends for the many wrongs I have committed."[39] God was calling, and Vaus, the expert wiretapper, was listening in. His religious awakening made national headlines.[40]

Vaus's conversion turned out to be crucial in Billy Graham's rise to prominence during the early 1950s.[41] (Coincidentally, so did the conversion of Jeanne Crain's co-star in *Pinky*, Ethel Waters.) Here was a well-known Hollywood gangster who had renounced a life of crime to be born again, all on account of a sermon at a Billy Graham tent revival. The story provided an opportunity to enhance Graham's national profile, and the Billy Graham Evangelical Association (BGEA) proved quick to exploit it.

In 1951, the Navigators, a BGEA-affiliated religious group, helped Vaus publish *Why I Quit Syndicated Crime: The Wiretapper's Own Story*

(1951), a book-length testimony of his journey from sin to salvation. Featuring a fulsome preface written by Mickey Cohen himself—certainly the only book in history to claim this honor—Vaus's autobiography proved successful among evangelical communities across the United States.[42] It also went through several foreign-language editions.[43] After Alice Vaus published a companion volume, *They Called My Husband a Gangster* (1952), the BGEA's Hollywood production division, World Wide Pictures, arranged to adapt the family's story for the silver screen. The film was eventually released to theaters as *Wiretapper* (1955), starring Bill Williams and Georgia Lee (see Figure 6.4).

Wiretapper is worth examining in some detail, even if the film does little more than recapitulate the most sensational episodes in Vaus's autobiography, from his ignominious stint in the military during World War II to his religious rebirth at Billy Graham's Canvas Cathedral in 1949. (Vaus's illegal exploits with the LAPD are omitted from the film altogether.) As with all of the movies that the BGEA backed in this period—notable titles include *Oiltown, U.S.A.* (1954), *Souls in Conflict* (1955), and *The Heart Is a Rebel* (1958)—*Wiretapper* was primarily the product of a proselytizing impulse.[44] World Wide Pictures suggested as much in an official press release: "'Wiretapper' in its entirety was never designed as an entertainment feature for Christians. Rather, it was produced with a single purpose . . . to reach the unreached, and bring them to Christ. . . . It was our conclusion that through this medium [cinema] we would reach the greatest number of people in the shortest possible time."[45] Georgia Lee, who co-starred as Alice Vaus in the film and went on to appear in three more BGEA / World Wide Pictures productions, later told an interviewer that *Wiretapper* sought to make the drama of divine redemption as relatable to mass audiences as possible, showing "how a man as stained with evil and sin as was Jim can be lifted back by the power of God."[46]

If the lesson of Vaus's religious transformation wasn't clear enough on its own, the film's climactic conversion sequence featured an appearance by Billy Graham himself, who delivers a version of the sermon that led to Vaus's awakening in November 1949 (see Figure 6.5). "There's a man somewhere in this audience who has heard this message many times before, and the spirit of God is striving mightily with him at this moment," Graham exhorts at the end of the film, speaking as much to Bill Williams's Jim as to the audience in the movie theater. "If he doesn't come to Christ now, he may never come. . . . There's still time for you to come. You come, and give your life to Christ." Jim gives his life, of course, and the film

Figure 6.4. Promotional poster for *Wiretapper* (World Wide Pictures, 1955), based on Jim Vaus's 1951 autobiography, *Why I Quit Syndicated Crime: The Wiretapper's Own Story*. Billy Graham Evangelistic Association.

Figure 6.5. Billy Graham delivers the climactic sermon in *Wiretapper*.
Wiretapper, World Wide Pictures, 1955.

concludes as he renounces all worldly ties. He breaks with his criminal
connections, surrenders his material possessions, and dedicates himself to
God's service.

As an artifact of American evangelical propaganda, *Wiretapper* doesn't
seem to warrant much in the way of analysis. But as a sensational episode in the public history of electronic eavesdropping, the film provides
much more interesting fodder. For one thing, *Wiretapper* represents a
notable link in a long chain of connections between eavesdropping and
American religious culture that runs from the 1916 wiretapping scandal
in New York City, which involved taps on five prominent Catholic priests,
to Francis Ford Coppola's thriller *The Conversation* (1974), which figures the morality of electronic surveillance as a question of spiritual faith.
More importantly, for our purposes, *Wiretapper*'s portrayal of midcentury tap-and-bug techniques turns out to have been curiously exacting.
Notwithstanding the BGEA's mawkish religious message, the film broke
new ground in exposing mass audiences to listening technologies that had
escaped the public's attention during the wiretapping controversies of the
early 1950s. The impulse to showcase eavesdropping devices is evident
from the film's opening title sequence, for instance, which begins with a

slow dolly shot into an open telephone junction box. As a flashlight shines down a row of terminal switches, a hand attaches a thin pair of wires to a single connection—one of the earliest authentic depictions of a direct wiretap in U.S. film history. The sequence is repeated later on in the movie, when the police catch Jim listening in on a Hollywood prostitute's telephone conversations (see Figure 6.6). "Tapping private wires is a serious offense," the arresting officer reminds him.

In another scene, Jim is shown searching for a miniature microphone hidden in the home of Charles Rumsden, the film's fictional stand-in for Mickey Cohen. The episode draws heavily on Vaus's autobiography, which early on offers a vivid description of the sorts of listening devices that American law enforcement agencies began using in the late 1940s: "These [hidden] microphones are difficult to find. They are very small, and are often buried between the two thicknesses of a wall. The wires that connect them are no larger than a human hair. Usually the wires are tucked into the cracks or crevices of the wood so that it is impossible to locate a microphone just by looking for it."[47] To accentuate the size of the device on screen, the camera once again dollies in for a close-up. As Jim works to dislodge the bug, we are afforded a better sense of scale: the microphone is only slightly bigger than his thumbnail. The film also includes similar shots of a working pen register, a wire recorder, and Vaus's trusted "talking cane."

Wiretapper's attention to technical detail gave the moviegoing public its first glimpses of the bugging devices that Vaus and other eavesdropping specialists had pioneered in the wake of the invention of the transistor. The technologies were part of the film's appeal. As one reviewer punned, although *Wiretapper*'s didactic storyline was likely to "short circuit" most moviegoers, the film's visual "fascination with machinery and gadgets" was more than enough to sustain audience interest.[48] It's crucial to note in this account, as in the film itself, that the electronic innovations showcased in *Wiretapper*—the miniature microphone, the pen register, and the "talking cane"—are figured as benign technological novelties, occasions for wonder rather than causes for alarm. These are ingenious devices that "defy known physical laws" and "hear the seemingly unhearable," as Vaus put it in his autobiography, and at crucial points in the movie they variously brand Jim as a "young Edison," a "whiz kid," and an electronic "magician."[49] Vaus, in short, is represented as a technological miracle worker throughout *Wiretapper*. His electronic expertise provides a quotidian analogue to Billy Graham's spiritual gift, which

Figure 6.6a,b. Jim Vaus (Bill Williams) uses a direct wiretap to listen to a telephone conversation in *Wiretapper*. *Wiretapper,* World Wide Pictures, 1955.

essentially allows Graham to overhear Jim's private crisis of conscience at the end of the film. Perhaps it isn't surprising, then, that Jim's melodramatic conversion from wayward wiretapper to born-again Christian seems plausible by the time the movie's credits roll. In *Wiretapper,* eavesdropping is a form of the Lord's work.

The idea that electronic eavesdropping devices were technological miracles, symptoms and symbols of God's hand in the world, would become the foundation of Vaus's crusade as an evangelical Christian. In the years that followed the publication of *Why I Quit Syndicated Crime* and the release of *Wiretapper,* Vaus traveled the country to deliver testimonies of his conversion experience to evangelical congregations (see Figure 6.7). His mission, according to one report on his exploits, was to "tap the souls of the sinful."[50]

Vaus's sermons in this period typically recounted the details of his life in organized crime and subsequent spiritual transformation. And true to his portrayal in *Wiretapper,* he frequently used electronic eavesdropping equipment to illustrate the Christian moral of his story.[51] As he explained the principle in a 1967 interview for the BGEA,

> I used the equipment as a means of perhaps attracting attention and as a good illustration. . . . Well, for example, I would take one evening and show some of the devices used for scientific investigation of crime. . . . Oh yes, I had about 2 or 3 tons of it and carted it in trucks and hired a fellow to truck it and set it up for me. . . . In fact I carried it all the way to Alaska. But one night I would use the equipment used in the scientific investigation of crime, show how it was possible to pick up conversations and record them without being anywhere near the location, and then end up with a thing like [a] black light and actually would allow somebody to take some money and try to hide it with the audience and then by using ultra violet light, locate where the money is hidden. The individual response—I'd go on many times to tie this in with a specific illustration. I had a talk I gave on 'No Place to Hide' pointing out . . . how difficult it is for man to hide from man, let alone man hiding from GOD.[52]

According to a 1953 article on a revival meeting in southern California, Vaus's "crackling electronics displays" were invaluable demonstrations of "points of religion. . . . It's a new idea, explaining God in terms of the amazing modernity of electronics and explaining electronics and science

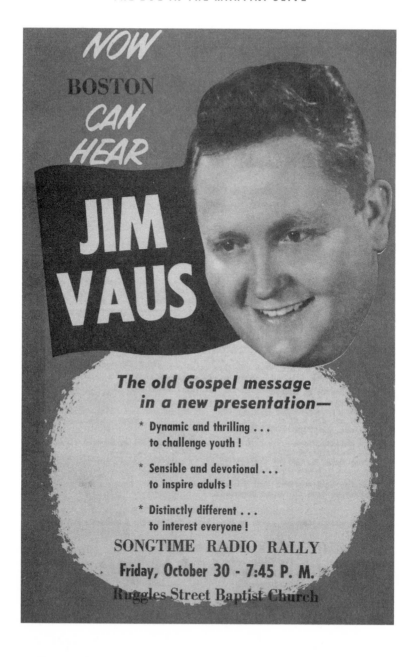

Figure 6.7a (above) and b (opposite). "Jim Vaus: The Old Gospel Message in a New Presentation" (1953). Papers of J. Arthur Vaus, Archives of the Billy Graham Center, Wheaton College.

JIM VAUS *Knows the answer!*

- to crime

- to juvenile delinquency

- to broken homes

- to every problem which confronts an individual today.

- The answer to the problems of 20th Century living . . . is the Gospel of Jesus Christ!

SEE . . . sermons illustrated with amazing electronics equipment!

Jim Vaus learned the answer the hard way . . . as a wire-tapper his electronic skill was put to use for Mobster Mickey Cohen. Four years ago in a Billy Graham meeting Jim accepted Christ as Saviour . . . today thousands from coast-to-coast have thrilled to his message.

Special word from Rev. Jim Vaus:

Recently Governor Arthur B. Langlie of Washington told me that "the great need in America today is a moral and spiritual awakening in the hearts of the American people." I am convinced that people everywhere are hungry to know God; to know the peace of mind and heart that comes with complete dedication and yieldedness to Jesus Christ. My story is an unusual one . . . but I give God all the glory. I tell it with the prayer that young people . . . adults too . . . might realize the emptiness and unhappiness outside of Christ.

in lay terms."[53] One of Vaus's devotees in upstate New York offered a more animated response to a 1956 sermon, once again connecting eavesdropping technology with the miraculous and divine: "He's . . . got a trunkful of gadgets with which he performs what look . . . like downright miracles!"[54] In real life, not just on screen, Vaus's eavesdropping devices seemed wondrous, almost godlike.

This was by no means a fringe assessment. During the 1950s, American news outlets frequently reported on eavesdropping devices as miraculous feats of technical ingenuity, particularly after the transistor began revolutionizing the field. In 1955, the same year *Wiretapper* appeared, *Science News Letter* ran an article titled "Tappers Called Ingenious," which marveled at the ability of professional eavesdroppers to "develop equipment for their trade so advanced . . . that some systems are similar to secret military devices."[55] *Newsweek* likewise represented the nation's growing population of bugging experts as "inspired tinkerers," citing Vaus's work directly and quoting an anonymous FBI official who clearly admired (and likely profited from) the creativity of men like him: "It's amazing what they'll come up with."[56]

Vaus's "No Place to Hide" theme also resonated in secular contexts. As Bernard Spindel explained in a tell-all article for *Collier's* in 1955, the discovery of the transistor made it next to impossible to imagine a private sanctuary beyond the reach of the wiretap and the bug:

> To be completely safe against electronic eavesdropping today, you would have to construct a special conference room deep in the interior of a large building. The room would have to be windowless to protect it from the parabolic microphone (whose dish-shaped 'ear' can pick up and amplify conversations through open windows from some distance away). The room would have to be encased in heavy metal to thwart any still-secret eavesdropping devices set up outside the room and to reduce the effectiveness of any radio transmitters hidden inside. Within the metal sheath there would have to be an inner shell of transparent plastic, separated from the metal by about six inches and resting on transparent plastic supports. This would foil contact microphones which, when placed against a resonating surface such as a nail driven through a wall, pick up conversations on the other side of the wall. Moreover, all the wiring in the room would be visible through the plastic shell; any tampering

with the lines would be immediately apparent. There would, of course, be no telephones.[57]

In the mid-1950s, Spindel's imaginary safe haven wasn't a nightmare prediction of a society without privacy. It was a thought experiment intended to enumerate what was technologically possible in the age of transistor miniaturization. The miracles of electronic engineering made it such that bugs could operate anywhere, free of detection. From the junction box to the pulpit, and from the pulpit to the silver screen, Vaus's career provided evidence of that seemingly improbable fact.

. . .

The rhetoric surrounding electronic eavesdropping shifted in the coming years—and no figure embodied that shift more than Vaus's successor in the public eye, Harold Lipset, a bugging pioneer whose exploits in the field of private investigation during the 1960s came to personify the frightening reach of concealed listening devices.

Born in Newark, New Jersey, and educated at the University of California, Berkeley, Lipset began his career as a private detective while serving in the U.S. Army's Criminal Investigation Division.[58] He earned a bronze star for investigating crimes committed by American soldiers during World War II, and in 1947 he returned to the Bay Area to open a licensed private investigation firm with his wife. By the mid-1950s, Lipset had already made his name as America's "super snooper," routinely working more than 500 cases a year. He cemented his reputation by using electronic eavesdropping devices to help solve them.

Unlike Vaus, Lipset didn't have a natural knack for electronics. He only began exploring the possibilities of hidden microphones when he realized that most Americans regarded private investigation as a disreputable line of work. "Private investigators were seen as easy pickings by opposing counsel in those days [the late 1940s]," Lipset later recalled. "I could claim until I was blue in the face that I would never lie to protect a client . . . but I wouldn't have any proof. So when recorders came along . . . here was incontrovertible proof that what we said and heard in the field was true. It was like hiring a witness for every investigation, every interview we conducted, only the witness wasn't human—it was electronic and so,

Figure 6.8. Geiss-America Wristwatch Transmitter (ca. 1962). Ralph Bertsche began designing devices like these for Harold Lipset in the 1950s. By the 1960s they were available for purchase on the open market. Division of Work and Industry, National Museum of American History, Smithsonian Institution.

we felt, more reliable."[59] When the transistor began transforming the field of audio surveillance, Lipset saw an opportunity to gain an investigative advantage and create valuable publicity for his services. He soon hired an in-house electrical engineer named Ralph Bertsche to design eavesdropping equipment for his cases.

It was Bertsche, more than anyone else, who was responsible for Lipset's fame. Working in the back room of a converted warehouse, Bertsche crafted a line of portable listening devices—bugged cigarettes, ink pens, wristwatches, and the like—for Lipset to use on the job, and his gadgets helped to solidify Lipset's reputation as a detective who could overhear any conversation, regardless of setting or circumstance (see Figure 6.8). In one of the pair's most imaginative collaborations, Lipset was tasked with recording a confidential exchange while sitting naked in a steam room at a Turkish bath house. With no obvious place to hide a bug, Bertsche fashioned a microphone for Lipset concealed in a bar of soap.

Throughout his career, Lipset took a populist's view of electronic eavesdropping, agitating for lax governmental restrictions on the private use of listening devices. According to Lipset, modern recording equipment had the potential to protect private citizens against a rapidly encroaching surveillance state. If police agents and government officials resorted to wiretaps and bugs regardless of the rule of law, the best course of action for the average American was to level the playing field: "The basic issue . . . is not whether some private eye is going to invade your privacy by bugging your home, but how well you can *protect* your rights against the invasion of privacy by Big Brother. . . . A recording of your voice or of someone making accusations could free you."[60] For Lipset, in other words, listening devices could exonerate as much as incriminate. It was a bug-or-be-bugged world.

Lipset first articulated this position while hawking his electronic wares before the Senate Subcommittee on Constitutional Rights in December 1959. "I am not defending the right of any outside party to intercept your conversation between yourself and the party you are talking to," he explained, with Bertsche and a table full of miniature listening devices beside him. "I am defending [your] right . . . to make a record of that conversation for your own protection."[61] Ever the showman, Lipset went on to demonstrate a range of state-of-the-art eavesdropping techniques for the Committee, stopping to play portions of his testimony back from a five-inch-long Minifon recorder hidden in his coat pocket. The performance struck a nerve. Missouri senator Thomas Hennings, who chaired the hearings, nervously interrupted Lipset's demonstration to compare him to Dick Tracy, the notorious comic strip detective.[62] In the coming weeks national reports ominously singled out Lipset's testimony as a sign that forces were mobilizing to "invade our lives and put to rout our innermost secrets."[63]

At the conclusion of the proceedings, Lipset read a prepared statement on the social value of listening devices: "[Bugging] equipment serves a useful social purpose. . . . I believe that the use of modern recordings is the greatest advance toward ascertaining the truth. . . . They [recordings] are a faithful and true reproduction of what was said, including the tone of voice used. Unlike the testimony of an individual, a recording is not subject to bias, prejudice, accuracy, intelligence, reliability, memory, and interpretation."[64] One year later, Lipset used this argument as the basis for his written response to *The Eavesdroppers* in the *Minnesota Law Review.*[65]

When Lipset was invited to return to Washington for a series of congressional privacy hearings in February 1965, the martini olive trick (see Figure 6.9) was to be the main attraction in a day dedicated to the testimony of eavesdropping industry luminaries. The gimmick was orchestrated well in advance of the proceedings. Missouri senator Edward V.

Figure 6.9. Harold Lipset's "Pry Martini," recreated for a 1966 *Life* magazine spread on electronic eavesdropping. Arthur Schatz / The LIFE Picture Collection / Getty Images.

Long, who agreed to let Lipset plant a hidden microphone in the Senate chambers and feign surprise when he played back his remarks at the end of his testimony, wanted the bugged martini to underscore the perils of continued congressional gridlock over privacy legislation.[66] Lipset, by contrast, envisioned it furthering his populist crusade. "It didn't seem all that unusual," he later recalled. "The martini glass . . . held a facsimile of an olive, which could hold a tiny transmitter, the pimento inside the olive, in which we could embed the microphone, and a toothpick, which could house a copper wire as an antenna. . . . Our point was that a host could wander through his own party, having drunk his own martini, and pick up the conversations that were directed at him. . . . We wanted to show the vast proliferation of this equipment, and the bug in the martini olive was one very feasible example of many."[67]

Yet both Long and Lipset miscalculated how drastically the revelations of the intervening years had raised the stakes of the eavesdropping debate. The national media ended up spinning Lipset's martini olive performance in the opposite direction. Papers around the country carried the story with an anxious mix of sarcasm and alarm. A tongue-in-cheek article on the front page of the *Washington Post* warned, "Don't Talk to a Martini, the Olive May Be Listening," and the *Baltimore Sun* sounded the same ironic note of caution: "Your Olive May Quote You if Snoopers Know Your Drink."[68] The *Chicago Tribune* lamented that "you can't even trust the olive in your martini these days."[69] The situation bordered on absurdity, a fact that wasn't lost on the syndicated humorist Russell Baker. "The big new anti-people thing is food with ears," Baker wrote in the aftermath of Lipset's appearance before Congress. "Olives that lurk under three inches of gin and listen. Parsley that eavesdrops. Peas that broadcast. . . . When [man] fears that private detectives are listening to broadcasting vegetables in his duodenum, he has lost more than his right to privacy and his right to indiscretion. He has lost his right to keep snoopers out of his esophagus. He can no longer tell himself: 'a man's stomach is his castle.'"[70]

Mocking Lipset's signature creation clearly helped to ease the troubling questions that the 1965 Senate hearings on electronic eavesdropping seemed to raise. For many, the "Pry Martini," as the device came to be known, epitomized the dystopian possibilities of the electronic age. If an olive could be transformed into a listening device, where could the average American citizen hope to find privacy?

The myth of the bugged martini olive grew in the coming years. Lipset's 1965 testimony was essentially an encore performance of his 1959 Senate bugging showcase—and his 1959 showcase itself clearly drew on the pageantry of Vaus's evangelical eavesdropping demonstrations of the early 1950s. But the martini olive gimmick would be remembered as the turn of events that hastened the national electronic surveillance reforms of the late 1960s, a narrative that was equal parts fact and fiction. Long, for instance, recalled Lipset recording his testimony on the floor of the Senate from the drink in his hand, making clear to all in attendance "that the growing market in transmitter devices . . . provided the eavesdropper with a wide-open field of operations."[71] (In truth, Lipset had used a microphone hidden in a nearby flower arrangement, since the martini olive bug wasn't reliable enough to use in a public setting.) An oft-cited cover story on eavesdropping in *Life* further stretched the tale, recreating a version of the "Pry Martini" in a glossy photo spread and claiming that the apparatus was already in widespread use.[72] (This, too, was misleading: the device that Lipset brought before Congress was merely a showpiece. Pouring gin in the glass would have caused the hidden microphone to short.)[73] Yet it was largely on the basis of accounts like these that in 1966 Lipset's home state of California passed a bill prohibiting the use of electronic eavesdropping equipment. The FCC soon followed suit, and when Congress finally prohibited the private manufacture and sale of eavesdropping devices in 1968, the "martini olive transmitter" was the first device it included on its list of banned items.[74] In response to the flurry of legislation, Lipset pronounced California a "police state" and warned that the federal government's bugging ban meant that American citizens could no longer protect themselves from the intrusions of government agencies.[75]

Whether or not the bugged martini olive was a callow public relations stunt, Lipset's testimony maintained a powerful hold on the national psyche well into the late 1960s. Like Vaus's miracle listening devices, the "Pry Martini" became a symbol of the possibilities of the electronic age, but its implications seemed more alarming than wondrous. Even professional eavesdroppers felt that Lipset's creation confirmed that Americans were at the mercy of technological forces beyond human control. According to Sholly Kagan, an independent manufacturer of listening devices who testified before Congress on the same day as Lipset, electronic eavesdropping was "another case of a human need developing a new technology which . . . is rather more potent than our means for controlling it," and the existence of the bugged martini was an obvious sign of the need to tighten federal eavesdropping regulations.[76] Bernard Spindel told the *Baltimore Sun* that the Americans were "defenseless" against the devices that men like Lipset were using—this, despite the growing number of U.S. security firms offering "debugging" services to the general public.[77]

A far more common response to the bugged martini came at the end of Robert M. Brown's *The Electronic Invasion* (1967), a best-selling eavesdropping industry exposé that went through three editions in the late 1960s. "The most effective anti-bugging device is you—so don't run off at the mouth!" Brown advised in the book's last chapter, resigning the nation to the terrifying future that Lipset's invention seemed to fore-shadow.[78] Richard Schwartz had come to the same conclusion a decade earlier, while finishing his research for *The Eavesdroppers*. The solution was simple: the most reliable defense against miniature listening devices—whether hidden in picture frames or cigarette packs, flower arrangements or martini olives—was to keep silent.[79]

. . .

If the 1960s marked the "beginnings of the end of privacy," as the historian Sarah E. Igo has argued, then there is good reason to regard *The Eavesdroppers* as the beginning of the beginnings.[80] Following the publication of Samuel Dash's book, a host of copycat studies flooded the popular market, all claiming to enumerate the death of privacy protections in the United States. Most of them took the form of paranoid jeremiads. For Vance Packard, the technological innovations of men like Lipset—and men like Vaus before him—represented an "insidious impingement upon our traditional rights as free citizens to live our own lives." The existence of wiretaps and bugs raised "somber questions about what the future holds for late twentieth-century society."[81] For Vice President Hubert Humphrey, who authored a preface to Edward V. Long's best-selling contribution to the genre, *The Intruders* (1966), the national response to the electronic listening invasion would "determine whether Americans in the last part of the twentieth century will be free or craven, independent or guarded."[82] Even Alan Westin's *Privacy and Freedom* (1967), the product of eight years of careful academic research, was rooted in the dystopian apprehension that the advancements of the electronic age had upended ancient social relationships, auguring a future-without-privacy to come.[83]

Not all pronouncements of the demise of privacy in the 1960s had to do with wiretapping. Newer, more intrusive forms of surveillance were beginning to lay claim to the nation's attention: closed-circuit television, data banks, lie detector tests, even (as Long suggested in *The Intruders*, with characteristic grandiosity) brainwashing and mind control.[84] Yet all of the decade's signature studies of the eavesdropping threat relied on Dash's research, and all shared his sense that the end of privacy in America—an end, it seems, that we have reached many times over—had come at the hands of a diverse set of actors, institutions, and technologies, none of which could be completely subsumed under the aegis of an Orwellian surveillance state. It seems especially telling that the device that eventually came to stand in for the intrusiveness of the age—Lipset's bug in the martini olive—conjures up images of sophisticates mingling at cocktail parties and singles flirting in dimly lit bars. In the public imagination, threats to communications privacy loomed well outside the realm of official state affairs. The shadowy government entities and secret surveillance programs that we know and fear today were only beginning to come into focus.

Even so, for all of its influence on the electronic surveillance debates of the 1960s, *The Eavesdroppers* would come to seem obsolete a decade later. In May 1973, a publishing house in New York contacted Dash for permission to reprint a cheap paperback version of the PBAE study, mostly in an effort to capitalize on a new bugging scandal brewing in the nation's capital: Watergate. William Sloane, Dash's editor at Rutgers University Press, begged off the request on Dash's behalf: "It was not the intention of the Pennsylvania Bar Association Endowment or the original funder [the Fund for the Republic] . . . to put out a sensational thing about wiretapping, but a serious study by competent people of the problem in our society. Actually, the book is more out of date than just about any other on our list because the whole emphasis both on the law and in the geography of the practice has changed. There isn't even a separate chapter on Washington in this book."[85]

What had happened to make the PBAE study, once the source of so much outrage, seem so outmoded?

The obvious answer is that the specter of government surveillance— "Washington," in Sloane's euphemistic parlance—had finally reared its head. When the breadth and audacity of the federal government's Cold War surveillance programs finally came to light in the mid-1970s, the revelations of *The Eavesdroppers* came to look quaint by comparison. But transformations in electronic surveillance policy had also hastened the PBAE study's obsolescence, and to understand them we need only turn to the final section of the book, written by the legal historian Robert Knowlton.

The primary lesson of Dash's research for the PBAE, supported in his testimony before Congress in the late 1950s and early 1960s, was that the "wiretapping-eavesdropping problem" would remain intractable until lawmakers found a way to align state and federal policies. Knowlton, for his part, shared Dash's sentiments. But he subtly pushed the argument for reform one step further, arguing that wiretapping and bugging policies needed to be aligned as well. As Knowlton explained at the outset of the book's final chapter, wiretapping was merely a "specialized form" of electronic eavesdropping, and the law's counterintuitive refusal to recognize that basic fact severely limited the government's power to regulate a rapidly changing field.[86] From *Goldman* and *Irvine* to *Benanti* and *Silverman,* the Supreme Court's historic rulings on electronic surveillance had largely upheld wiretapping and bugging as separate technological spheres. Knowlton implied that order would only come when a legal case brought them back together again.

In 1965 that case came along. A bookie in Los Angeles named Charles Katz was using a public pay phone to transmit wagers to an East Coast gambling syndicate, and federal agents had arranged to record his conversations without a warrant. Unlike most electronic surveillance operations, however, the investigating officers didn't have to tap any lines to hear what Katz was saying. On the outside of the phone booth was a miniature listening device. They used the bug to listen in instead.

Part III

THE LISTENING AGE

Chapter 7

Title III

PERHAPS PREDICTABLY, THE MOST FAMOUS MOVIE about electronic eavesdropping ever made, Francis Ford Coppola's *The Conversation* (1974), ends with a telephone call.

After discovering that he has facilitated the murder of a high-powered corporate executive, professional wiretapper Harry Caul (Gene Hackman) sits alone in his apartment, playing his saxophone along with the jazz recording that blares from his stereo. Harry's number is unlisted, so when the sound of a telephone interrupts his performance he hesitates to pick up the receiver. At first no one responds on the other end of the line. But the phone rings a second time a few moments later, and the high-pitched sound of rewinding audiotape answers Harry's reluctant greeting. A familiar voice cuts in—"*We know that you know, Mr. Caul. . . . For your own sake don't get involved any further. . . . We'll be listening to you*"— and to underscore the threat, we hear a recording of the tune that Harry has just played on his horn. Realizing he has been bugged, Harry abruptly hangs up the phone.

In the film's bleak final act, Harry dismantles his apartment in search of the hidden recording device (see Figure 7.1). He starts by examining locations that were known to be the favorites of the men who pioneered the business of bugging. He inspects a picture frame and studies the inner workings of an electrical switch. He peers into an air vent and checks the curtains that hang from his windows. Still on the hunt, he scrutinizes the knickknacks on his bookcase, resting his eyes on a small, plastic statuette of the Virgin Mary. Too scandalized to proceed, Harry turns his attention to a light fixture and a telephone. But the camera pans back to the statuette. Could the microphone really be inside? Harry takes one final look— pausing, as if to ask for forgiveness—and smashes the idol with his fist, sending the bookcase tumbling to the floor. Still nothing.

Figure 7.1. Harry Caul (Gene Hackman) checks his telephone for an eavesdropping device at the end of Francis Ford Coppola's *The Conversation*. *The Conversation,* Paramount Pictures, 1974.

Coppola's final shots capture Harry in desolation. He peels back the wallpaper and pulls up the floorboards, but there's still no sign of the bug. While the film's credits roll, the camera pans back and forth across the ruined expanse of Harry's apartment, simulating the slow sweep of closed-circuit surveillance feed.

Every time I teach *The Conversation,* my students want to talk about this scene. Perhaps it's because the smashing of the Virgin Mary provides a rare moment of clarity in a film otherwise clouded by narrative ambiguities. The heavy-handed imagery wasn't lost on the critics who first reviewed the movie when it was released in the spring of 1974. For Nora Sayre of the *New York Times,* the final scene of *The Conversation* showed how easily the modern-day eavesdropper might "wind up a victim of his craft."[1] A critic in the *Wall Street Journal,* convinced that Coppola wanted to make a modern-day technological "horror story," took Sayre's observation one step further: in an age of pervasive electronic surveillance, the average American, much like Coppola's Harry, would find sanctuary only by "destroying [the] private self," obliterating the ideals, both secular and sacred, on which the myth of man's private castle once stood.[2] For some reviewers, the image of the shattered Virgin Mary stood out because

Coppola made Harry's devout Catholicism a special point of emphasis throughout the film.

"Like the dismembered telephone that reveals nothing but circuitry," Lawrence Shaffer noted in a 1974 article on *The Conversation* for *Film Quarterly*, "the icon too is empty." In a world where technologies of all sorts invisibly trespass on the sanctity of the private individual, the Virgin Mary "hold[s] no answers."[3] Harry has lost faith. But why?

For many viewers in 1974, and for many in the decades since, current events would offer a convenient explanation. Almost as soon as *The Conversation* premiered, filmgoers and critics began enumerating a laundry list of chance connections between the film's plot and the Watergate scandal, another affair of hidden bugs and empty icons that had shaken the faith of so many Americans in the early 1970s. Some of them seemed impossible to chalk up to coincidence.

Much of *The Conversation* follows Harry as he reconstructs surveillance recordings on an Uher 5000 model reel-to-reel deck, the same machine that Richard Nixon's long-time assistant, Rose Mary Woods, used to transcribe the president's Oval Office conversations. And like Nixon himself, Harry's final descent into paranoia and paralysis comes as a result of recordings of his own making, a fictional turn of events that recalled *Time* magazine's controversial April 30, 1973 cover, which showed the principals of the Watergate conspiracy tangled in a web of incriminating audiotape (see Figure 7.2). A more direct connection between *The Conversation* and the political crisis in Washington involved Harold Lipset. At the exact same time that Lipset served as an investigator for the Senate Watergate Committee, following leads for the group's chief counsel, Samuel Dash, the San Francisco private eye volunteered as a "technical advisor" on the set of Coppola's film. (Lipset's name appears in the credits that overlay the movie's celebrated opening zoom shot.) Lipset was forced to resign from the Watergate investigation when his lone criminal conviction for electronic eavesdropping came under public scrutiny.[4] Nevertheless, news outlets around the country relied on his dismissive assessment of the Oval Office recording setup ("adequate, but not professional") as the drama in Washington came to a close.[5]

In hindsight, the final sequence of *The Conversation* can itself be read as a darkly ironic riff on a refrain that Nixon was known to have repeated to White House staffers when rationalizing his administration's record of wiretap abuse: "For Christ's sake, everybody bugs everybody else."[6] The president wasn't alone in succumbing to cynicism about the pervasiveness

Figure 7.2. "Watergate Breaks Wide Open." The April 30, 1973 cover of *Time* magazine, which lampoons the Nixon administration as caught in a web of incriminating audiotape. © 1973 TIME USA LLC. All rights reserved. Used under license.

of electronic surveillance. One month before *The Conversation* appeared in theaters, the media theorist Marshall McLuhan remarked that the "whole matter of privacy is suspect. . . . The Watergate affair makes it quite plain that the entire planet has become a whispering gallery, with a large portion of mankind engaged in making its living by keeping the rest of mankind under surveillance."[7] Both McLuhan and Nixon appear to have known what Harry somehow doesn't, at least until *The Conversation*'s dramatic concluding scene: even the buggers were bugged.

Sentiments like these were widespread by 1974. Contrary to popular memory, Watergate didn't so much spark as corroborate the belief that electronic surveillance was *everywhere* in American society, a permanent fixture of the paranoid order of things. Coppola, for his part, wrote a treatment for *The Conversation* as early as 1966, six years before the break-in at the Democratic National Committee headquarters in Washington set off the Watergate scandal.[8] By the time Coppola managed to shoot the film, which he described to reporters in 1972 as a meditation on the "nightmarish situation that has developed in our society, a system that employs all the sophisticated electronic tools that are available to intrude upon our private lives," popular images of wiretapping and bugging had come to stand in for cultural crises of all sorts.[9] They would continue to serve that function long after Nixon's downfall. Hollywood films such as Alan Pakula's *Klute* (1971) and Sydney Lumet's *The Anderson Tapes* (1971) prominently featured wiretapping in an effort to dramatize quotidian assaults on individual privacy. Others—*The French Connection* (1971), *The Day of the Jackal* (1973), *The Parallax View* (1974), *Three Days of the Condor* (1975)—used wiretaps and bugs as throwaway plot devices, tools that fictional characters could rely on to survive in the corrupt worlds they inhabit.

On stage, the playwright Arthur Miller made the ubiquity of electronic surveillance the basis for *The Archbishop's Ceiling* (1977), a drama set in eastern Europe that revolves around a group of writers who discuss their dissident political views in a bugged sitting room. In the preface to a revised edition of the play produced in 1984 (the year of its re-release was no accident), Miller went so far as to characterize the late 1960s and early 1970s as the "era of the listening device." He compared his experience in the United States during those years to that of a subject living under totalitarian rule behind the Iron Curtain. In both places, Miller explained, ordinary people went about their daily business assuming that "power's ear is most probably overhead."[10]

American culture took its paranoid turn at something of a paradoxical moment. As a number of historians have pointed out, popular anxieties about the pervasiveness of wiretapping and electronic eavesdropping reached a fever pitch at the same time that efforts to regulate privacy began making concrete gains on the national stage. On one hand, the twin issues of privacy and surveillance gained a "sudden visibility" in some of the most consequential judicial decisions of the sixties and seventies.[11] In a series of rulings that probed the limits of the government's right to wiretap—*Berger v. New York* (1967), *Katz v. United States* (1967), and *United States v. United States District Court* (1972)—the U.S. Supreme Court brought order to the chaos of the nation's eavesdropping laws and established new standards for protecting privacy in the electronic age.[12] On the heels of *Griswold v. Connecticut* (1965), a historic case involving the use of contraceptives by married couples, the Court's wiretapping decisions of the late 1960s and early 1970s transformed privacy from a civic ideal to a constitutional rule, creating the modern "right to privacy" that we know and invoke today.[13]

On the other hand, privacy and surveillance also gained visibility in the halls of Congress. While the Supreme Court worked to lay the constitutional foundations for modern privacy rights, lawmakers on both sides of the aisle commissioned high-profile studies of surveillance abuse, campaigned on promises of privacy reform, and opened the legislative gates for a flood of new laws intended to protect private life against the incursions of government and technology. In 1968, Congress passed Title III of the Omnibus Crime Control and Safe Streets Act, a major piece of wiretap legislation designed to protect the "privacy of innocent persons" while legalizing police wiretapping under judicial supervision.[14] A host of rights-minded privacy laws would follow in the coming years: the Fair Credit Reporting Act (1970), the Family Educational Rights and Privacy Act (1974), and the Privacy Act (1974). Later, in response to the Watergate scandal and the Church Committee intelligence investigations, the passage of the Foreign Intelligence Surveillance Act (1978) ended more than four decades of uncertainty surrounding the legality of national security wiretapping, establishing protocols to check the government powers that had led to the warrantless surveillance of countless American artists, activists, and political organizations: from Abby Hoffman and Alan Ginsburg to the Socialist Workers Party and the Southern Christian Leadership Conference. A wholesale revolution in privacy and surveillance oversight seemed to have unfolded over the course of a decade. At least in theory.

In practice—as Coppola, who tapped his first telephone at the age of thirteen, doubtless understood—the revolution was one in name only.[15] Visibility could obscure as much as it revealed. Even though the late sixties and early seventies seemed to herald a newfound national commitment to protecting privacy and limiting surveillance, the era's signature pieces of electronic surveillance legislation had the ironic effect of normalizing taps and bugs in areas of American life that would have seemed unthinkable only a decade prior. It's here, at long last, that we can arrive at a viable interpretation of Harry Caul's crisis of faith at the end of *The Conversation*. As much as the upheavals of the sixties and seventies shed frightening new light on threats to individual privacy in America, the basic political response to the period's surveillance scandals was to enshrine government eavesdropping into law, scrubbing the "dirty business" of wiretapping clean in the process. At the time, and in the decades since, that paradox mostly escaped public notice, hidden as it was in explosive national debates about race and crime, dissent and conformity, law and order.

Perhaps, then, the reality of the "era of the listening device," like Harry's missing bug, was never really visible in the first place.

. . .

No public figure did more to shape the national conversation about wiretapping and electronic eavesdropping in the sixties and seventies than Edward V. Long. We last encountered Long in Chapter 6—he was the Missouri senator who engineered Harold Lipset's bug-in-the-martini-olive stunt, an event that triggered some of the decade's most frenzied pronouncements about the death of privacy. Long was much more than a political showman, however. He was the most formidable and embattled electronic privacy advocate of the period. As Democratic chairman of the U.S. Senate Subcommittee on Administrative Practice and Procedure, an entity charged with investigating government surveillance in the mid 1960s, Long played a starring role in a series of public hearings that blew the lid off wiretapping in Washington. He summarized the committee's findings in *The Intruders: The Invasion of Privacy by Government and*

Industry (1966), a best-selling book intended to drum up support for electronic surveillance reform, and he went on to float a sweeping Senate bill, backed by President Lyndon B. Johnson, that would have wiped out government eavesdropping for good.

Yet the world that Long made was far different from the one he intended. Over time, his public campaign against wiretapping fizzled, and the defeat of his regulatory proposals opened the door for the passage of Title III of the Omnibus Crime Control and Safe Streets Act (1968), a groundbreaking piece of federal legislation that sanctioned government eavesdropping at the very moment when the modern right to privacy was born. Title III of the Safe Streets Act—also known as the Federal Wiretap Act, the Wiretap Act of 1968, or, most commonly, "Title III"—still authorizes police wiretapping in America today. To follow the arc of Long's career is to track the origins of our society of normalized government surveillance, years before the controversies over Watergate and the Church Committee erupted.

Edward Long was an unlikely candidate to take up the mantle of electronic surveillance reform. After attending Culver-Stockton College and the University of Missouri, Long cut his political teeth as a prosecuting attorney for Pike County, Missouri, a small-town district located a little more than ninety miles northwest of St. Louis. The job would almost certainly have given him contact with the practice of law enforcement wiretapping. At the time, the state of Missouri lacked a statute prohibiting electronic surveillance. Long may well have had the opportunity to authorize police wiretaps and even to make use of wiretap evidence in court. (Missouri held out as a "virgin jurisdiction" until 1989, when the state legislature passed a law that made it legal for police to tap under the supervision of the Missouri attorney general.)[16] Long was elected as a state senator in 1945, and he worked his way up through the ranks of the Missouri State House in the decade that followed. In 1956 he was appointed lieutenant governor.

Long came to Washington in November 1960, after winning a special election to fill the Senate vacancy created by the death of Missouri lawmaker Thomas Hennings Jr. A stalwart of the Democratic establishment, Hennings had chaired, in the late 1950s, a much-ballyhooed Senate Subcommittee on Constitutional Rights that worked to expose illegal wiretapping among state police and private citizens. Almost as soon as Long inherited Hennings's seat, he picked up where his predecessor left off, writing editorials for Missouri news outlets that made grave predictions about the erosion of constitutional rights in the electronic age. The histrionic titles of his published articles—"Big Brother Is Listening," "Big Brother Is Watching You," and, later on, "How to Kill Big Brother"—say a great deal about the nature of his political stance at the time.[17] Long believed that American citizens of the 1960s were subject to a technological assault on privacy and security that exceeded even novelist George Orwell's dystopian predictions in *Nineteen Eighty-Four.* Public awareness was the only bulwark against creeping totalitarianism.

"It may well take us until 1984 to destroy him, and we should expect to lose a few battles along the way," Long mused in 1967, relying on the Orwellian metaphors that were seemingly everywhere at the time. "With the help of an enlightened and aroused American public, Big Brother may finally have met his master."[18]

In all likelihood, Long's commitment to civil liberties emerged out of his interest in aligning his political platform with the policies that had cemented his predecessor's electoral success. An effusive appraisal of Hennings's legacy that Long wrote in a November 1961 issue of the *Missouri Law Review* suggests as much—not to mention the hundreds of constituent letters that Long received in support of electronic privacy reform, now collected among his personal papers at the State Historical Society of Missouri.[19] Prospective changes to federal law enforcement policy provided another, more immediate cause for Long's activism. In 1961, U.S. Attorney General Robert F. Kennedy revived the Justice Department's moribund Organized Crime and Racketeering Division to combat mob activities in the United States. To aid the anticrime operation, one of Kennedy's first moves was to lobby for legalized police wiretapping. "Wiretapping often may be the only way of getting evidence or of getting the necessary leads to break up major criminal activities," Kennedy wrote in an editorial for the *New York Times* in June 1962. "The solution is a coherent law which, with stringent safeguards, permits the gathering of evidence by wiretapping in vital cases but at the same time effectively forbids other wiretapping, public or private."[20] Kennedy's proposal, which raked the coals of the bitter debate over Section 605 of the 1934 Federal Communications Act, spurred Long to action.[21]

After assuming control of the Senate Subcommittee on Administrative Practice and Procedure, Long took up government wiretapping as his pet political cause. Inspired by a March 1964 article in *Time* magazine titled "Bug Thy Neighbor," which alleged that the U.S. government was spending more than $20 million per year on electronic surveillance operations, the

QUESTIONNAIRE RELATING TO INVASIONS OF PRIVACY

Part I. Use of Telephone System

1. (a) How many listening-in circuits and/or telephone transmitter cutoff buttons (i.e., devices to permit a third person to monitor telephone conversations without being heard on the line) were installed on telephones of your agency in the Washington area as of June 30, 1964?

(b) What was the total cost for the installation of such devices during each of the fiscal years 1959–64?

(c) What was the total cost for use and maintenance of such devices during each of the fiscal years 1959–64?

(d) Are there restrictions either upon who may use these devices or who may authorize their use?

(e) When personnel listen in on telephone conversations (either by means of transmitter cutoffs, listening-in circuits or telephone extensions), are there specific agency rules or regulations making it mandatory to reveal that a third party is listening in? (Please see pt. VII, question 21(a).)

2. (a) Is the mechanical or electronic recording of telephone conversations prohibited?

(b) If not, is it mandatory to inform the other party that their conversation is being recorded?

3. (a) How many recording devices, which are equipped to record telephone conversations, were either owned by or leased to your agency and either installed or available for use on agency telephones in the Washington, D.C., area as of June 30, 1964?

(b) What was the total cost for the purchase or lease of such devices during each of the fiscal years 1959–64?

(c) How many of these recorders are equipped with "beeping" devices?

(d) What was the total cost for the lease of the beeping devices during each of the fiscal years 1959–64?

(e) Are any of the beeping devices of the type that may be disconnected, bypassed, or turned off during the recording of a conversation and, if so, how many are of this type?

(f) Does your agency permit the recording by employees of their telephone conversations under any circumstances by means of noninstalled, nonbeeping recorders, and if so, under what circumstances?

4. (a) Has your agency, or anyone on its behalf, ever surreptitiously tapped or monitored a telephone?

QUESTIONNAIRE RELATING TO INVASIONS OF PRIVACY

BY THE

SUBCOMMITTEE ON ADMINISTRATIVE PRACTICE AND PROCEDURE

OF THE

COMMITTEE ON THE JUDICIARY
UNITED STATES SENATE

SEPTEMBER 1964

Figure 7.3. U.S. Senate Subcommittee on Administrative Practice and Procedure, "Questionnaire Relating to Invasions of Privacy" (September 1964). Missouri senator Edward V. Long sent this document, containing twenty-two questions about government use of electronic surveillance, to thirty-three federal agencies in the fall of 1964. Many refused to respond. U.S. Government Printing Office.

committee sent an exploratory questionnaire to thirty-three government agencies whose dealings, on the surface, had little to do with the enforcement of criminal law or the protection of national security: from the Civil Aeronautics Board and the Department of Agriculture to the Securities and Exchange Commission, the Tennessee Valley Authority, and the U.S. Post Office (see Figure 7.3).[22]

Most of the questions in the survey made direct inquiries into the purchase and use of electronic eavesdropping equipment: "Has your agency, or anyone on its behalf, ever surreptitiously tapped or monitored a telephone?" "In the fiscal years 1959–1964 has your agency purchased any miniature (i.e., under 5-pounds weight) wire recorders?" "What was the unit and total purchase cost?" Other questions probed the "dirty" surveillance and security tricks often associated with wiretapping: "Has your agency purchased any closed-circuit TV equipment, infrared photographic equipment, one-way glass, or two-way mirrors?" "Has your agency sought the placement of 'mail covers' by the Post Office Department, its officers, or employees?" "Does your agency have an internal 'security force' which consists of persons assigned to check on the loyalty and security of employees?"[23]

The goal, as Long explained it in an internal committee memorandum, was to ascertain "'how much' and 'what kind' of snooping is done" in Washington.[24] Along the way, he hoped to convince the American public that corrective legislation against rampant government surveillance was decades overdue.

Wary of ruffling powerful political feathers, the Long Committee opted to steer clear of the federal intelligence establishment in the opening phase of its investigation. In late March 1964, the author of the salacious *Time* article euphemistically advised the Long Committee's lead counsel, Bernard Fensterwald, not to "mess with the CIA thing," and the ranking senators in the group all agreed—at least initially.[25] But the inquiry encountered resistance even outside of the secretive corridors of the FBI, the CIA, and the Pentagon. Many of the thirty-three federal agencies refused to respond to the questionnaire. A few included ambiguous and misleading answers, or stalled for months before replying.[26]

When political pressure began to mount in the spring of 1965, Long managed to convene a series of public hearings that revealed much of what the thirty-three federal agencies wanted to hide. In March 1965, the committee questioned officials from the U.S. Post Office, who promptly divulged that their agency was in the process of monitoring the correspondence of some 24,000 American citizens. One month later, representatives from the Food and Drug Administration and the Department of Health, Education, and Welfare admitted to dozens of cases of illegal wiretapping in the investigation of grocery stores and pharmaceutical corporations, including one involving a chain of supermarkets in Long's home state of Missouri. The first round of hearings came to an explosive climax in July–August 1965, when top officials from the Internal Revenue Service, an agency long suspected of illicit investigative activities, testified about the unreported use of illegal wiretaps in thousands of tax fraud and racketeering cases. One of the most sensational details of the IRS testimony was the disclosure of a "Technical Investigative Aids School"—in essence, a wiretap training program—for federal agents, located four blocks from the Treasury Department's headquarters in downtown Washington. More than seventy government employees had attended the school between 1959 and 1964. Course offerings included "Amplifiers and Recorders," "Microphone Installation," and "Surreptitious Entry."[27]

Federal officials complained bitterly about the political effects of the Long Committee's work during the summer of 1965. When the embarrassment of the hearings forced the IRS to scale back its electronic surveillance activities, the *New York Times* ran a front-page article claiming that Long's efforts had "crippled" the Justice Department's Organized Crime and Racketeering Division. Because tax investigations produced more than 60 percent of the government's syndicated crime convictions, the story went, Long's checks to the Treasury Department's wiretapping powers had forced federal prosecutors to relinquish their most reliable investigative tool in Mafia cases. "This will kill us," warned one anonymous Justice Department attorney.[28] More conservative news outlets dismissed the Long Committee hearings as publicity stunts that similarly hurt the cause of law enforcement. The *Washington Star* pilloried Long for "helping the criminals by demoralizing the men who pursue them. . . . In a period when the nation is alarmed and confused by many aspects of the crime problem, Congress should produce something more constructive and definitive than the sideshows staged by Long."[29] On the eve of Long Committee hearings held in Kansas City, the *Kansas City Star* ran a cartoon that depicted Long as a media-hungry inquisitor: while a paunchy Long threatens an imprisoned IRS agent with instruments of torture, spotlights and television cameras are trained on the lurid spectacle, following a sign that points the way to "Senator Long's gala Kansas City wire tapping investigation" (see Figure 7.4).[30]

The early critiques of the Long Committee, many of which were directed at Long himself, fed a familiar political narrative: that opposing electronic surveillance was tantamount to siding with the criminals in the nation's ongoing war on crime. That narrative gained steam in December 1965, when a federal corruption case involving a well-connected Washington lobbyist, Fred B. Black, was revealed to involve the unwarranted use of a concealed listening device. Prominent officials at the FBI and the Justice Department—including J. Edgar Hoover and Robert Kennedy—subsequently took turns blaming each other for the offense.[31] In what proved to be a fateful move, Long responded to the controversy by signaling a willingness to broaden the scope of his inquiry to include potential surveillance abuses at the FBI. "We have been extremely reluctant to call officials from the FBI and Justice Department," Long told the

Figure 7.4. "Come On—Admit You've Been Using Harassment and Intimidation Methods!" (October 1965). Edward V. Long Papers, State Historical Society of Missouri.

Washington Post in 1966. "But now that some of the principal participants have opened up these matters we feel that an on-the-record hearing is necessary."[32]

The day never came. Behind closed doors, Hoover summoned his deputy associate, Cartha DeLoach, to pay several off-the-record visits to Long's office on Capitol Hill. DeLoach implored Long not to "open Pandora's box," underscoring that hearings on the FBI's electronic surveillance operations would endanger national security and give criminal defendants a leg up in pending federal cases. DeLoach went so far as to draft a fabricated letter in Long's name exonerating the FBI of any wrong-doing. He never released it to the press, but the back-channel pressure worked. DeLoach wrote to his supervisors in January 1966 to say that he had "neutralized the threat of being embarrassed by the Long Committee" for the time being. "Keep on this situation at all times," DeLoach warned.[33] Long had made himself an enemy of Hoover's FBI.

Fearing political reprisal, Long backed away from his promise to hold hearings on the FBI. But he soon found other avenues to pursue in his campaign against electronic surveillance, in part building on alliances he was cultivating inside a sympathetic White House.

Lyndon Johnson had expressed reservations about government wire-tapping early in his presidency. In a June 1965 memorandum—a document crafted on the eve of the Long Committee's IRS hearings, doubtless anticipating the revelations to come—Johnson characterized wiretapping as a "highly offensive practice" that leads to "serious abuses and invasions of privacy."[34] He went on to order the heads of federal agencies to curb the use of electronic surveillance except in cases involving the protection of national security. Johnson's attorney general, Nicholas Katzenbach, spent much of his two-year tenure at the Justice Department quietly instituting Johnson's guidelines, and the historical record suggests that the work of the Long Committee gave teeth to his efforts. Notwithstanding DeLoach's dealings with Long, for instance, Hoover capitulated to Johnson's demands in 1965–1966, citing the "present atmosphere" of "congressional and public alarm" as justification for cutting back on active FBI taps and bugs.[35] The White House later made explicit its tacit support for safeguards against electronic surveillance when Long published *The Intruders* (1966), a book-length commentary on the Administrative Practice and Procedure Committee's findings. For the first edition of the study, Johnson's vice president, Hubert Humphrey, wrote a foreword in praise of Long's commitment to protecting civil liberties.[36] *The Intruders* went through three editions in early 1967.

The White House's opposition to electronic surveillance gave Long a narrow window of political opportunity. His break came when Johnson used his January 1967 State of the Union Address to call for an end to "all wiretapping, public and private, wherever and whenever it occurs, except when the security of the nation itself is at stake."[37] The statement was what Long had been waiting for. Three weeks later, Long introduced a bill on the Senate floor designed to enact Johnson's proposed framework for surveillance control, banning all police wiretapping and criminalizing the manufacture, use, and advertisement of technical equipment "primarily useful for the purpose of wire interception or eavesdropping."[38] Long went out of his way to exclude national security cases from his proposal. But in keeping with the procedural protections against government wiretapping first established in the *Nardone* decisions of the late 1930s, the bill also included a blanket prohibition against the use of wiretap evidence in court.

Long gave the bill a title lofty enough to match the ambition of his campaign: the Right of Privacy Act of 1967. As winter turned to spring, all signs seemed to point to its safe legislative passage. On top of the growing chorus of support Long had received from the Johnson administration, the U.S. Supreme Court was in the process of deliberating in *Berger v. New York* (1967), a case that legal experts deemed a wholesale referendum on the constitutionality of wiretapping and bugging in law enforcement investigations.

The case involved a public relations consultant named Ralph Berger, who was convicted of conspiring to bribe the chairman of the New York State Liquor Authority. The evidence against Berger rested on conversations that the New York Police Department had bugged over the course of several months in accordance with New York's 1938 eavesdropping law, which authorized state police to use electronic surveillance under limited judicial oversight. The issue, for the Court, was whether the bug on Berger lacked sufficient "particularization" to be considered valid— that is, whether the New York law warranted law enforcement eavesdropping in violation of basic constitutional protections against general searches, conducted in dragnet fashion for extended periods of time. Reversing the conviction spelled doom for New York's wiretapping law, if not for the entire cause of government wiretap legalization. "[The] petitioner [Berger], in effect, argues that it is impossible to set up a court authorized eavesdropping system consistent with the Constitution," a legal scholar at the University of Florida explained in a memorandum to the Long Committee in April 1967. "If the New York statute is struck down, it may well mean that electronic eavesdropping will be out completely as a source of judicial evidence."[39]

Two months later, a 6–3 ruling did just that: reversing Berger's conviction and striking down the Empire State's eavesdropping statute. The political fallout was swift, particularly because lawmakers in Washington had for decades considered New York's system for court-authorized surveillance as a prototype for federal law. In an outraged dissent, Justice

Byron White claimed that the Court's decision in *Berger v. New York* indirectly rendered all forms of electronic surveillance unconstitutional: "Today's majority does not, in so many words, hold that all wiretapping and eavesdropping are constitutionally impermissible. But, by transparent indirection, it achieves practically the same result by . . . imposing a series of requirements for legalized electronic surveillance that will be almost impossible to satisfy. In so doing, the Court ignores or discounts the need for wiretapping authority."[40] The White House agreed that the *Berger* ruling all but snuffed out the prospect of legalized government surveillance. "To the extent the Court left open a crack [for federal wiretap authority], it's a narrow one—maybe not wide enough to make any law worthwhile," a Johnson administration official told the *Wall Street Journal* two weeks after the Supreme Court handed down the *Berger* decision.[41]

Reading the writing on the wall, Katzenbach's successor at the Justice Department, Ramsey Clark, issued stringent new eavesdropping guidelines to federal agencies the following month (see Figure 7.5).[42] At the FBI, the number of active wiretap and bug installations soon dropped to its lowest level since 1940.[43]

• • •

For Long, the outcome of the *Berger* case was cause for celebration. He felt that his efforts to expose government surveillance abuse had moved the judicial needle in the direction of privacy. He believed that the Supreme Court had vindicated his civil libertarian crusade.

A few days after Clark instituted the Justice Department's new restrictions on wiretapping, Long wrote a confidential memorandum to President Johnson to report that his investigations with the Senate Subcommittee on Administrative Practice and Procedure were "beginning to bear real fruit." Long struck an optimistic tone throughout the message. He assumed that the White House would continue to stand behind the Right of Privacy Act, and that the *Berger* ruling paved the way for its passage in Congress the following term. "Now we have our first major victory," he wrote to Johnson, with tellingly inclusive rhetorical emphasis. "With the significant assist of the Berger case, it is time for us to consolidate gains and push on to strengthen privacy."[44]

Long's optimism would prove to be short-lived. The political winds were already shifting. The Long Committee surveillance inquiry; the

Figure 7.5. "Sorry, But You'll Have to Start Working Again"
(ca. July 1967). Attorney General Ramsey Clark ordered a halt to
hundreds of active federal eavesdropping operations following
the U.S. Supreme Court's ruling in *Berger v. New York* (1967). At
the time, advocates for electronic privacy criticized government
wiretapping as both intrusive and lazy. Edward V. Long Papers,
State Historical Society of Missouri.

Johnson administration's effort to curb taps and bugs; the unveiling of
the Right of Privacy Act; and the civil liberties "victory" of *Berger v. New
York:* all of these developments—major signposts in the history of wiretapping and electronic eavesdropping in the 1960s—unfolded against a
backdrop of cataclysmic unrest in American cities, chaos both real and

imagined. The Long Committee's hearings on the IRS, for instance, shared headlines with the Watts rebellion of August 1965. While Long's chief counsel, Bernard Fensterwald, grilled IRS officials about the Treasury Department's wiretap training program, African American residents of south Los Angeles clashed with state police. The conflict left thirty-four dead and a stunned city in ashes. The turmoil only intensified in the coming years. While Hoover discontinued hundreds of active FBI wiretaps in compliance with the Justice Department's new electronic surveillance restrictions—and while Long put the finishing touches on *The Intruders*—student activists began burning their draft cards, eventually taking to the streets to protest the war in Vietnam. The Supreme Court handed down the *Berger* ruling on June 13, 1967, in the opening stretch of the "long, hot summer" of racial violence that flared from Atlanta and Buffalo to Milwaukee and Tampa, in total reaching almost 170 cities across thirty-four states. One month later, neighborhoods in Newark and Detroit were in flames.

Convinced that the nation was "rapidly approaching a state of anarchy," as one prominent Republican put it, conservative lawmakers responded to the unrest by disparaging the promises of Lyndon Johnson's Great Society.[45] They saw civil rights protests and race riots as the product of an incipient culture of lawlessness, and they cited soaring national crime rates to promote the idea that a more punitive approach to government would make the country whole again. In Washington, Republicans and southern Democrats offered a variety of explanations for the disorder in America's cities: liberal social policies that mistakenly attempted to ameliorate crime's "root causes"; Supreme Court decisions that protected the rights of criminal defendants; and, most importantly, the watershed civil rights legislation of the 1960s, which seemed to reward civil disobedience and upset the racial status quo. Yet in the conservative mind the solution to the problem of race riots and street crime, now linked, always remained the same: the reinstatement of *law and order*.[46]

By the end of the decade, that turn of phrase had emerged as an electoral rallying cry and a partisan legislative agenda. A powerful coalition of conservatives wielded it to endorse a range of repressive law enforcement policies, most of which were directed at the urban communities of color that seemed to be at the center of the epidemic of crime and disorder. From the beginning, the legalization of government surveillance was a key facet of the law-and-order platform. The Right of Privacy Act would eventually fall victim to its ascendance.

It was President Johnson's own response to the law-and-order agenda that set the stage for Long's political undoing. In July 1965, Johnson made his first major move to combat what conservatives had identified as a national "crime epidemic" by appointing a task force of nineteen experts to study crime and policing across the United States.[47] After conducting two years of exhaustive research, the President's Commission on Law Enforcement and Administration of Justice, as it was known, submitted its eagerly anticipated final report, *The Challenge of Crime in a Free Society*, in late February 1967. The study called for sweeping changes to the workings of American law enforcement. Among them—buried 200 pages into the document, and flying in the face of Johnson's own public statements—was a recommendation to grant municipal, state, and federal police the authority to use electronic surveillance in criminal investigations.

Citing little more than the collective opinion of a "great majority of law enforcement officials," the Johnson Crime Commission declared wiretapping a "necessary step in the evidence-gathering process" in complex criminal investigations. Without it, the report continued, the nation's police agencies were powerless to prosecute high-level cases. The only sensible solution was to legalize law enforcement wiretapping. "Congress should enact legislation dealing specifically with wiretapping and bugging," the Commission concluded. "All private use of electronic surveillance should be placed under rigid control, or it should be outlawed . . . [and] legislation should be enacted granting carefully circumscribed authority for electronic surveillance to law enforcement officers." In a shot across Long's bow, the report also implied that the Right of Privacy Act, unveiled on the Senate floor just two weeks prior, was "thoroughly confused."[48] Amid the clamor about wiretap abuse and invasions of privacy, *The Challenge of Crime in a Free Society* sent a clear message: electronic surveillance was key to controlling crime and protecting the social order.

That was February 1967. What Long described as a "victory" for privacy the following July—on the heels of *Berger v. New York,* in the middle of the long, hot summer—was thus merely the opening battle of a protracted war over the future of crime control in America, a war that ended up pitting civil libertarians like Long against conservative hawks for law and order, many of whom believed that liberal defenses of constitutional rights were behind the nation's crime problem in the first place.[49]

By then Long was fighting uphill. In late May 1967, a sensationalized article in *Life* magazine raised troubling questions about the personal motivations behind Long's political crusade. Likely the product of a Hoover smear, the story was a variation on the tried-and-true law-and-order theme of helping the criminals by limiting the government's investigative powers. Long, sources alleged, had embarked on his campaign against electronic surveillance in order to discredit the FBI's case against Jimmy Hoffa, the notorious Teamsters president then in the process of appealing a federal conviction on grounds of illegal wiretapping.[50] An ethics investigation later cleared Long of any wrongdoing, but the damage was done. The Missouri senator found himself politically radioactive. The death-knell for his cause sounded in December 1967 when the Supreme Court handed down a ruling in its second major wiretapping case of the year: *Katz v. United States* (1967). The *Katz* decision gave the law-and-order camp a coherent legal language to argue against the Right of Privacy Act. It also set the conservative crime control agenda on a collision course with the White House.

• • •

We encountered the basics of the *Katz* story at the end of Chapter 6. Charles Katz was a prominent Los Angeles bookie who was caught illegally transmitting college basketball wagers from a public pay phone located a few blocks from his apartment.[51] Like *Berger,* the evidence in the case rested on incriminating conversations that the government recorded via concealed microphone. This time, the bug was affixed to the outside of the glass-enclosed phone booth.

Unlike *Berger,* however, the investigating officers in the *Katz* case took every precaution to minimize the intrusiveness of their eavesdropping operation. They didn't plant the listening device until they first secured material evidence to show that Katz was using the pay phone to conduct illegal activities. They switched off the microphone whenever they overheard conversations unrelated to bookmaking. And they took special precautions to avoid listening in on other callers. In total, the government obtained just six surveillance recordings, around three minutes apiece, over the course of the investigation—hardly a dragnet operation of the sort that had occupied the Court in *Berger v. New York.* The one problem was that the officers had neglected to secure a warrant to monitor the phone booth beforehand.

For the Court, the *Katz* case boiled down to two key questions: Was the bugging of a public pay phone a "search and seizure" by Fourth Amendment standards? And if so, did the government's use of electronic surveillance in the investigation violate constitutional guarantees? Court-watchers on both sides of the political spectrum expected the final decision to answer both questions in the vein of *Berger,* sending yet another warning to Congress about the legitimacy of wiretapping and bugging.[52] Instead the Court did something more complicated, so much so that generations of legal scholars have wrestled with its meaning ever since.[53]

In answering the first question, the majority reasoned that electronic surveillance constitutes a Fourth Amendment search because it involves intruding on conversations that an individual in Katz's situation would reasonably assume to have the privilege of carrying on in private. In the words of the opinion, "What [Katz] sought to exclude when he entered the booth was not the intruding eye—it was the uninvited ear. . . . No less than an individual in a business office, in a friend's apartment, or in a taxicab, a person in a telephone booth may rely upon the protection of the Fourth Amendment. One who occupies it, shuts the door behind him, and pays the toll that permits him to place a call is surely entitled to assume that the words he utters into the mouthpiece will not be broadcast to the world. To read the Constitution more narrowly is to ignore the vital role that the public telephone has come to play in private communication."[54]

Privacy, in this formulation, isn't a physical condition, solely dependent on protected spatial areas into which the government might or might not trespass. It is an *expectation* that individuals carry with them, a state of mind to be respected regardless of one's location in society. This was a lofty and somewhat nebulous legal conclusion—an argument that departed from decades of Fourth Amendment jurisprudence. But it yielded immediately tangible results for Katz: If electronic surveillance counts as a Fourth Amendment search, then it stands to reason that the government is constitutionally bound to secure a warrant before initiating it. The phone booth bug in the Katz case was thus invalid—above all else, because it lacked judicial preapproval. Had the investigating officers first secured a warrant, Katz would presumably have had little cause to appeal his conviction.

Katz v. United States is now regarded as one of the most consequential cases in the history of Fourth Amendment law. For better or worse, we still look to the language of the ruling, more than fifty years later, to determine what counts as a criminal search, and whether that search violates the constitutional rights that protect individuals.[55] On its face, *Katz* was a win for privacy. The decision redefined the Fourth Amendment to protect "people, not places," sweeping aside the more limited, property-based understanding of trespassory searches that *Olmstead v. United States* established four decades prior.[56] Henceforth, the state's use of phone taps and bugs would have to accord with the standard principles of search and seizure. The yawning legal loopholes that had enabled decades of police and government wiretap abuse were seemingly closed.

The move to bring electronic surveillance under Fourth Amendment protections came with an important trade-off, however. Beneath its civil libertarian surface, *Katz v. United States* essentially functioned as a political compromise—a concession, both implicit and explicit, to those who feared that the *Berger* ruling had rendered all forms of electronic surveillance unconstitutional. While *Katz* came down on the side of individual rights—and while the language of the opinion codified norms for protecting them against government intrusion—the decision also articulated the conditions under which wiretapping and electronic eavesdropping could be construed as permissible: when the investigating parties first acquire a warrant, signed by a neutral judicial authority; when the subsequent use of electronic surveillance is limited in duration and scope, avoiding an open-ended search for mere evidence; and finally, when the federal investigation in question touches on a matter of national security.[57] (In a much-debated footnote to the majority opinion, the Court deemed this last stipulation beyond the purview of the case, thereby leaving national security wiretapping an open field.)[58] In a single stroke the Supreme Court had neutralized the effects of the *Berger* ruling and relegitimized the wiretap.[59] In the process, the decision halted the momentum of Long's controversial Right of Privacy Act. The political implications were so drastic, so unexpected, that contemporary observers characterized the Court's decision as a "flip-flop."[60] What *Berger* killed, *Katz* resurrected.

The stage was now set for electronic surveillance—once considered an extraordinary investigative technique of last resort, used only in the defense of national security—to become an ordinary tool of law and order. All Congress needed to do to create a federal wiretap law consistent with the Constitution was to follow the general requirements for lawful police eavesdropping that the Supreme Court had outlined, threading the needle between the *Berger* and *Katz* decisions. The unabated national panic over riots and street crime afforded electronic surveillance advocates in the law-and-order camp the perfect opportunity to do it.

Months earlier, a small coalition of Republicans and southern Democrats, backed by the national law enforcement lobby, had discreetly proposed adding a wiretap authorization clause to the White House's "omnibus" crime control bill, then called the Safe Streets and Crime Control Act of 1967.[61] After *Katz,* they increased the volume of their calls, and by spring 1968 they managed to yoke the passage of Johnson's crime control measures—a critical component of the Democratic party's electoral platform that year—to the statutory authorization of law enforcement wiretapping. Forecasting an imminent social crisis, the rank and file of the law-and-order coalition relied on boilerplate ideas about "assisting law enforcement" and "stopping street crime" to rationalize their efforts. Taps and bugs became the technical weapons the government needed to keep America's streets safe in a time of unrest.

The politician to lead the charge for wiretap authorization was John L. McClellan, an Arkansas Democrat who styled himself as the Senate's "fiercest crime-buster."[62] Like many law-and-order conservatives of the 1960s, McClellan believed that the nation's crime problem was the result of a justice system gone soft.[63] Rights-minded Supreme Court decisions— from *Mallory v. United States* (1957) and *Miranda v. Arizona* (1966), which protected criminal suspects against self-incrimination, to *Berger v. New York,* which questioned the legality of electronic surveillance—had forced police to abandon the tactics that put crooks behind bars. The liberalization of the nation's crime laws had mollycoddled a new generation of criminals. According to McClellan, sentimental pieties about constitutional rights and civil liberties, newly ubiquitous in the age of the Warren Court, had the effect of handcuffing the law enforcers rather than the lawbreakers. Overemphasis on fairness and due process had ushered in a "golden age for criminals," who were free to "rob, rape, murder, and mug day and night with . . . impunity."[64] The future of America hung in the balance. "Everywhere the traditional agencies of social control are breaking up," McClellan asserted in a 1968 jeremiad on crime, typical of the apocalyptic views he espoused throughout the period. "We have now begun to reap the grim harvest."[65]

For McClellan, the only way to stave off the forces of lawlessness was to defy the courts and allow the government to make use of its full arsenal of crime-fighting tools, wiretaps and bugs especially. "The criminal element in this country . . . makes tremendous use of the telephone," he mused in 1967. "They use it as a weapon, as an instrument to further their own nefarious trade. Shall we say law enforcement shall not be permitted to use the same weapon?"[66] McClellan asked versions of that question time and again in letters, in editorials, and in public remarks delivered on the floor of the Senate: "Are we to permit . . . the hoodlums, the underworld, to continue to use [the telephone] but not combat it with the same effort, the same instrumentalities?" Almost without fail, he offered the same answer—"the end justifies the means"—and throughout the late 1960s he cannily used his leadership position in the Senate Judiciary Committee to hold hearings on crime and policing that helped amplify his ideas about the importance of government surveillance power.[67] Out of his efforts grew a suite of legislative proposals for legalizing police wiretapping that gained traction as the election of 1968 approached.[68]

McClellan was also a hard-line segregationist, and he believed that keeping special electronic tabs on African American civil rights organizations was essential to maintaining the social order.[69] (It's unclear whether he was aware that the FBI's secret Counter Intelligence Program, known as COINTELPRO, designed to disrupt the "subversive" activities of civil rights organizations and antiwar groups, was already doing just that.) The Johnson Crime Commission had limited its endorsement of police wiretapping to federal cases involving organized crime.[70] McClellan, by contrast, wanted much more, advocating for legalized wiretap use in "street crime" cases involving drugs, rape, kidnapping, murder, and—most important of all for the law-and-order crowd—incitements to riot. After *Katz* and the tumult of 1967–1968, he began exploiting race-baiting tactics to make his case. When McClellan proposed appending surveillance provisions to the Safe Streets Act in May 1968, for example, he openly played on reactionary anxieties about Black nationalist militancy to support a broadening of the government's wiretap mandate:

> Each crime . . . for which an electronic surveillance order may be obtained has been selected [in this bill] because it is either serious in itself or characteristic of organized crime or subversives. . . . You could bug a room or a hall in which [Stokely] Carmichael was meeting, in which [H.] Rap Brown was meeting, where they were inciting to riot, telling people to get their guns, 'Go get whitey,' and do this and do that. Do you want to take that out of the bill? . . .
>
> I do not know whether we are free from future riots. I do not say that we are going to have them. No one knows. I do say that every

effort, every protection, every instrumentality that is within the Constitution should be made available to law enforcement officers to try to prevent the planning for rioting if any such thing is going on and to try to prevent rioting from happening.[71]

Long took to writing articles in specialized technical journals to point out the flaws in McClellan's logic. "Eavesdropping techniques when used by law enforcement officers frequently do not improve effective law enforcement efforts," he remarked in an essay on crime control for the *Forensic Quarterly.* "There are many crimes which normally are not susceptible to eavesdropping no matter how sophisticated his gear may be. Rape, most murders, purse snatching, mugging, and other street crimes fall into this category."[72] So, too, did rioting. But in the wake of the uprisings in Detroit and Newark, McClellan's dog-whistle calls for wiretapping as a riot-prevention tool drowned out the civil libertarian sentiments that were dominant only a year prior. Electronic privacy advocates stood little chance to stem the political tide.

The Safe Streets Act left the Senate on May 23, 1968, with McClellan's expanded provisions for police wiretapping intact. Tragedy curtailed the House's much-anticipated debate on the amended version of the bill two weeks later. Shortly after midnight on June 5, an assassin's bullet struck Robert F. Kennedy in the head as he left a presidential campaign event celebrating his victory in the California Democratic primary. As the nation mourned Kennedy, the latest casualty in an explosive cycle of political violence, few lawmakers saw the wisdom in impeding the most sweeping public safety measure to come out of Washington in decades.[73] In the weeks that followed, the White House found itself caught in the middle. Looking back on several years' worth of campus protests, political assassinations, and race riots—and looking ahead at the Democratic Party's upcoming election fight against a presidential challenger, Richard Nixon, who had made crime control a defining issue of his campaign—Johnson had no choice but to heed the call of law and order. He caved to McClellan's demands. The Omnibus Crime Control and Safe Streets Act of 1968 was the law that resulted from the ensuing compromise.

Titles I and IV of the Safe Streets Act followed through on the most important recommendations of the Johnson Crime Commission: providing federal grants to the states to improve policing and public safety, and curtailing the interstate trade in handguns. In the middle were Titles

II and III, the triumphs of the conservative crime control push. Title II contained provisions designed to kneecap the legal protections for criminal suspects created by the Supreme Court's rulings in *Mallory v. United States* and *Miranda v. Arizona*. And then there was Title III, which established the first comprehensive system of electronic surveillance oversight in the nation's history.

*   *   *

In its final form, Title III of the Omnibus Crime Control and Safe Streets Act of 1968 was a contradictory mix of *Katz*-inspired privacy protections and law-and-order assertiveness. So it remains today. Aside from a few notable changes, the current structure of the law still looks much as it did in 1968.

Under the guise of safeguarding the "privacy of innocent persons," as the language of the act described it, Title III criminalized both private-sector wiretapping and the unlicensed manufacture, distribution, or advertisement of electronic eavesdropping equipment.[74] This was a crucial, if long overdue, step in the direction of privacy, codifying a prevailing consensus about the evils of the tap-and-bug trade that had been in place since the days of the 55th Street wiretap nest scandal. But the bulk of the statute was given to authorizing police wiretapping in cases that exceeded even Long's worst fears about the possible scope of state-sanctioned surveillance: espionage, sabotage, treason, riot incitement, kidnapping, robbery, murder, bribery, corruption, extortion, fraud, racketeering, drug trafficking, gambling, and counterfeiting—a "shopping list" of major crimes, in the words of Hiram Fong, one of Long's chief allies in the Senate.[75] Per *Katz*, the one crucial prerequisite to a lawful Title III wiretap was a warrant signed by a federal judge. The act also encouraged state police, when authorized to tap under state law, to obtain similar surveillance orders when investigating any crime "dangerous to life, limb, or property" punishable by more than one year in prison.[76]

The final shape of the Safe Streets Act wasn't a total loss for Americans who were concerned about the surveillance abuses that the Long Committee had uncovered. The law took significant steps toward constitutional fairness and public accountability. For a police official seeking an electronic surveillance order, Title III required showing probable cause to the presiding judge, and demonstrating an exhaustion of all other available methods of gathering evidence. This meant that law enforcement could only turn to taps and bugs as tools of last resort. Title III also placed strict limits on the duration and scope of police surveillance operations. Before Title III, a wire investigation might drag on for months at a time, trawling all manner of innocent communications in the hunt for incriminating information—this was the main reason the Supreme Court had overturned New York's warranted wiretap law, recall. After the Safe Streets Act, wiretaps and bugs couldn't be active for more than thirty days. Anything longer required another round of judicial approval. Moreover, according to Title III's critical "minimization" provisions, throughout the process police also had to take measures to avoid eavesdropping on conversations unrelated to criminal activity.

Perhaps most importantly, the Title III system created a paper trail—a bane of overzealous police efforts from time immemorial. In addition to subjecting each wiretap order to a rigorous application process, accessible to discovery at trial, the law required police to notify the targets of electronic surveillance within ninety days of a tap's discontinuation. It also required state and federal law enforcement to provide Congress with an annual accounting of all wiretap and microphone operations. The Administrative Office of the U.S. Courts still publishes its Title III-mandated *Wiretap Report* annually.[77] The American public's awareness of the expansion of law enforcement surveillance in the years since 1968 owes no small debt to the paperwork that Title III generated.

Still, the Safe Streets Act wasn't a triumph for privacy, at least not by the lofty standards that Long had set over the course of the 1960s. Other portions of the bill worked to undermine the constitutional protections that the court-order system appeared to establish. In a last-minute bargaining coup that outraged congressional liberals, McClellan managed to smuggle in a special provision authorizing police to tap and bug for forty-eight hours without a warrant in any "emergency situation." Even more controversial was an executive exception clause that allowed the White House to authorize warrantless surveillance "to protect the Nation against actual or potential attack or other hostile acts of a foreign power" or "to protect the United States against the overthrow of the Government by force or other unlawful means."[78] Crafted with the crucial *Katz* footnote in mind, likely at J. Edgar Hoover's suggestion, the provision appeared to grant the White House a unilateral right to monitor any individual or group deemed dangerous to national security.[79] After twenty-eight years of secrecy, confusion, and abuse, the Roosevelt doctrine was finally legal.

The law's apparent acquiescence to the executive branch's surveillance powers was perhaps the most obvious measure of how far the goalposts of electronic privacy seemed to have moved in the short time since the "victory" of *Berger v. New York*. Title III made government wiretapping permissible. Reversing decades of mainstream political consensus, taps and bugs now had the force of the law behind them.

On June 19, 1968, a doleful Johnson signed the Omnibus Crime Control and Safe Streets Act. A prominent White House official captured the mood of defeat among the liberal establishment when he wrote to Johnson to say that the bill's wiretapping provisions "may do more to turn the country into a police state than any law we have ever enacted."[80] Playing on Long's lost-cause legislation, the Democrats who opposed the inclusion of Title III referred to the final version of the bill as the "End of Privacy Act."[81]

In a statement presented to reporters at the bill's signing, Johnson remarked that the Safe Streets Act contained "more good than bad. . . . [I]t is in America's interest that I sign this law today." But he also underscored that legalizing police wiretapping was an "unwise and dangerous step," a concession to conservative hysteria over race and crime. Johnson was well aware that the law-and-order faction had achieved what was unthinkable only a few years prior. He closed by making a dramatic appeal for corrective action:

> I call upon the Congress immediately to reconsider the unwise provisions of Title III and take steps to repeal them. If we are not very careful and cautious in our planning these legislative provisions could result in producing a nation of snoopers bending through the keyholes of the homes and offices of America, spying on our neighbors. No conversation in the sanctity of the bedroom or relayed over a copper telephone wire would be free of eavesdropping by those who say they want to ferret out crime. . . . We need not surrender our privacy to win the war on crime.[82]

Long, for his part, had made an eleventh-hour motion to strike Title III from the text of the Safe Streets Act. He was defeated by a vote of 68–12. Less than two months later he lost his bid for reelection in the Missouri Democratic primary.

Chapter 8

Big Brother, Where Art Thou?

EDWARD V. LONG'S RIGHT OF PRIVACY ACT represents a fascinating road not taken in the history of wiretapping and electronic eavesdropping in the United States. In hindsight, the demise of Long's daring legislative proposal, now largely forgotten, marks the onset of a tectonic shift in the politics of electronic surveillance in America—one that still reverberates today. Almost overnight, what was embattled became mundane. Rendered constitutional after *Katz,* and rendered procedurally routine after the passage of the Safe Streets Act, police wiretapping proliferated during the late 1960s and early 1970s. Between 1968 and 1973, the earliest years the federal government kept public records of its own electronic surveillance activities, warranted wiretap and bug installations increased fivefold at both the state and federal level.[1] The vast majority were approved in gambling investigations. As the 1970s wore on, Title III surveillance moved increasingly into the domain of narcotics enforcement, a transition we'll explore more fully in Chapter 10. In all cases, it's likely that Title III surveillance rates were in fact much higher than the early editions of the federal *Wiretap Report* indicate. Government officials routinely indulged in sly "numbers games" to minimize the amount of warranted taps on the public record.[2]

It's hard to comprehend just how improbable this turn of events was at the time. The use of phone taps and bugs in criminal investigations now arouses little in the way of public comment, much less criticism or concern. Most of the nation's fears about eavesdropping seem to focus on the forms of electronic surveillance that the federal government conducts in the name of national security, not crime control. (To be sure, calling a wiretap a "crime control device," as law-and-order advocates did throughout the sixties and seventies, sounds tellingly foreign to our ears today.) But the disorienting fact of history is that it wasn't always this way. Conservative policymakers had worked tirelessly to pass a law like Title III in the forty years between *Olmstead v. United States* and the Safe Streets Act. More than fifty wiretap authorization bills made it to the floor of Congress in that chaotic interval—so many, and so often, that Long, writing early in his tenure in the Senate, could compare the predictability of their appearance in Washington to that of the seven-year locust.[3]

Yet legalized electronic surveillance remained a nonstarter for all that time on account of a basic political fact: most Americans didn't like wiretapping. It still seemed illicit and immoral, a "dirty business" to be conducted only in the most extreme cases. The turmoil of the late 1960s changed all of that. The panic over urban disorder made wiretapping seem like a palatable solution to federal crimes of all sorts.[4] If the *Katz* case opened the door for legalized government wiretapping, the politics of law and order—which is to say, the politics of race—pushed it through. The nation was left with a wiretap policy that a coterie of federal lawmakers hurried through Congress in the spirit of opportunism.

More disorienting still, no one seemed to care.[5] One of the most peculiar dimensions of the dispute over wiretapping and street crime is how quickly it dissipated. Given the outpouring of grassroots support the Long Committee received in the 1960s—and given the fact that the issue of electronic privacy became "good politics" for liberals and conservatives alike in the 1970s, particularly in the wake of the Watergate scandal—we might expect privacy-minded factions in Congress to have revisited Lyndon Johnson's desperate calls to repeal Title III.[6] By 1969, however, signs were already pointing to an end to the great debate over the use of electronic surveillance in criminal investigations. Just eleven months after the passage of the Safe Streets Act, a New York state law enforcement official told the House Select Committee on Crime that "it is about time that the argument that wiretaps and electronic eavesdropping do not materially contribute to criminal investigations . . . should finally be put to bed."[7] Despite the fact that the group on the dais consisted of several representatives with ideological leanings similar to Long's, no one so much as batted an eyelash. By April 1970 John McClellan, the primary sponsor of Title III in the Senate, could claim that eighteen months' worth of legalized law enforcement wiretapping had vindicated the wisdom of the new system.

"Never again should we hear expressed a doubt that wiretapping is needed to break the back of organized crime in America," McClellan wrote in the pages of the *American Legion Magazine*.[8] What he didn't mention was that only a handful of warranted wiretaps were in operation at the time. Fewer had yielded tangible results in court.

Acceptance of the new state of affairs in Washington only seemed to deepen in the early 1970s. The compromise that produced the final text of the Safe Streets Act stipulated that Congress convene a special task force to assess the effects of Title III after a period of five years. The National Wiretap Commission (NWC), as the group came to be known, published its final report in 1976. After visiting forty-six states, conducting hundreds of interviews, and holding seventeen days of public hearings, the NWC issued official conclusions that merely "reaffirmed the finding[s] of Congress in 1968," claiming that "electronic surveillance is an indispensable aid to law enforcement" and asserting that "the interception of a wire or oral communication . . . should be allowed only when authorized by a court of competent jurisdiction."[9] In an irony that we'll return to at the end of this chapter, the NWC's one major recommendation was to *expand* the list of crimes for which federal law enforcement agencies could employ Title III surveillance. Here again, the push to enlarge the purview of police wiretapping encountered scant resistance. By that time, almost 70 percent of Americans condoned the use of electronic surveillance by state and federal law enforcement agencies when properly warranted—a noteworthy jump from the meager 46 percent of Americans who felt that way at the moment the Safe Streets Act became law.[10]

Calling Title III the "End of Privacy Act" was thus right in a way that the civil liberties advocates of Long's generation would never have dreamed. The law's passage marked the end of the decades-long era in which political opposition to police wiretapping was mainstream. Whether they knew it or not, a growing majority of Americans had come to accept the wisdom of the new order. The legitimization of the wiretap was complete.

. . .

Over time, the consensus surrounding the merits of Title III helped to sideline nagging questions about the warrant system's ability to safeguard the "privacy of innocent persons," per the law's stated objective. In the fifteen years following the passage of the Safe Streets Act, state and federal law enforcement agencies submitted more than 8,600 applications for Title III wiretap installations. Only 20 were denied.[11] The high rate of approval seemed to confirm a prediction that Long and his supporters made early on: that the court-order process would merely serve a ceremonial function, enabling judges to rubber-stamp most any wiretap application that crossed their desks. Law enforcement officials met even higher rates of approval for extensions of thirty-day wiretap installations, raising concerns about the law's ability to minimize the scope and duration of any given surveillance operation in accordance with Fourth Amendment standards.[12] A practice that came to be known as "judge shopping" also emerged as an endemic problem in the early 1970s. In many jurisdictions, police and prosecutors had the leeway to seek out judges who were known to be favorably disposed to signing Title III wiretap orders. This enabled them to circumvent more cautious judges who might find cause to reject a proposed surveillance operation.[13]

Despite these procedural challenges, by the mid-1970s most experts believed that the Title III system worked. Where its success remained uncertain was in its role as a deterrent against illegal wiretapping, outside the realm of law enforcement. Beyond the ambiguous goals of controlling crime and maintaining social order, eradicating the private-sector eavesdropping industry was a key impulse behind the wiretapping provisions of the "ominous omnibus bill," as one interested onlooker dubbed the Safe Streets Act.[14] If there was one point of agreement between liberals and conservatives on the question of electronic surveillance reform in the 1960s, this was it. Yet Title III's record in the fight against the scourge of the bugging business would prove uneven at best. A brief look at its history in this period shows just how little the Safe Streets Act did to eradicate the phone tap and the bug from the areas of American society where it had thrived for more than a century.

It's difficult to find reliable statistics testifying to the amount of illegal wiretapping in America before and after the passage of the Safe Streets Act. Studies conducted in the 1960s vary widely in their estimates, suggesting that individuals and firms privately offering bugging and debugging services did anywhere between $1.5 million and $1 billion worth of business annually.[15] An oft-cited 1964 report published in *Time*—the article that launched the Long Committee investigation—claimed that the number was closer to $20 million.[16] Whatever the actual size of the industry, all of the studies of the period agreed that the lion's share of its dealings occurred outside of official channels. The Long Committee came to the same conclusion. Most electronic surveillance in America was happening off the government's books, more often than not in the same legal gray areas that had driven the private eavesdropping market for decades.[17]

Despite the gradual liberalization of the nation's divorce laws, marital wiretapping remained common throughout the 1960s.[18] Corporate uses of electronic surveillance were equally pervasive. A 1966 ABC News report suggested that as many as one out of five American businesses used miniature listening devices to monitor their employees and competitors, a rate that corroborated a prominent Harvard Business School study from the previous decade.[19] Even if the numbers were inflated—a measure less of the vitality of the bugging business than of the intensity of national anxieties surrounding the "invasion" of transistorized listening devices, per Chapter 6—accounts from industry insiders suggested that corporate use of electronic surveillance equipment was both widespread and routine in the years leading up to the passage of the Safe Streets Act.

In his tell-all book *The Electronic Invasion* (1967), for example, the electronics engineer Robert M. Brown divulged a list of twenty-four prominent corporate entities that had purchased audio surveillance equipment from Consolidated Acoustics of Hoboken, New Jersey, a middle-of-the-road retailer of listening devices. The list included household names like American Airlines and Avis Rent-a-Car; Chevron Oil, Chrysler, and Coca Cola; Encyclopedia Americana and Prudential Insurance; Walt Disney, Western Union, and Westinghouse Electric.[20] What uses a rental car agency or an insurance provider—much less an encyclopedia publisher—might have had for a hidden microphone we must leave to the imagination. The diversity of the clientele is telling in itself.

The passage of the Safe Streets Act in 1968 led to all manner of official pronouncements about the end of illegal eavesdropping in America. By most accounts, the new regime was a resounding success. In 1971, Attorney General John Mitchell testified to Congress that private-sector wiretappers were "extinct" as a result of Title III. "Because of the penalties involved," Mitchell reported, "the force of the federal law has practically run them out."[21] The NWC was less grandiose in its assessments five years later. But the commission's majority came to the same general conclusion about the criminalization of unsanctioned electronic surveillance, stating that "Title III has reduced the incidence of . . . illegal interceptions through its controls on the manufacture, sale, and advertising of surreptitious devices and its criminal sanctions for their use."[22]

The official narrative of the slow demise of the bugging business both reflected and reinforced a gathering sense that the "private ear"—the great bugaboo of the American privacy scares of the fifties and sixties—had become something of an obsolete American professional, left behind by new policies and changing times. In February 1971, freelance wiretapper Bernard Spindel, the real-life inspiration behind *The Conversation*'s Harry Caul, died in prison while waiting to appeal a Title III conviction, the first criminal charge against him to stick after more than three decades of work on both sides of the law. Spindel's obituaries would marvel at the breadth of his clientele and his mastery of the field, more often than not in a vein that can only be described as nostalgia. The *New York Times* memorialized Spindel as "no. 1 big-league freelance eavesdropper and wiretapper in the United States," as though reporting on the death of a baseball star.[23] A few months later the *Times* announced the passing of William Mellin, a retired Treasury Department investigator known affectionately as the "dean of wiretappers," whose specialized skills had secretly helped to bring down the old-time bosses of syndicated crime: Legs Diamond, Lucky Luciano, Dutch Schultz.[24] The peculiar wistfulness of such accounts betrays an important assumption about the world that Title III appeared to have wrought. By the early 1970s, the reign of the private ear seemed distant enough to have become the stuff of legend. Tapping for hire was romanticized as a colorful pastime of a bygone era, perhaps another reason Coppola's *Conversation* struck a nerve when it did.

And yet, digging deeper into the historical record reveals a slightly different story about illegal wiretapping and the bugging business in the 1970s. Aside from a few high-profile national cases, like that of Spindel—or like that of the Watergate burglars, who were themselves brought to trial for the unlicensed possession of electronic eavesdropping devices—violations of Title III's illegal wiretapping clause in fact proved as difficult as ever to investigate and prosecute. Between 1969 and 1974, American Telephone & Telegraph discovered 1,457 eavesdropping devices unlawfully installed on the wires and phones of its subscribers throughout the Bell System. A mere 27 of them, less than 2 percent, resulted in an arrest.[25] Even fewer led to criminal charges. The NWC dismissed statistics like these as drops in the telecommunications ocean: U.S. phone companies received an average of 10,000 "tap check" requests each year, but the percentage that yielded real, live wiretaps was infinitesimally small compared to the number of telephones in use on a daily basis.[26] "The average citizen's fears that he might be the victim of electronic surveillance are mainly unjustified," the NWC concluded in 1976.[27]

Yet less-heralded government studies conducted around the same time showed that the business of illegal wiretapping still flourished, despite the period's economic woes. When a special subcommittee of the Senate Committee on the Judiciary surveyed 115 detective firms around the country in 1976, as many as 42 admitted to offering clandestine wiretap services.[28] "I won't do it, but, sure, the other guys will," one private investigator explained. "But you can bet your life they'll never admit it. The biggest demand is still from the husband who's curious about what the wife's up to back at the house."[29] Another "basement operator" (a colloquial term for a professional wiretapper that came into use in the postwar period) admitted to the American Civil Liberties Union that business was so good "it was like Title III never happened."[30] Media attention to the Watergate scandal only enhanced the industry's profile—and perhaps even its profits.[31]

One reason the illegal eavesdropping business remained unchanged in the 1970s was that the federal government had limited resources for pursuing potential Title III violations.[32] Another was creative rebranding. After 1968, private manufacturers and retailers of electronic surveillance equipment began courting unexpected markets for their services. Hoping to exploit potential loopholes in Title's III's illegal wiretapping clause, many moved to repackage the tools of the eavesdropping trade for the American consumer. Wiretap devices were frequently advertised in newspapers and magazines as "answering machines" for homes and offices. Bugs were often billed in trade catalogs as "wireless microphones" for concert halls and television studios. The transistor radio—long employed in remote eavesdropping operations and given to government informants wired for sound—led an improbable double life on the open market as a "baby monitor" throughout this period.[33] The same technology used to spy on powerful mobsters and crooked politicians could also help parents tell if their children were sleeping.

The attempt to market eavesdropping technologies by other means was a somewhat transparent twist on a tried-and-true story about electronic surveillance that first took root in the age of the transistor: if a household item could be transformed into a listening device, so, too, could a listening device be transformed into a household item. The gimmick kept many manufacturers afloat throughout the 1970s, despite its painfully obvious absurdities. In 1976, a supplementary government report on surveillance technologies contracted by the NWC found that state-of-the-art listening devices known variously as "harmonica bugs" and "infinity transmitters," designed to turn home telephones into remote room amplifiers, were often listed as "home security systems" in electronics equipment catalogs:

> Now you can check your premises anytime from anywhere. This system consists of the monitor unit which plugs into standard telephone jacks supplied with each system. This unit is placed at the site to be monitored. The remote activator will allow you to activate the monitor unit from any dial telephone. The receiver amplifier unit which sits by your bedside telephone continuously monitors your premises even while you are asleep. The telephone never rings. You won't alert a burglar by a ringing phone. The telephone in your office or business won't make the slightest sound. It will sit there just as innocently as ever while you safely monitor your premises.[34]

Advertisements for such devices typically included ostentatious legal disclaimers—"WARNING: THIS IS NOT A BUG"—to prevent authorities from regulating their distribution and sale under Title III.[35] Engineering firms that sold electrical equipment to educators and hobbyists took advantage of the same ambiguities in the law. Edmund Scientific, one of the largest suppliers of educational instruments in the United States, included a "bionic ear" in its trade catalogs well into the early 1990s. The company scarcely made an effort to obscure the device's original use: "Designed with personal security in mind. . . . this portable electronic listening tool weighs under 1–1 / 2 pounds with the headphones included. It is the size of a flashlight, and is powered by one common 9 volt battery. . . . . Applications include: bird watching, security law enforcement, hunting, and hearing impaired."[36]

Creative marketing wasn't the only reason the eavesdropping industry managed to survive the Safe Streets Act. Far more important was the fact that Title III made it legal for manufacturers and retailers of electronic surveillance devices to sell their equipment to the government. If in the 1950s and 1960s the bugging business primarily catered to aggrieved spouses and anxious executives, in the 1970s the most reliable customers were police detectives, at long last armed with the authorization to use taps and bugs in routine criminal cases. State and federal agencies had of course worked with surveillance hardware firms for decades, purchasing equipment for eavesdropping operations that delicately toed the line of the law. In fact, one of the Long Committee's first moves was to command the public testimony of three of the government's biggest suppliers: Criminal Research Products, based in Conshohocken, Pennsylvania; Mosler Research Products, based in Danbury, Connecticut; and Fargo Corporation, with branches in San Francisco and Washington, D.C.[37]

Yet even as Title III worked to limit the illegal eavesdropping trade in other contexts, the law had the unintended effect of sanctioning the clandestine relationships between government agencies and wiretap device firms that had developed in the postwar period. The owner of Spy Shop, a notorious surveillance technology retail store located in Washington, D.C., explained the paradox in a 1971 interview: "When a customer walks in, we would tell him that we don't sell bugs or wiretap equipment. If he identified himself properly as being from government or law enforcement, we'll sell bugging equipment with a contract. What we sell to [them] is legal."[38]

The legitimization of law enforcement surveillance, in other words, meant the legitimization of the sale of surveillance technologies.[39] Title III formalized an economy that was otherwise informal. Private-sector firms that managed to secure contracts with state and federal agencies reaped the biggest rewards.

The unintended consequences of Title III were never more apparent than in the summer of 1974, when a trade fair held in Moscow, of all places, created an unusual public controversy over the newly sanctioned reach of the American eavesdropping industry. The spat centered on *Krimtehnika,* an annual Soviet police convention that showcased surveillance and crime-control technologies. Such events were common enough in the United States after the Safe Streets Act legalized police surveillance. (The long middle section of *The Conversation,* which follows the wiretapper Harry as he attends a surveillance convention, was in fact shot on location at a real law enforcement technology showcase in San Francisco.) The difference, in the case of *Krimtehnika,* was that the event took place in the Soviet Union, then widely known for its repressive police tactics. What's more, for the 1974 meeting the Russians had taken the unprecedented step of inviting American firms to participate.

As early as November 1973, specialized research and development entities across the United States began receiving written requests from the Soviet Ministry of Internal Affairs to display new devices in Moscow the following summer. According to documents that Congress subpoenaed in the wake of the affair, the desired technical categories included "Equipment for Crime Scene Investigations," "Laboratory Equipment and Instruments," "Registration Equipment," "Mobile Crime Scene Laboratories," "Searching Devices," "Photo Cinematographic Equipment and Fittings," "Systems and Devices for Video and Sound Recording," and "Communication Facilities."[40] The Soviets also expressed a special interest in walk-through metal detectors, tear-gas guns, and—in an unsurprising twist—cutting-edge eavesdropping devices.

In total, ninety-four American firms received an invitation to participate in *Krimtehnika '74*. The vast majority declined the opportunity outright. But a few expressed a preliminary interest in courting the Soviet law enforcement market. By June 1974, two private firms had signed official agreements to set up display booths in Moscow. The first was Voice Identification, Inc., an audio device company that had received an ungrammatical request the previous fall to "demonstrate maximum instruments" at the convention.[41] The Soviet ministry had a special interest in displaying Voice Identification's Series 700 Sound Spectrograph, a primitive version of what we now call "voiceprint" technology, which helps to identify voices captured on surveillance recordings.[42] The second firm was Criminal Research Products, whose work supplying the federal government with wiretap equipment had attracted the attention of the Long Committee a decade earlier. With Title III permitting the sale of electronic surveillance devices to law enforcement agencies at home—and without a law on the books explicitly barring the distribution of those same devices to similar entities abroad—both firms decided to proceed. American eavesdropping technologies were heading to Moscow.

News of the *Krimtehnika* contracts broke on July 7, 1974, just two and a half weeks before the Supreme Court ordered President Nixon to relinquish his embattled Oval Office tapes, a decision that initiated the Watergate affair's explosive final phase.[43] Notwithstanding more pressing political concerns, the stir on Capitol Hill was immediate. Representatives of both parties rushed to denounce the exportation of eavesdropping devices to the Soviet Union. In an emergency meeting of the Senate Committee on Government Operations, an enraged member of Congress claimed that the "sale of police equipment to a police state, where it will almost certainly be used to increase the efficiency of repression and intimidation, is so alien to our traditions and so contrary to our values that it ought not to be licensed by this government."[44] Another senator likened sending eavesdropping devices to *Krimtehnika* to "exporting gas chambers to Hitler."[45]

"I'd hate to have weapons of oppression in the Soviet Union or any other country carry the mark 'Made in America,'" he continued. "I'm for détente and I'm for trade, but this equipment could be used for the suppression of dissent and minorities."[46]

In one of his last official acts before Congress voted to impeach him—a supreme irony of history, all things considered—Nixon successfully negotiated a prohibition on the advertisement and sale of American-made eavesdropping devices across the Iron Curtain. American participation in *Krimtehnika '74* was scuttled in the process. But there was hypocrisy in the deal, a contradiction in terms that went well beyond the obvious fact that illegal wiretapping was at the top of the Nixon administration's long list of crimes. Officials from both Voice Identification, Inc., and Criminal Research Products were quick to point it out.

"The equipment [going to *Krimtehnika*] was the *same equipment* used by police departments in the United States," a sales representative from Criminal Research Products complained to the *Los Angeles Times* when the Soviet trade show opened its doors in August 1974, his ticket to Moscow canceled. "It is not equipment used for oppressive things."[47]

Why were American eavesdropping firms prohibited from hawking their wares to police and security agencies abroad when it was already legal to sell the same devices to state and federal law enforcement at home? Couldn't government wiretapping in the United States amount to "repression and intimidation" and "suppression of dissent" too?

The questions made a perverse sort of sense. There were major differences between government surveillance in the United States and government surveillance in the Soviet Union. After 1968, Americans had legal mechanisms in place for controlling wiretap abuse and protecting individual rights. The Russians did not. All the same, the *Krimtehnika* controversy illustrated, if only for a fleeting moment, the contradictions and failures of Title III in its regulatory infancy. In the summer of 1974—six years into the life of the Safe Streets Act, as the nation steeled itself for a final showdown with a president who seemed to have bugged everyone, including himself—illegal wiretappers still went unprosecuted and eavesdropping technologies still went unregulated. Private-sector surveillance still operated according to the principles of the market rather than the rules of law. One government study concluded in 1976 that electronic eavesdropping, both sanctioned and not, remained "uncontrollable in a meaningful democratic sense," even with a law like the Safe Streets Act finally on the books.[48] Proclaiming Title III an unqualified victory for privacy was thus inaccurate, if not risky.

In December 2018, I sat down for an interview at the home of Ralph V. Ward, a former sales executive at Mosler Research Products. During the 1950s and 1960s, Mosler was one of the largest suppliers of eavesdropping equipment in America. Now in his nineties, Ward was eager to share his experiences in the surveillance industry. He even went so far as to type out a short memoir of his seventeen years of work for Mosler, which started in 1955 when he answered a classified advertisement in a D.C. city paper. At the time Mosler was known simply as Research Products, Inc.; the firm had offices overlooking the Potomac River in Washington's Georgetown neighborhood.

After exchanging pleasantries over tea, Ward drew attention to a small leather carrying case perched in the chair next to him. It was an unassuming item that resembled an old-fashioned doctor's bag. This was a "Schmidt Kit," a portable toolbox for electronic surveillance technicians designed by Ward's business partner, Kenneth H. Schmidt, one of the pioneers in the tap-and-bug field. Everything inside Ward's Schmidt Kit was secured in a snug compartment: a tone generator for testing electrical wires and pliers for stripping lines; alligator clips for connecting taps and headphones for listening in. Ward sold hundreds of them to government agencies over the years—it was Mosler's best-selling item. He received six different security clearances so that he could travel between federal buildings and make sales pitches for various devices in the Mosler line. This was his main job until the corporate electronics giant Westinghouse acquired the firm in the early 1970s, at which point he went off on his own to pursue more lucrative avenues in the private security field.

When I asked Ward about Title III's effect on the American eavesdropping industry, he seemed nonplussed. He eventually told me that the measures the government put in place to curtail the private manufacture, sale, and use of wiretapping and bugging equipment made little difference to Mosler's day-to-day operations.

"The law didn't help, the law didn't hurt," Ward told me. "It really didn't change anything."[49] With a broad smile, he shrugged his shoulders. Our conversation moved on.

. . .

Electronic baby monitors and Soviet police conventions aren't what Americans typically talk about when they talk about wiretapping in the sixties and seventies. Nor, for that matter, are Edward V. Long and the Safe Streets Act.

Thus far we haven't encountered the crimes that the FBI and other government agencies committed in the name of national security throughout this period: the bugging of the Warren Court; the tapping of the homes and offices of members of Congress; the systematic effort to monitor student protestors and antiwar activists, "communists" and "subversives," artists and writers with dissenting beliefs of various sorts.[50] We haven't discussed the relentless surveillance of African American political leaders and civil rights organizations during the 1960s: Martin Luther King Jr., Malcolm X, Elijah Muhammad, and Stokely Carmichael; the Southern Christian Leadership Conference, the Student Non-Violent Coordinating Committee, the Nation of Islam, and the Black Panthers.[51]

The National Security Agency falls outside of my account here, despite the fact that its most highly classified eavesdropping programs—codenamed SHAMROCK and MINARET—provided critical background information to many of the U.S. government's political surveillance operations.[52] (In those days, the joke was that the capacities of American signals intelligence, which already dwarfed those of the FBI and the CIA, were so secret that the acronym "NSA" stood for "No Such Agency" or "Never Say Anything.")[53] And then we've barely scratched the surface of the Nixon administration itself. The wiretaps, planted by White House operatives, that monitored the telephones of journalists and National Security Council staffers. The tiny microphones discovered in a White House office safe, wired for sound inside tubes of Chap Stick (see Figure 8.1).[54] The fact that a sitting president's conspiracy to obstruct justice wound up on tape for the world to hear.

These are well-known wiretaps—some of the most discussed in American history. Their renown, a direct result of the fallout from the Watergate and Church Committee investigations, would help to make the 1970s synonymous with government eavesdropping. Yet both then and now, our attention to the doings of the national security state has tended to obscure the historical forces that ended up underwriting more prosaic forms of state surveillance. The eavesdropping scandals of the period confirmed

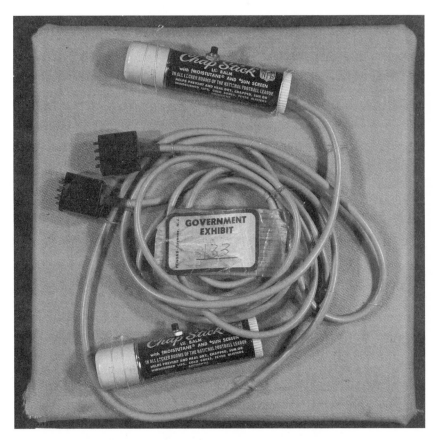

Figure 8.1. "Chap Stick Tubes with Hidden Microphones" (1972). Government Exhibit 133, *United States v. G. Gordon Liddy, Eugenio Martinez, Frank Sturgis, E. Howard Hunt, James McCord, Bernard Barker, and Virgilio Gonzalez* (1972). National Archives and Records Center, 304967.

what Americans had long suspected about the perils of consolidated executive power and excessive government secrecy; both Congress and the courts took a momentous series of steps to combat them. But Title III slipped into the background of American life along the way. Police wiretapping, long disparaged as an affront to civil liberties, became *normal*.

Regardless of how one assesses the validity and value of Title III surveillance today—and there is no doubt that the wiretap continues to serve as a vital law enforcement tool when employed responsibly under the law's baroque guidelines—the fact that police wiretapping remains normal shows just how much the terms of the national debate have shifted over time. While public acceptance of the Title III system has stayed more or less constant since the early 1970s, government wiretapping for "national security" has occupied a near-perennial state of political crisis. This contradiction is the final story of wiretapping and electronic eavesdropping in the sixties and seventies, the story perhaps most frequently overlooked. It begins in the months after the Safe Streets Act became law, when officials at the Justice Department lobbied to indict a group of political activists for conspiring to incite a riot at the 1968 Democratic National Convention in Chicago.[55]

The defendants in the case—popularly known as the "Chicago Eight," a veritable who's who of sixties radicalism—were wiretapped without a warrant in the months leading up to their arrest. But to stave off procedural complications at trial, John Mitchell, then newly appointed as attorney general, devised a preemptive legal strategy to justify the government's actions. In a thirty-four-page memorandum that the historian Jeff A. Hale has characterized as "one of the most sweeping assertions of Executive power in the history of the United States," Mitchell declared that Title III's national security provision empowered the president to "authorize [wiretap] interceptions in certain specified instances involving the security of the nation."[56] The argument became known as the "Mitchell doctrine," a repackaged version of Roosevelt's secret 1940 national security directive, which, by the late 1960s, stood behind more than 9,200 warrantless wiretap and microphone installations on alleged enemies of the state.[57] The idea was that it was legal for the White House, via the attorney general, to use electronic surveillance on anyone who posed a threat to domestic order. No oversight necessary.

The Nixon Justice Department unveiled the Mitchell doctrine at a pretrial hearing for the Chicago Eight held on June 13, 1969. Mitchell's aggressive interpretation of Title III came under fire almost immediately. Two weeks later, the American Civil Liberties Union filed a countersuit against the Justice Department on behalf of the Chicago defendants and nine other organizations associated with the civil rights and antiwar movements.[58] Several other legal challenges to the warrantless wiretapping of radical activists made inroads in 1970–1971.[59] Yet even as the cases against the Mitchell doctrine made their way through the courts, officials at the White House and the Justice Department considered the issue of national security surveillance settled. On the shaky ground of Mitchell's argument, the Nixon administration adopted increasingly audacious electronic measures to protect its political interests. It was in this period, for instance, that Nixon and Secretary of State Henry Kissinger, working with Hoover, hatched a plan to wiretap fourteen NSC staffers and three journalists in order to discover the source of White House news leaks.[60]

It wasn't until 1971, in a trial involving a radical leftist group charged with detonating a bomb outside a CIA office in Michigan, that the Justice Department's strategy finally began to crumble.[61] In that case—*United States v. Sinclair, Plamondon, and Forrest* (1971)—federal authorities had made extensive use of wiretaps and bugs with Mitchell's explicit stamp of approval. When the prosecuting attorneys again cited Title III's national security provision to prevent the court from excluding the wiretap evidence at trial, the presiding district judge—Damon J. Keith, a figure whose name would eventually become synonymous with the episode—pushed back, authoring an impassioned opinion that rejected warrantless wiretapping as an abuse of executive power. Keith did little to hide his contempt for the Nixon administration's attempt to use claims about "national security," perennially in the eye of the political beholder, to circumvent the authority of the Fourth Amendment.

"We are a country of laws and not men," Keith wrote. "Such power held by one individual [the president] was never contemplated by the framers of our Constitution and cannot be tolerated today."[62] The statement would prove prophetic.

So began the high-stakes legal battle that became *United States v. United States District Court* (1972), also known as the Keith case, the Supreme Court's third major electronic surveillance ruling of the period. The Nixon administration appealed Judge Keith's decision in June 1971, and the Mitchell doctrine made its way to Washington the following year. In a unanimous decision handed down on June 19, 1972—just two days after the Watergate break-in—the Court declared the warrantless wiretapping of U.S. citizens unconstitutional, even in cases involving perceived threats to national security. The ruling was categorical about the dangers of the Mitchell doctrine:

> History abundantly documents the tendency of Government—however benevolent and benign its motive—to view with suspicion those who most fervently dispute its policies. Fourth Amendment protections become the more necessary when the targets of official surveillance may be those suspected of unorthodoxy in their political beliefs. The danger to political dissent is acute where the

Government attempts to act under so vague a concept as the power to protect "domestic security." Given the difficulty of defining the domestic security interest, the danger of abuse in acting to protect that interest becomes apparent.

The price of lawful public dissent must not be a dread of subjection to an unchecked surveillance power. Nor must the fear of unauthorized official eavesdropping deter vigorous citizen dissent and discussion of Government action in private conversation. For private dissent, no less than open public discourse, is essential to our free society.[63]

The soaring rhetoric of the Supreme Court's decision led the press to extol the outcome of the Keith case as a triumph of the rule of law over an attempt to usurp deliberative democracy.[64] The assessment wasn't without cause. The Keith ruling amounted to a stunning rebuke to the Nixon administration, particularly given that the Court's newest member, Lewis F. Powell, authored the opinion to the case. Powell was a Nixon appointee with a long record of support for government wiretapping.[65] He had even aligned himself with the executive branch's expansive national security claims during his confirmation hearings the previous fall.[66]

The Keith case provided the opening act of an extended political drama over wiretapping and national security that would unfold in fits and starts over the course of the 1970s. The final installment came six years later, in 1978, when Congress passed the Foreign Intelligence Surveillance Act (FISA). The law, at long last, established an elaborate system of oversight for the use of electronic surveillance in national security investigations. After 1978, national security wiretaps were to obey all of the warrant procedures outlined in Title III. The key difference was that a secret court assumed the responsibility of judicial administration, a capitulation to the necessary confidentiality of foreign intelligence operations.[67] At its signing, FISA was praised as a triumph of bipartisan cooperation: a law to heal the wounds of political trauma, and a system to prevent future lapses in good governance. The bill's primary architect called it a "landmark in the development of effective safeguards for constitutional rights" and a "triumph for our constitutional system of checks and balances."[68] Order seemed to be restored.

Historians of electronic surveillance typically recount the period from the Keith case to FISA—from the Watergate hearings of 1973–1974 to the Church Committee investigations of 1975–1976—as a period of *revelation* and *response*. These were years when state surveillance secrets were exposed, critiqued, and corrected; years when Americans' longstanding suspicions about illegal government wiretapping were confirmed on the record. FISA, in this narrative, stands out among the tide-turning legislative actions that Congress took in the wake of the political scandals of the day. Responding to the discretionary abuses of both the White House and the intelligence community, lawmakers in Washington seemed to have moved to restore the confidence of the American people. The paranoid drama of the "age of the listening device" appeared to end on a positive note.

We now know that another drama was already beginning to unfold backstage. As with many of the intelligence oversight laws that emerged out of Washington's volatile season of inquiry, FISA's check on the federal government's surveillance powers would quickly prove hollow.[69] By 1980, just two years after the triumphant passage of FISA, it was already clear that the secrecy of the Foreign Intelligence Surveillance Court's proceedings had the potential to shield the intelligence establishment from meaningful public accountability.[70] And in the meantime, Title III expanded its foothold in an invisible political give-and-take that the panic over national security surveillance had only served to facilitate.

In February 1974, for instance, the Supreme Court handed down a minor ruling that ran counter to much of the pro-privacy politics of the day. Now largely relegated to the footnotes of doctrinal history, the case of *United States v. Kahn* (1974) established that wiretap intercepts could legally incriminate not just the named target of a Title III investigation but anyone the police happened to overhear on the line along the way.[71] From the perspective of the civil liberties advocates of Edward V. Long's generation, the outcome would have signified an egregious affront to constitutional rights. Amid the furor over Watergate's "dismal catalog of abuse," as the *Washington Post*'s editorial board described it, *Kahn* came and went without comment.[72]

The National Wiretap Commission's final recommendation to expand the investigative purview of Title III was met with a similar degree of public indifference. In a telling coincidence—low-hanging fruit, once again, for the paranoid historian—the NWC's study of Title III wiretapping was published on April 30, 1976, one day after the Church Committee delivered its final report on illegal activities at the FBI and the CIA. While the Church Committee proposed new laws to curtail spying in the federal intelligence community ("Too many people have been spied upon by too many Government agencies and too much information has been collected"), the NWC advocated for increased wiretap authority among state and federal police ("Congress should carefully consider . . . [an] expansion of the list of crimes for which electronic [surveillance] orders may be obtained").[73] At the time, few noticed that the two reports seemed to work at cross-purposes, or indeed that they had anything to do with each other. The *New York Times* buried its coverage of the NWC's recommendations, which would have been scandalous fare in Long's day, on page 20 of the paper.[74]

Small wonder that Americans embraced the bleak message of Hollywood films like *The Conversation* in the years when lawmakers finally moved to bring government wiretapping under official control. In reality, the explosive skirmishes over privacy and surveillance of the 1970s were taking place on receding territory. As the nation came to uneasy terms with the age-old problem of national security wiretapping, far more mundane forms of electronic surveillance—wiretapping in criminal investigations, wiretapping in the private sector—were becoming more entrenched, and sometimes even expanding their reach. By the end of the decade, many Americans were left with a sinking sense that the battle over wiretapping was finished. The defenders of civil liberties had come up short.

"We lost . . . when they passed Title III," remarked Herman Schwartz, a legal scholar who emerged as one of the nation's most vocal opponents of electronic surveillance after Edward V. Long was voted from office. "Wiretapping is here to stay whether we like it or not. We ought to concentrate on making sure it's used properly and sparingly. . . . If we don't take care of our remaining enclaves of privacy, we soon won't have any."[75]

For Schwartz, and for many others, 1968 was the point of no return, regardless of what ended up transpiring in the chaotic decade that followed. It was enough to make steadfast believers in privacy lose faith.

Chapter 9

Limited Assistance Necessary

BETWEEN 1976 AND 1980, federal judges across the United States approved 3,001 separate wiretap orders under Title III of the Omnibus Crime Control and Safe Streets Act. Of these, 1,544—more than half—were carried out in just two jurisdictions: New York and New Jersey. Year after year, the two states vied for the ignominious title of America's police surveillance capital, with New Jersey pulling ahead by decade's end. During this period, police in the Garden State tapped more telephones than all federal law enforcement agencies combined.[1]

Edward J. Tomas was one of the many technicians who installed and monitored those wiretaps, working on the front lines of New Jersey's electronic war on crime.[2] Tomas grew up outside New Brunswick. Like many wiretap experts before him—like Jim Vaus, Bernard Spindel, and even Francis Ford Coppola—he exhibited an interest in electronics at an early age. Tomas's father had worked at Western Electric, AT&T's long-time manufacturing subsidiary. As a teenager, the young Tomas spent much of his spare time tinkering with the telephone hardware and testing equipment that occasionally found its way into the family home.

"They [telephones] were like toys to me," Tomas recalled when we spoke in 2018. "I wanted to figure things out, how they worked. When something didn't make sense, I could contact other people and they could help me work through the problems I was having. We were almost like hackers are today."

After enlisting briefly in the Marine Corps Officer Candidates School (OCS), Tomas took a position with law enforcement in New Jersey, initially working odd jobs for the Somerset County state prosecutor's office. It was there, in 1975, that he met Leonard Arnold, an up-and-coming assistant prosecutor who had worked for Bell Laboratories, AT&T's celebrated research and development subsidiary, after attending law school.

An electronics enthusiast himself, Arnold encouraged Tomas to learn more about wiretapping. His advice was prescient. New Jersey's statewide effort to employ electronic surveillance in high-level criminal investigations was already starting to gain momentum. New Jersey police installed 138 wiretaps that year, 19 percent of the national total.[3] By 1980, the state's share would climb to 30 percent.[4]

Tomas first learned the art of wiretapping with the New Jersey State Police Electronic Surveillance Unit, a small but dedicated team of troopers who worked out of an unassuming red-brick building in West Trenton. Later on, he took a brief leave of absence from the force to learn more advanced surveillance techniques at Audio Intelligence Devices (AID), then the nation's largest licensed supplier of eavesdropping equipment. Founded in the early 1970s, AID boasted an exclusive government contract and a catalog of specialized instruments that ran more than a hundred pages. The firm also held regular training seminars for state and federal law enforcement officers at its headquarters near Fort Lauderdale, Florida. (Now known as Law Enforcement Associates, Inc., the firm still sells equipment and holds training seminars today.)[5] Tomas remembers paying his own way to attend a multiweek course at AID sometime in the winter of 1978. When he returned to New Jersey, he established himself as one of the most trusted wiretap technicians in the state. "Ed was *the* investigator you wanted on your file," remembered one of the prosecutors with whom he worked in the early days. "He was so highly skilled. He showed everyone, including members of the state police, just how important electronics could be. I certainly didn't have the knowledge, or the fortitude, to do what Ed did."[6]

At the time, tapping a telephone was simple enough, notwithstanding the training required to do it. Technically speaking, little had changed since the days of Prohibition, when federal agents first began soldering extension wires onto copper telephone lines, waiting to overhear a conversation that might bring down a bootlegging syndicate. As soon as Tomas located a suspect's cable and pair appearances—proprietary information that the phone company typically provided—all he needed to eavesdrop on a phone call was a terminal wrench, a set of alligator clips, and an automatic tape recorder. Monitoring a wiretap for weeks or months on end was punishing work. "I'm never going to get that time of my life back," laughs Tomas, recalling the tedious days and nights he spent listening to live audio surveillance recordings. But more often than not the tedium paid dividends. A wiretap order in a narcotics or corruption investigation almost always led to indictments—and eventually convictions.[7]

According to Tomas, the one true challenge of the job was climbing the occasional telephone pole to install or inspect a wiretap. This was a treacherous business. Utility company linemen didn't climb for service checks unless they were specially trained, and protective union contracts prevented them from going up in foul weather. Police wiretaps, by contrast, didn't wait for sunshine. Tomas remembers losing his footing while climbing a telephone pole on one unlucky occasion. Tweezing dozens of stubborn, creosote-covered splinters from your forearms and thighs isn't something you're likely to forget.

·　·　·

Throughout the 1970s and 1980s—for most of the twentieth century, really—the key to a successful wire investigation was the cooperation of the phone company. In New Jersey, Tomas recalls, service providers typically staffed one or two law enforcement "liaisons" whose primary job was to provide technical assistance to the police.[8] Their efforts to broker wiretaps made most electronic surveillance operations run like clockwork. It was the law enforcement liaison who fielded the initial police request for a suspect's cable and pair appearances. This was the crucial first step in the process. It cleared the way for detectives to locate the line and install a pen register, an electronic device that tracks the numbers dialed from a target telephone. Down the road, after the detectives on the case had exhausted all other angles of investigation, the law enforcement liaison established the "leased line" wiretap setups that entities like the New Jersey Electronic Surveillance Unit favored.[9] This officially got the police up on a wire, as the saying went. The rest was a matter of waiting and listening.

According to Tomas, leased-line wiretaps were more common than direct wiretaps during this period. Their convenience made telephone surveillance something of an armchair sport. After presenting a wiretap order to a service provider, law enforcement agencies would "lease" a vacant line near that of the target telephone, enabling the provider to route all incoming and outgoing signals on the target line to an external listening post. This gave the police the ability to listen to calls from the security of a remote location. It also allowed them to monitor multiple wiretaps at once. In cases involving short-term or emergency wire surveillance, investigating officers often bypassed the telephone company's leased-line arrangement, instead opting to work with a handy wiretap device known, somewhat suggestively, as a "telephone slave." The slave served as an electronic bridge between the target telephone and any nearby vacant line. The police could then listen in after dialing the vacant connection.

A few independent telephone providers—"small-town types," Tomas labels them—had the reputation of resisting overtures to install leased lines. Sometimes their employees tattled to criminal suspects that the cops were in the process of tapping their phones. But the big providers, like New Jersey Bell, were more professional, more reliable. Tomas remembers that phone company contacts occasionally risked their jobs to offer him uniforms, decals, and other official-looking accessories to provide cover in the event of a blown operation. Under-the-table dealings like these were strictly forbidden, however. Tomas told me he never took them up during his time on the force.

The one downside to the police's reliance on telephone carrier assistance, at least from the perspective of the police, was cost. True to their name, leased lines were in fact leased: phone companies charged law enforcement agencies regular rates to use them. This made budgets for surveillance operations hard to keep under control. "Imagine what your monthly bill would look like if you were using the phone twenty-four hours a day, seven days a week," Tomas explains. "That's what we were doing when we leased a line to set up a wiretap, and the government paid dearly for it." Compounding this problem was the fact that state and federal agencies also had to provide overtime pay to the officers who monitored wiretap recordings around the clock. In the event of a lengthy or complex wire investigation, many of those same officers would have to be pulled off other ongoing cases, initiating yet another drain on official resources.

More than anything else—more than the minimization guidelines established in the Safe Streets Act, more than the judges charged with overseeing Title III operations, more than genuine concerns for privacy or due process—bottom-line considerations of manpower and cost were what prevented states like New Jersey from tapping more lines during the 1970s and 1980s. Wiretap investigations exacted a heavy toll on law enforcement agencies. They still do today. The consequence, Tomas explains, is that "at every stage someone, somewhere wants to shut you down." Throughout the first two decades of legalized police wiretapping in America, however, the telecommunications industry wasn't one of them.

. . .

In early 1980, Tomas accepted a job at an intelligence agency in Washington. He spent the better part of the next decade tapping phones, planting bugs, and conducting photographic and aerial surveillance in locations far more exotic than New Jersey—a good story, he assured me, for another time, perhaps for another book. What's important for us is what happened when he returned home.

In 1989, Tomas left Washington and resumed working wire investigations for the New Jersey Division of Criminal Justice. In many respects his job was much the same as when he left it. Wiretaps were still vital to the work of law enforcement. In fact, new investigative strategies in the nation's War on Drugs meant that police in the Garden State were tapping more lines than ever. Yet Tomas also began noticing occasional hiccups in his work. The process of tapping a phone to advance a criminal investigation—perfected with the telecommunication industry's help, both on and off the record, over the course of the twentieth century—wasn't as seamless as it used to be. New developments were getting in the way.

In the first place, Tomas's investigative targets were using more than just the telephone to communicate with each other. The telecommunications industry had changed a great deal in the years between 1980 and 1989. In January 1982, the U.S. Department of Justice forced AT&T to settle in its landmark antitrust case against the Bell System monopoly. Two years later, AT&T restructured its operations and divested its holdings. The breakup of the Bell System ushered in an era of unprecedented competition, creativity, and innovation in the field of telecommunications. A host of new devices and features soon began flooding the marketplace: call forwarding, call waiting, and caller ID; alphanumeric pagers, fax machines, and mobile phones.

For the average American consumer, the new telephone services made communicating at a distance more instantaneous, more reliable, and more convenient. But for Tomas they caused frequent headaches because many of them were resistant to established methods of eavesdropping. "We didn't exactly have advance notice," Tomas recalls of the newfangled communications technologies that rolled out, one after another, over the course of the 1980s and 1990s. "We were always forced to react after the devices were out. . . . A lot of times there would be a year or two lag before effective law enforcement surveillance could happen." As those lags got longer and more frequent, as the logistical problems piled up, police in New Jersey found themselves playing the wiretap game from behind. For a short period of time, a tech-savvy criminal using, say, a pager or a fax machine had the ability to foil a wiretap investigation altogether.

Another problem Tomas encountered—a problem that seemed to bode much more ominously for the future—stemmed from an apparent sea change in the nature of electronic information itself. By the late 1980s, telecom carriers were starting to replace the copper wires of the old electromechanical switching system with fiber optic cables and computerized exchanges. The two-way traffic of the analog age was gradually giving way to something new and unfamiliar: a digital communications network, defined by flexible data flows and multiplatform use (see Figure 9.1).

The transformation from analog to digital appeared to pose an existential threat to the work of law enforcement eavesdropping. Fiber optic cables were faster and more secure than their electrical predecessors. Tapping them directly yielded an unintelligible muddle of noise. Making matters worse, the technical workarounds that digital carriers began offering to police were cumbersome, even inadequate. Because early digital switching offices lacked the bandwidth to support multiple wiretap operations at one time, Tomas and his colleagues often found themselves waiting for higher-priority federal taps to run their course before they could initiate their own wire investigations.

The combination of accelerated technological change and operational gridlock made Tomas begin to wonder whether the digital age might eventually render wiretap surveillance obsolete. "When they [service providers] went from analog to digital, that was a huge, huge problem," he remembers. "For a while there, a lot of people had completely secure communications, unbeknownst to them. Those of us who were in law enforcement were totally blind."

Ed Tomas's story isn't unique. Throughout the late 1980s and early 1990s, as global telecommunications firms leaped headfirst into the digital age, law enforcement agencies across the United States encountered many of the same logistical problems. The time-tested process of securing a Title III order, contacting a service provider, and installing a wiretap— none of it seemed to work in the new communications environment, at

Figure 1.—The traditional telecommunication network

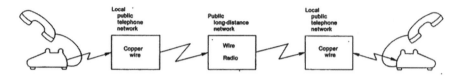

Figure 2.—A View of the Future Telecommunications System

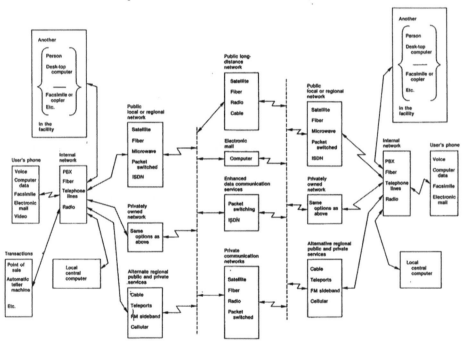

Figure 9.1. "The Traditional Telephone Network/A View of the Future Telecommunications System" (1985). The U.S. Office of Technology Assessment created these two diagrams to illustrate the differences between analog and digital communications systems. Lawmakers referred to them in the congressional hearings that preceded the passage of the Electronic Communications Privacy Act (1986). Robert Kastenmeier Papers, Wisconsin Historical Society.

least not in the way that it had. Behind closed doors, lawmakers and police officials began considering the grim possibility that the government might end up "going dark" by the turn of the twenty-first century. Anything seemed possible in those days. The march of technological progress was moving America off the wire in several ways at once.

As we'll see in this chapter (and in Chapter 10), the U.S. government responded to the looming wiretap crisis by doing what it always did when faced with a glitch in its electronic surveillance operations: it called on the telephone industry for help. But after reaping the rewards of more than a century of quiet cooperation, the state suddenly found a much less welcoming voice on the other end of the line. In the early 1990s, telecommunications companies began pushing back against the imperatives of law enforcement surveillance. Wary of consumer backlash in an increasingly competitive marketplace, common carriers of all sizes challenged the government's pleas for wiretap assistance. Telephone industry organizations began embracing the popular cause of privacy.

The ensuing political struggle would come to define the history of wiretapping and electronic eavesdropping at the close of the twentieth century. It would also raise key questions that Americans continue to ask about their technological life here in the twenty-first: How much should corporate communications firms work to protect our privacy, and how much should they yield to the prying eyes and ears of law enforcement? Should surveillance capabilities be built into the hardware of communications technologies, or should that hardware afford us the right to be let alone?

None of these questions were new. But to understand why Americans came to ask them with such urgency in the 1980s and 1990s—and to understand why they played such a critical role in the debates over privacy and surveillance that ensued—we need to return, one final time, to the origins of law enforcement wiretapping itself. Needless to say, those origins are impossible to imagine without the cooperation of the telecommunications industry, always standing by.

．　．　．

Throughout the second half of the nineteenth century, as we have already seen, wiretaps were primarily the province of common crooks and confidence artists. Aside from the Civil War operators who transformed the practice of wiretapping into a military art, official agents of the state seldom deigned to tap telegraph lines. Among law enforcement officers and investigative agencies, reported the *New York Times* in 1880, there was "little temptation or opportunity to intercept telegrams, and no practice of doing it." The government's indifference to wiretapping was one of many reasons state and federal lawmakers left the issue of electronic privacy unregulated for so long. "If, indeed, [the] Government were addicted to tapping the wires, stationing spies in offices to read the clicking of the instrument, or stopping boys on their way to deliver messages, in the hope of detecting 'treasons, stratagems, or spoils,'" the *Times* continued, "a law saying that dispatches should be respected like letters in the mail would have some practical meaning. But there is no such abuse."[10]

The government's disinterest in wiretapping was a function of the telegraph's patterns of use. As a convenience to customers, telegraph carriers typically saved a hard copy of every message sent or received. It was much easier for officials to acquire the duplicates—filed in most telegraph offices for at least six months, and accessible by subpoena—than to intercept electronic signals in real time. In this way, the data retention practices of early U.S. telecom firms laid the operational groundwork for a rule of government surveillance that still holds true today: communications "at rest" are more efficient to monitor than communications "in motion."[11]

In the early years, telegraph carriers readily complied with official requests for duplicate messages. During the Civil War, for example, Western Union gave the U.S. War Department unfettered access to stored telegrams in order to help uncover Confederate plots against the government. Later on, when lawmakers began investigating cases of political corruption during Reconstruction, the company surrendered thousands of messages to Congress to assist in the impeachment trial of President Andrew Johnson.

Western Union would adopt a less cooperative policy once the press began criticizing rival telegraph firms for submitting to similar dragnet searches.[12] In 1873, the company instructed its employees not to comply with subpoenas for duplicate telegrams. And in 1876–1877, Western Union made headlines by refusing to turn over messages that would help determine the rightful winner of the contested Hayes-Tilden presidential election. According to Western Union president William Orton, who testified before Congress at the height of the controversy, telegraph messages were "strictly private and confidential" whether they were in transit over the wires or stored in company offices.[13] He rejected the assumption that the government had a right to view their contents. After a protracted legal dispute, Western Union eventually relinquished more than 30,000 messages to Congress, in the process laying bare the company's questionable lobbying efforts during the run-up to the 1876 election. Yet other telegraph providers would choose to adopt Western Union's policy of noncompliance in the coming years. In a competitive marketplace, protecting the "inviolability" of the telegraph was good public relations.

Early telephone carriers proved much more amenable to state cooperation. The New York Police Department (NYPD) began tapping lines through the New York Telephone Company's central exchange in 1895. At the time, telephone employees referred to the custom of backdoor police eavesdropping as "bridging over" or "censoring."[14] New York Telephone's arrangement with the NYPD didn't become public until the Thompson Committee hearings of 1916 (see Chapter 2). By that time, New York Telephone was running tappable lines directly into the NYPD wiretap squad's unmarked headquarters at 50 Church Street, in the heart of lower Manhattan. The setup gave the wiretap squad the ability to eavesdrop on almost every connection in the city.

The fallout from the Thompson Committee hearings of 1916 underscored the political perils of carrier cooperation. From one angle, New York Telephone's relationship with the police suggested that the company had a mission beyond the mere maximization of revenue. In public testimony before the committee, New York Telephone officials repeatedly insisted that carrier-assisted wiretaps were "of great aid to the police department and of benefit to the general public"—an important narrative to cultivate at a time when the popular press routinely caricatured telephone providers as insatiable corporate monopolies, tentacular in their reach.[15] But from another angle, New York Telephone's partnership with the NYPD was entirely consistent with that same sinister image. At every turn in the 1916 hearings, telephone company representatives felt compelled to underscore that carrier-assisted wiretaps were neither profitable nor common. "The New York Telephone Company never derived any benefit, financial or otherwise, from tapping wires," one corporate executive testified. Police surveillance was a "serious disturbance" to day-to-day operations, an onerous responsibility that "we [New York Telephone] would like to be relieved of."[16]

All the same, the NYPD admitted to installing more than 350 wiretaps in the months between October 1914 and May 1916. Every single one went up with New York Telephone's help, each time after company officials received the same boilerplate notice signed by NYPD commissioner Arthur Woods: "I have reason to believe that the following telephone is being used for criminal purposes, and respectfully request the co-operation of your company in detecting this matter."[17]

In the years that followed, telecommunications firms across the country worked to tread the line between compliance and resistance. For the most part, telephone carriers cooperated with government wiretap requests on an off-the-record, case-by-case basis. During Prohibition, for example, federal law enforcement wiretaps were often installed with the cooperation of AT&T's operating companies.[18] When they weren't, it was usually because the investigating Prohibition agents were fortunate enough to have a willing informant at the central exchange, or—better yet—a former telephone lineman among their ranks who had knowledge of system infrastructure. At AT&T, officials justified such work on the grounds that the telephone had always provided a valuable "public service." Assisting the police was part and parcel of the company's baseline commitment to operating in the national interest. As the historian Alan Stone has shown, this was ironically the very same logic that AT&T used to justify its monopoly over the telephone industry and stave off federal regulation.[19] Helping with wiretaps may well have been a way to stay in the government's good graces.

As in the case of New York Telephone, however, AT&T's unofficial policy of government cooperation came with the risk of customer backlash. Any news about wiretapping, whether by private actors or by agents of the state, was certain to shake the public's faith in the security of telephone service. Bell System officials thus kept their work with the federal government under wraps, resting uneasily on the assumption that the Prohibition Bureau's wiretapping operations were seldom extensive enough to make headlines on their own. It was a familiar logic of customer relations, borrowed from the government that the telephone industry felt compelled to serve: what you don't know can't hurt you.

The networks were right—at least until Roy Olmstead came along (see Chapter 1). And yet even with telephone tapping at the center of the largest criminal case in the history of Prohibition, AT&T at first seemed content to stay on the sidelines. Bell System officials barely commented on the Olmstead affair when the legendary "Case of the Whispering Wires" went to trial in January 1926. It was only in the fall of 1927, when the U.S. Supreme Court finally agreed to hear Olmstead's appeal, that the company opted to break its silence on the subject. From the perspective of AT&T executives, the Court's decision to grant certiorari in *Olmstead v. United States* had brought the Bell System to the brink of a public relations catastrophe. A ruling against Olmstead, in favor of the government, had the potential to sow customer distrust, possibly threatening the company's ability to expand its services. What's more, the arguments before the Court all but guaranteed that the most salacious details of the federal investigation—such as the fact that the government's primary eavesdropper, Richard Fryant, had monitored customer lines for the New York Telephone Company before joining the Seattle Prohibition Bureau—would reach a national audience.

Recognizing the gravity of the situation, AT&T sprung into preventative action. In October 1927, the company banded together with three other telephone industry organizations to submit an extended *amicus* brief in support of Olmstead's complaint. The document laid out a case against wiretapping that would become the telephone industry's party line about privacy and surveillance for much of the twentieth century.

"The function of a telephone system in our modern economy is, so far as reasonably practicable, to enable any two persons at a distance to converse privately with each other as they might do if both were personally present in the privacy of the home or office of either one," the AT&T group's *amicus* brief began. The everyday workings of the system, the argument went, depended on the carrier's guarantee of security and the subscriber's expectation of privacy. Any conversation that took place on a telephone thus belonged to the callers themselves: "When the lines of two 'parties' are connected at the central office, they are intended to be devoted to the exclusive use, and in that sense to be turned over to the exclusive possession, of the parties. A third person who taps the lines violates the property rights of both persons then using the telephone, and of the telephone company as well. It is of the very nature of the telephone service that it shall be private. . . . The wire tapper destroys this privacy."[20]

This is, to say the least, a counterintuitive interpretation of the relationship between electronic communication and private property—one that differs from the canonical formulation that the *Olmstead* majority would eventually adopt. In the industry's view, eavesdropping on a telephone call was tantamount to "stealing" an oral conversation, owned, like any other private possession, by the callers on the line. Moreover, the physical act of tapping the wires involved tampering with the telephone company's equipment, an offense already prohibited by law in twenty-eight states, including Olmstead's home state of Washington. Taken together, those two facts alone suggested that wiretapping was an illegal search, whether or not a government eavesdropper had in fact trespassed on a protected area, such as a home or an office, in the process of carrying out the disputed act. By those standards, the Supreme Court had no choice but to rule in Olmstead's favor.

"The telephone companies deplore the use of their facilities in the furtherance of any criminal or wrongful enterprise," the AT&T brief concluded. "But . . . it is better that a few criminals escape than that the privacies of life of all the people be exposed to the agents of the government, who will act at their own discretion, the honest and the dishonest, unauthorized and unrestrained by the courts."[21] Justice Louis Brandeis drew attention to this last formulation when he penned his celebrated dissent to *Olmstead v. United States* the following summer.[22]

As we saw in Chapter 1, the Supreme Court eventually rejected Olmstead's claim of government misconduct, and with it the telephone industry's unorthodox legal argument about wiretapping and private property. But AT&T officials still used the ruling as an opportunity to burnish the Bell System's reputation as a defender of privacy. Three days after the Supreme Court handed down the *Olmstead* decision, while national news outlets skewered the majority opinion, AT&T president Walter S. Gifford sent an extended telegram to Bell System executives. He wanted to outline the official position that AT&T's operating companies were to take on the outcome of the case, and to underscore that the telephone industry would remain steadfast in its opposition to government wiretapping.

"It has always been the policy of the Bell System companies to respect and protect the privacy of telephone conversations," Gifford wrote. "We are disappointed in the decision, but it will not make any change in our policy. While the Supreme Court decided that evidence obtained by federal agents by wire-tapping is admissible in a criminal prosecution, it is a mistake to suppose that the Supreme Court has approved wire-tapping itself. Tapping or otherwise tampering with telephone lines is an unlawful trespass on the property of the [operating] companies which they will continue to resist; and intercepting and divulging telephone conversations is an invasion of the rights of telephone users which the criminal statutes of most of the states forbid under drastic penalty."[23]

This was the basic stance AT&T executives took on government wiretapping for most of the twentieth century. The Supreme Court, through *Olmstead,* had permitted the use of wiretap evidence in court. But it hadn't

Figure 9.2. "Your Respect for Secrecy of Communications Is the Basis of Good Service" (ca. 1969). Courtesy of AT&T Archives and History Center.

endorsed wiretapping itself, at least not directly, and the state laws that already prohibited the practice remained intact. In an attempt to counter the impression that the *Olmstead* decision had fatally compromised the workings of the telephone system, several Bell operating companies soon announced that they would reject future overtures from the police to install wiretaps in criminal investigations.[24] "Privacy" and "secrecy" became system-wide watchwords. In the years that followed, AT&T would use those two terms interchangeably.[25] Telephone offices around the country displayed posters illustrating the Bell System's official privacy policy, which borrowed from the language of the 1934 Federal Communications Act to stipulate that "except as it may be necessary to inform your supervisors of service situations or conditions requiring their attention . . . you are not permitted, in any manner, directly or indirectly, to divulge to anyone the existence, contents, or nature of any communication" (see Figure 9.2).[26] The rhetoric remained intact well into the 1960s, even as AT&T withstood accusations that it wasn't doing enough to safeguard its networks against illegal wiretapping.[27]

"Privacy of communications is a basic concept in our business," AT&T vice president Hubert Kertz told Edward V. Long's Senate Subcommittee in September 1966. "The public has an inherent right to feel that they can use the telephone with confidence. . . . Any undermining of this confidence would seriously impair the usefulness and value of telephone communications."[28]

The intended message was simple, and AT&T would revert to it frequently: *Olmstead* or not, customers were entitled to expect that they could use the telephone without someone listening in, just as they would expect when talking face to face.

. . .

As with Western Union's policy of government noncompliance in the 1870s and 1880s, it's crucial to separate rhetoric from reality, the Bell System's customer relations strategies from its internal policies in action.[29] It turns out that much of the talk about "privacy" and "secrecy" was just that: talk. Although AT&T operating companies took visible steps to oppose wiretapping in response to the legal expansion of the state's surveillance powers, as in the case of *Olmstead v. United States*—and although Bell officials paid frequent lip service to the company's commitment to protecting the intangible rights of its subscribers—the American telecommunications industry actually dealt with the dirty business of government eavesdropping at something closer than arm's length. The arrangements varied from state to state, and from carrier to carrier, but technical cooperation was the norm rather than the exception. To borrow a formulation that civil liberties advocates had adopted by the late 1960s, "Ma Bell" was Big Brother's handmaid.

Over the course of the twentieth century, government cooperation took on two main forms at AT&T, neither of which reliably showed up in official records or congressional testimonies. The first was what we might call *direct assistance,* which involved providing law enforcement agencies with equipment, hardware, and even centralized network connections in the service of ongoing wiretap investigations. The aforementioned "leased line" arrangement, which took off in the 1940s, was perhaps the most common form of direct assistance available, allowing state and federal investigators to monitor telephone calls from the comfort and security of a remote location. The only difference between leased lines and the wiretap setups that carriers like New York Telephone had offered the NYPD in the early 1900s was that leased lines came with a price tag.

The second form of carrier assistance was more *indirect,* involving the exchange of proprietary information. This was much easier to hide. Throughout the twentieth century, Bell operating companies—and smaller, regional providers too—often disclosed cable and appearance locations in order to help detectives pinpoint the correct wires to tap. Local exchanges also readily coughed up dialed number registries and other forms of customer calling data, collected in the course of regular business to ensure accurate billing. In some jurisdictions, maintaining friendly telephone company contacts came to be regarded as essential to regular police work. Citing hundreds of off-the-record interviews across a dozen different U.S. cities, Samuel Dash told a Senate subcommittee in 1959 that "almost every police wiretapper, whether he be a plainclothesman or assistant district attorney, if he is assigned to supervise wiretapping . . . has a contact in the telephone company. . . . [T]his is condoned by the company itself, but expected to be kept quiet and not to be admitted."[30] Despite their stated opposition to wiretapping, private detective firms like the Pinkerton Agency encouraged their operatives to develop "informants" at local telegraph and telephone companies as well.[31]

Revelations in the 1960s vindicated Dash's allegations. The Long Committee's earliest hearings in Kansas City (see Chapter 7) showed that Southwestern Bell frequently offered leased lines and call data to the Kansas City Police Department. The state of Missouri lacked a wiretap law at the time, so cooperation between the two sides wasn't illegal in and of itself. But when the Kansas City Police partnered with the Internal Revenue Service to wiretap organized crime syndicates, a frequent occurrence in the period, all of the parties involved—phone company employees, local police officers, federal agents—were in theory exposed to prosecution under Section 605 of the Federal Communications Act.[32]

Cooperation between telephone providers and law enforcement agencies was especially common in large cities. In Washington, D.C., the FBI maintained a working relationship for several decades with the Chesapeake and Potomac Telephone Company. By the late 1960s, the Bureau was annually leasing more than 450 lines from the C&P Telephone central exchange, most of which fed into a secret listening post housed in the capitol city's old Post Office Building—now the site of the Trump International Hotel.[33] At the time, C&P Telephone filtered the FBI's wiretap requests through so many supervisory levels that many of the employees who activated them had no idea of their intended use. Those who did nicknamed the backdoor wiretaps "Horace Hampton circuits," in honor of the corporate executive who served as the FBI's primary contact at C&P Telephone for more than twenty-two years.[34]

Whether AT&T's assistance to law enforcement was direct or indirect, on the record or off, the ratification of the Omnibus Crime Control and Safe Streets Act in 1968 only served to deepen the contradictions inherent in the company's public stance. Much like the outcome of *Olmstead v. United States* four decades earlier, the passage of Title III was a bitter pill for the telecommunications industry to swallow. Even though the measure created a long-overdue system of regulatory guidelines for law enforcement wiretaps, it also gave the embattled practice of government eavesdropping the force of law. Once again, AT&T responded to an apparent expansion of the state's electronic surveillance authority by affirming its commitments to privacy and secrecy. But in so doing the company seemed to set itself up for a public reckoning.

On June 19, 1968—the same day President Lyndon Johnson signed the Safe Streets Act into law—AT&T circulated a memorandum about wiretapping in its internal *Management Report,* intending to give Bell System officials company-approved answers to common questions about the effects of Title III. On the issue of illegal wiretapping, AT&T instructed operating company managers to assure customers that they had little cause for anxiety or fear:

> *How extensive is wiretapping?*
> The amount is very small. Nevertheless, it is enough to give cause for concern, particularly since new, more sophisticated methods have been developed in recent years. We believe, however, that the new federal crime control act [the Safe Streets Act] will help control the problem of unauthorized wiretapping.

On the issue of network security, AT&T recommended citing the Bell System's long-standing efforts to create technical deterrents to third-party eavesdropping:

> *What precautions does the Bell System take against wiretaps?*
> Our plant is designed to provide a high degree of security against wiretapping. Most telephone cables, for example, are equipped with automatic alarms, which, among other things, warn us of tampering. In addition, employees working on our facilities are constantly on the alert for signs of unauthorized connections or tampering. When there is reason to suspect a wiretap, detailed tests and physical inspections can disclose if a tap is in operation.

And on the critical issue of law enforcement cooperation, AT&T encouraged its managers to rehearse an updated version of the narrative that the company had touted in the decades since the *Olmstead* ruling:

> *Does the Bell System wiretap for law enforcement officials? For anyone else?*
>
> No. We do not wiretap for law enforcement officials or any other agency or person, even if they should ask us. However, under the new law wiretaps can be made by federal law enforcement officials and by state law enforcement officials in jurisdictions where authorized by state law.

This last explanation wasn't exactly a lie. Technically speaking, neither the leasing of lines (direct assistance) nor the divulgence of customer data (indirect assistance) were equivalent to installing a physical tap on a customer's phone. But the story wasn't exactly the truth either. The ambiguity seemed to signal the beginnings of another customer relations crisis. Now that police wiretaps were a matter of public record, it seemed certain that the spotlight would turn to the role that the telephone industry had been playing behind the scenes. The June 19 AT&T memorandum noted that Title III "should be of special interest to managers."[35] In hindsight, the potted corporate language reads like a warning.

. . .

The contradictions in AT&T's position on wiretapping resolved sooner than anyone could have anticipated. It was the result of a lucky break, set in motion by a competitor. In the early months of 1969, the Central Telephone Company of Nevada—a regional provider unaffiliated with the Bell System—refused to comply with a request to furnish equipment to the FBI in the service of a wiretap order. The motivations behind Centel Nevada's obstinacy remain mysterious. Perhaps it was the result of high-minded corporate resistance to Title III, still criticized in civil libertarian circles as the "End of Privacy Act." Or perhaps it was the result of less innocent factors. Either way, the U.S. Court of Appeals for the Ninth Circuit ruled in the company's favor when the FBI applied for relief. The court's decision, handed down in May 1970, hinged on an odd counterfactual, so simple as to seem almost absurd. Title III of the Safe Streets Act was "rather extensive, containing ten sections, some quite lengthy," the three-judge panel noted.[36] Given the comprehensiveness of the law's treatment of the mechanics of court-ordered surveillance, Congress would almost certainly have included a provision about carrier assistance if carrier assistance was necessary. There was no such provision. Ergo: carrier assistance wasn't necessary.

Congress took decisive action in response to the Centel Nevada ruling—surprisingly, given the lingering rancor over the inclusion of Title III in the Safe Streets Act. In July 1970, two months after the Ninth Circuit's decision, a bipartisan contingent in the House and Senate ratified an amendment to Title III that clarified the telecommunications industry's role in the work of court-ordered surveillance. With little fanfare or debate, the measure initiated a policy that became known among telephone carriers as *limited assistance*. When presented with a wiretap order, phone companies were now required to provide the "information, facilities, and technical assistance necessary to accomplish the [wiretap] interception unobtrusively and with minimum disruption of service."[37] But they weren't required to place a physical tap on the line itself. In the near-unanimous opinion of Congress, that was the job of the state.

For AT&T, this was a welcome development. The limited assistance amendment to Title III was beneficial for two main reasons. First, it provided cover for the forms of technical support that the network's operating companies had been offering to law enforcement agencies all along. By amending the law, Congress had placed the final onus for wiretap installations on the government, giving the telephone industry a convenient sort of deniability in the public eye. As William P. Mullane, AT&T's press relations director, was quick to point out, the policy of limited assistance determined what it was that providers *weren't* responsible for—and that negative line of demarcation could prove politically useful in the event of wiretap abuse. "Law enforcement agencies will need certain limited assistance to accomplish court-authorized wiretapping," Mullane explained in a letter to the Bell System's press corps. "But in no instance does the telephone company make that actual wiretap or place any bug, nor do we send a telephone employee along. Nor do we trace incoming calls to a suspect's line. Nor do we furnish telephone company trucks, tools, equipment, uniforms, employee identification cards, or training to law enforcement agents. . . . [P]rivacy of communications must still be safeguarded."[38] Such a "hands-off" policy toward government wiretaps, sanctioned by the Title III amendment, would enable AT&T to escape censure at the height of the Watergate and Church Committee investigations of the mid-1970s. In fact, anxieties about the Bell System's role in telephone surveillance throughout the paranoid "era of the listening device" came to focus much more on the network's unrelated internal practice of "service monitoring": that is, randomly recording calls in order to thwart toll fraud.[39]

Second, and just as important for the health of the telecommunications industry as a whole, Congress' 1970 amendment to Title III contained financial incentives. For their trouble—and offering limited wiretap assistance wasn't much trouble, clearly—telephone companies were to receive full compensation for expenses incurred in the regular course of cooperating with the government.[40] At least in theory, this meant that court-ordered wiretaps wouldn't hurt the industry's bottom line. By 1976, the National Wiretap Commission reported that telephone companies around the country were happily complying with the arrangement: common carriers were supporting Title III operations "no more and no less than required," and wiretaps were neither a burden on service nor a drain on resources.[41] The system seemed to work for both sides.

In what appeared to be the final act of the decades-long drama behind the relationship between law enforcement and the telecommunications industry, the U.S. Supreme Court redoubled the force of the limited assistance principle in *United States v. New York Telephone Co.* (1977), a case that involved a set of circumstances similar to the Centel Nevada complaint. In a 5–4 decision, the Court determined that the government was within its rights to force telecommunications providers to furnish "any assistance necessary" to accomplish a warranted surveillance operation.[42] The ruling hinged on a seldom-cited legal precedent: the All Writs Act of 1789, which gives judges the power to compel third parties to take actions that are necessary to uphold the law. The decision would have powerful consequences despite the obscurity of its legal foundations. In 1978, Congress pounced on the *New York Telephone Co.* case to justify a proposal to expand the telecom industry's role in cooperating with national security surveillance operations. AT&T offered its customary public pushback, with several other industry organizations joining in opposition to the proposed measure.[43] But an expanded definition of limited assistance was eventually written into the final version of the Foreign Intelligence Surveillance Act (1978), the law that still governs the protocols for national security wiretaps today.

New York Telephone Co. and the All Writs Act, the doctrinal lynchpins of limited assistance, have made headlines in recent years. Pay close attention to the debates over technology and privacy that are perennially in the news and you'll notice how often they come up. When federal agencies petition to bypass the security features that global technology companies have built into our communications devices, they almost always lean on *New York Telephone Co.* and the All Writs Act.[44] That isn't merely the result of the symbiotic relationship between telephone carriers and the state that blossomed over the course of the twentieth century, the very same relationship that received the Supreme Court's endorsement in 1977. It's also the product of the battles over electronic privacy that Americans waged throughout the 1980s and 1990s. These would take place on more contentious ground.

Chapter 10

Off the Wire

IN LIGHT OF EVERYTHING WE'VE LEARNED in recent years about the tele-communications industry's role in the project of state surveillance, the story of AT&T's decades-long partnership with American law enforcement isn't exactly breaking news. Nor is it surprising that AT&T's willingness to assist police wiretaps contradicted both the company's internal security policies and the Bell System's highly choreographed image as a defender of customer privacy.

We now know, for example, that global communications firms were integral to the clandestine surveillance programs that the National Security Agency (NSA) oversaw throughout the Cold War. "All the big international carriers were involved, but none of 'em ever got a nickel for what they did," explained one official familiar with the process, which entailed carriers such as RCA, ITT, and Western Union forwarding copies of foreign cables to Washington for intelligence analysis.[1] In the years since the September 11, 2001, attacks, stories of similar forms of cooperation between the telecommunications industry and the U.S. intelligence apparatus have surfaced with predictable regularity. In 2006, an engineer at AT&T leaked information about a secret intercept facility in the company's San Francisco offices: room 641a of the SBC Communications switching center at 611 Folsom Street. Inside, NSA monitoring devices were busy siphoning reams of data from AT&T servers.[2]

More recently, the digital giants of Silicon Valley have followed in the footsteps of their analog predecessors. Companies such as Microsoft, Google, Facebook, and Apple have all offered the NSA access to network traffic through backdoor channels in the Internet's architecture.[3] Today it isn't simply that telephone companies function as "Big Brother's hand-maid," on scales that scarcely seem fathomable from the perspective of the twentieth century. It's that almost every other corporate entity that touches our private information seems willing to perform the very same role.

We can understand why a company like AT&T would participate in the project of state surveillance, however questionable, if we recognize that the telecommunications industry has always conceived of its services as a "public good": vital to economic growth, essential to the functioning of democracy, and (most important of all) best delivered through corporate practices that border on the monopolistic. As the writer and legal scholar Tim Wu has shown, AT&T and its subsidiaries essentially functioned as a "branch of government" for most of the twentieth century.[4] Capitulating to the demands of crime control and national security—working in the "national interest," broadly construed—was a small, pragmatic price to pay for the company to stay in the good graces of antitrust regulators. The research that scientists conducted at Bell Laboratories, which over the years was used to improve everything from personal computing power to disability access, was one side of AT&T's attempt to fashion its private interests in the image of the public good. The other side, a much darker side, was electronic surveillance.

Whatever the true motivations behind AT&T's amenability to wiretap assistance, the industry and the government had developed a close relationship, a kind of symbiotic accord, by the late 1970s. The state got wiretaps whenever and wherever it wanted. In turn, AT&T was allowed to protect the industry status quo and expand its market reach. Bell operating companies even got the opportunity to profit by leasing lines and otherwise following the letter of the limited assistance principle, reinforced by the U.S. Supreme Court's ruling in *United States v. New York Telephone Co.* (1977). This was why Ed Tomas, the electronic surveillance expert whom we met at the beginning of Chapter 9, had such an easy time on the job before he left to work in Washington. For the average surveillance technician, working a wire was simple because the telephone network was simple. It was also simple because it was in the financial interest of the telephone companies to lend a helping hand.

"Wiretaps were a money-making venture at the phone company, there's no doubt about it," Tomas told me.[5] He repeated this sentiment several times throughout our conversations. He was right, perhaps more than he realized.

If the telecommunication industry's policy of wiretap assistance was a byproduct of AT&T's desire to protect its status as a sanctioned monopoly, it stands to reason that the government's effort to dismantle the Bell

System would have upset the accord over electronic surveillance power that had evolved over the course of the twentieth century. Why give the government what it wants, when and where it wants it, if its interests are hostile to your own? Why help with wiretaps—why violate customer trust—if doing so gives your competitors an advantage in a cutthroat market?

Such questions make sense in the abstract. The historical record, however, tells a more complicated story. There are several factors that put telecommunications carriers and the U.S. government on different sides of the issue of law enforcement wiretapping during the 1980s and 1990s. The breakup of the Bell System provides the backdrop for some of it. But the gradual fraying of the partnership between service providers and the police has more direct roots in two concurrent developments.

The first development was the Reagan administration's redoubled efforts in the War on Drugs, which led to a dramatic increase in the cost and reach of police wiretaps nationwide. The second, somewhat more difficult to pin down, was an unsettled popular conception of the proper American balance between privacy and surveillance. The new network services that flooded the marketplace in the wake of the Bell System's demise, from cordless phones to electronic mail, played an important role on both fronts. By the early 1990s, it was clear that policymakers needed to broker a new electronic surveillance accord between service providers and law enforcement—one that recognized just how dramatically innovations in communications technology were reshaping American life on the eve of the twenty-first century.

• • •

Until the late 1970s, American law enforcement agencies typically employed Title III surveillance to disrupt the gambling industry. The case that brought *United States v. New York Telephone* to the Supreme Court had itself started with the investigation of a Manhattan bookmaking syndicate.[6] At the time, law enforcement officials saw gambling as the cornerstone of organized crime in the United States. Estimates suggested that the industry generated illicit profits of more than $60 billion annually, most of which flowed into the coffers of a coterie of crime families whose violent affairs seemed to stretch into every corner of the country.[7] From its inception, Title III appeared to give a lifeline to law enforcement agencies struggling to disrupt the gambling industry using traditional investigative methods. Whether electronic surveillance tactics could make meaningful inroads into the criminal organizations behind the illegal gaming rackets, however, remained an open question.

At face value, the numbers looked favorable. Between 1968 (when Title III of the Safe Streets Act first went into effect) and 1974 (when the National Wiretap Commission completed its preliminary study of legalized law enforcement surveillance), state and federal law enforcement agencies installed 4,334 separate wiretaps and bugs under court order.[8] Investigators overheard more than three and a half million conversations in that six-year span. The voices of a quarter of a million American citizens were involved, and 9,210 of them ended up getting convicted using evidence from what they said on the line.[9] But upon closer inspection, the vast majority of America's warranted wiretapping operations—as much as 72 percent on the federal level—turned out to be directed at individuals suspected of committing low-level gambling offenses. Those offenses carried short prison sentences: on the order of months, not years.[10] Most of the gambling convictions produced by Title III wiretaps between 1968 and 1974 resulted in petty fines and probation.[11] None of that, needless to say, was likely to disturb the workings of a complex criminal organization.

A representative milestone in the government's early efforts to combat gambling via Title III was Project Anvil, a massive multistate wiretap investigation initiated by the U.S. Department of Justice in 1971. The Anvil operation lasted for more than eighteen months and involved the coordination of dozens of wire technicians working in multiple jurisdictions.[12] It culminated in the indictments of 250 small-time bookmakers around the country, most of whom made their money by taking bets on professional football games. Yet after thousands of hours on the wire, the agencies running Project Anvil never found the opportunity to move beyond the bookies in the chain of criminal command. What's more, in many of the resulting indictments evidence unrelated to Title III recordings turned out to be sufficient for making a conviction.[13] Faced with the same nationwide mandate to combat the gambling industry, state law enforcement agencies encountered similar outcomes in wiretapping cases as the decade wore on.[14]

So were wiretaps worth the trouble? A small but vocal chorus of skeptics believed that they weren't: the numbers in the Title III gambling crusade simply failed to add up.

Herman Schwartz was among the most prominent of this group. A graduate of Harvard Law School, Schwartz joined the campaign against law enforcement wiretapping in the early 1960s after reading a copy of *The Eavesdroppers*—yet another measure of the influence of Samuel Dash's landmark 1959 study (see Chapter 5). One of Schwartz's earliest professional positions was with the New York Bar Association Committee on Civil Rights and Civil Liberties, a job that gave him the opportunity to conduct research on electronic surveillance in the nation's eavesdropping capital. In 1962, while teaching at the University of Buffalo, Schwartz authored a widely circulated field report for the American Civil Liberties Union (ACLU) titled "The Wiretapping Problem Today," the first of several pamphlets on the subject he would publish in the coming years.[15] When I sat down with Schwartz in 2018, he talked about his early work agitating against electronic surveillance with the ACLU as though he'd been tasked with reining in the lawlessness of the old West. As our conversation progressed, I found myself picturing him in the vein of Ranse Stoddard, the idealistic young attorney Jimmy Stewart plays in John Ford's *The Man Who Shot Liberty Valance* (1962), charged with bringing the rule of law to an unruly American frontier.

"A few of us took a wash-rag and tried to clean up the dirty business of wiretapping," Schwartz joked from his office at American University in Washington, D.C., where he has worked as a professor of constitutional law for the last four decades. "We didn't end up doing so well."[16]

For Schwartz, the dependable arguments about wiretapping as a violation of constitutional guarantees, handed down from Louis Brandeis's infamous dissent in *Olmstead v. United States,* seemed to have lost their political edge by the 1970s. The law-and-order victories that produced Title III had rendered the civil libertarian case against electronic surveillance a dead letter. The best claim that Schwartz had at his disposal was much more pragmatic. After scrutinizing several years' worth of data from the government's annual *Wiretap Report,* he saw that wiretapping was an obvious drain on state resources: a time-consuming manpower killer and an overpriced threat to the government's bottom line. Electronic surveillance tactics were too costly to justify.

The allegation was hard to ignore. At $27,000 a pop—a figure that omitted the time that judges and attorneys spent filing orders, reviewing paperwork, and processing legal challenges—the average law enforcement wiretap was an expensive proposition amid the fiscal shocks of the 1970s.[17] And nowhere were the costs more obvious, and more difficult to rationalize, than in the government's campaign against gambling. On top of their obvious drain on police agencies with limited budgets, gambling wiretaps also tended to draw investigators off of higher-priority cases, overload the courts with protracted legal battles, and in the end yield negligible inroads into syndicated crime. Even the National Wiretap Commission, an entity otherwise sympathetic to Title III's scope, deemed operations like Project Anvil "disruptive to law enforcement more than the gambling industry."[18] The point was that there wasn't much to validate such a massive investment in official resources, particularly if the result seldom incriminated more than the average bookie or go-between.

As Schwartz concluded in a 1977 report for the Field Foundation, published a few months before the Supreme Court handed down its ruling in the *New York Telephone* case, "Title III has borne out what its opponents feared: electronic eavesdropping has become an almost routine investigative tool for use in minor cases." Among the arsenal of crime-fighting tools that U.S. law enforcement had at its disposal, wiretaps were "no more effective than the conventional non-electronic methods, such as 'buy and bust' and 'turning' witnesses."[19]

By the end of the decade, law enforcement officials were starting to see the wisdom of Schwartz's argument. The costs of using wiretaps to fight gambling were taking their toll nationwide. "Inflation is a factor—everything is more costly than it was," Manhattan district attorney Robert M. Morgenthau told the *New York Times* in 1980. "There are a lot of investigative techniques which are not being used because of the lack of resources.'"[20] Morgenthau's office had run twenty-eight telephone taps the previous year. At a total cost of $433,057, the results in court were meager returns on the state's investment. Across the board, the U.S. government's *Wiretap Report* for 1979 listed its lowest totals for Title III surveillance in years, a decline of 36 percent since the height of the gambling crusade in 1973.[21] Many officials went on the record to explain the decrease in wiretap usage as the product of resource constraints and poor results, especially in gambling cases. Some also singled out large-scale wiretap operations outside of the push against gambling as exorbitant and ostentatious.

The *coup de grâce* was a bribery investigation run out of the FBI's Las Vegas field office from January 1979 to January 1980. The target was Teamsters president Roy L. Williams. Later described in court as the most extensive wiretap sting in the history of telephonic interception, the Williams investigation produced more than 2,000 reels of tape that captured 2,013 individuals speaking on 30,416 separate telephone calls. When the case went to trial, however, prosecutors were forced to reveal that just 1,400 of those 30,416 calls—a paltry 4.5 percent—could be considered incriminating. In the end, Williams's conviction rested mostly on the testimony of cooperating witnesses, not on the fruits of the Title III intercepts. The bill for the wiretaps: $1,004,110. In an update on the Williams case published in 1983, a reporter for the *New York Times* somewhat gleefully noted that the surveillance operation's million-dollar price tag didn't include the cost of the team of engineers hired to "electronically enhance" hundreds of otherwise inaudible wiretap recordings for the jury to hear at trial.[22] The phone calls of more than two thousand Americans, most of them innocent of any wrongdoing, were monitored by the FBI for no reason.

With a record tarnished by toothless anti-gambling stings (like Project Anvil) and expensive dragnet cases (like that of the Teamster Roy Williams), it was obvious why state and federal agencies were opting to scale back on wiretaps by the end of the 1970s. Title III seemed to have lost its way, just as Schwartz predicted. What the government needed in the new decade was a more dependable national target for electronic surveillance—a target whose political and economic street value could match the steep costs, in both dollars and manpower, involved in any given wiretap investigation. Drugs proved just what the doctor ordered.

. . .

Lawmakers recognized the wiretap's potential utility as a weapon against narcotics trafficking almost as soon as President Richard Nixon declared the War on Drugs in 1971.[23] Yet drug wiretaps would lag as a distant second behind gambling wiretaps throughout the decade. In 1976, the National Wiretap Commission reported that telephone intercepts in narcotics cases led to conviction rates that were slightly higher than those found in gambling cases. The average time on the line was shorter, too. The problem was that the agencies tasked with waging the drug war—most notably Nixon's new Drug Enforcement Administration (DEA), founded in 1973—were unequipped to mount more than a few wiretaps at a time.

"A major concern about the use of Title III surveillance, especially by the DEA, is the impact on the agency's manpower," the National Wiretap Commission warned in 1976. "In some offices it becomes necessary to set aside all other business for the duration of the surveillance; elsewhere, up to one-third of the office's agents may be required for a single surveillance."[24]

In place of wiretaps, the DEA tended to rely on more conventional police methods, such as foot patrol stops (i.e., "stop and frisk"), buy-bust ploys, and the long-term cooperation of witnesses. At this point it should go without saying that investigative tactics like these had disastrous effects on the Black and brown communities living on the front lines of America's drug war during the 1970s.[25] So it has gone in the decades since.

Only in New Jersey, where Ed Tomas was busy climbing telephone poles and installing slave lines, would police find a way to make narcotics wiretaps worth the inevitable drain on resources. From the mid-1970s onward, electronic surveillance in the Garden State yielded a 90 percent conviction rate. Most of the cases involving telephone taps and bugs were drug-related, carrying extended prison sentences that chipped away at trafficking organizations from the margins.[26] According to Tomas, the statistics for electronic surveillance in this period were actually more favorable than the official records showed. Some of the state's data on conviction rates in Title III cases failed to account for the fact that many police wiretaps—most, in Tomas's own experience—ended with plea deals. They seldom went to trial.[27]

Observers skeptical of New Jersey's early experiments with drug enforcement wiretaps dismissed the state's successes as the result of favorable environmental factors. Call it the fish-in-a-barrel theory: because the Garden State was a hotbed for crime families who made their money in narcotics, common sense suggested that the police could go up on a wire in the average street-level drug case and come away with a mafia lieutenant on the other end. In any event, it was only a matter of time before dealers and suppliers stopped using the phone to do business. But in a 1983 interview, G. Robert Blakey—a professor of law at Notre Dame who, fifteen years earlier, had worked with conservative firebrand John McClellan to draft the language of Title III—explained that New Jersey's early wiretapping victories were the result of "aggressive" training in surveillance tactics and criminal procedure.[28] The attention to detail on the ground helped the Garden State become the leading edge of a new electronic strategy in the drug war, a catalyst for the reinvention of the wiretap in the 1980s and 1990s.

"Wiretapping is concentrated in certain areas because there are many areas that don't have well-trained police or prosecutors," Blakey observed. "They're not sophisticated enough to use wiretaps, they're not integrated in their repertoire." Michael Bozza, a prosecutor for the New Jersey Division of Criminal Justice's Organized Crime Section, agreed with Blakey's assessment: "You can't tell me that New Jersey is the only place where there's large-scale organized crime. . . . Any place you have that type of problem, then electronic surveillance is appropriate. I don't understand why other states haven't caught up with us."[29]

Such claims may well have been true for wiretap work on the level of state law enforcement. But the feds were already starting to follow the New Jersey model. As soon as President Ronald Reagan took office and recommitted the nation to fighting the drug war ("We're taking down the surrender flag that has flown over so many drug efforts," Reagan famously announced from the Rose Garden. "We're running up a battle flag"), Title III wiretap use skyrocketed nationwide.[30] In 1981, Reagan's first year in Washington, federal telephone intercepts increased by 31 percent.[31] The following year, the number of Title III investigations jumped another 23 percent nationwide, and then another 40 percent in 1984.[32] The average length of a telephone tap extended from twenty-four hours to twenty-six days during this period.[33] DEA narcotics cases were behind the ballooning numbers at every turn.

Bob A. Ricks, the DEA's general counsel, defended his agency's shift toward electronic surveillance as a cutting-edge technological response to the mechanics of sophisticated trafficking organizations. Wiretaps gave the government an advantage in the drug war, Ricks explained, because the world of illegal narcotics had come to depend on the phone system: "The telephone is the drug dealers' Achilles heel. . . . Their organizations are spread out all over the place, and they have to communicate by phone. Not all of their business can be in person."[34]

The logic would have been familiar to anyone with a cursory knowledge of the evolution of policing and crime in the twentieth century. This wasn't the first time in the nation's history when the traffic of controlled substances seemed to depend on the medium of the telephone. Both in the mechanisms that enabled their distribution schemes, and in the hierarchical organizations that provided the capital behind them, the drug dealers of the 1980s resembled the bootleggers of the 1920s.[35] By turning to wiretaps to combat the drug trade, the DEA was effectively running the old Prohibition Bureau playbook. The rest of the nation's law enforcement agencies followed suit in the coming years, oblivious to a lesson that an earlier generation of crime warriors had learned the hard way. Employing electronic surveillance may well be an effective way to raise the political profile of a crime war: a way to make criminal investigations appear more efficient and scientific, a way to combat modern criminal organizations using state-of-the-art technologies and methods. But electronic surveillance doesn't necessarily end up making a crime war more winnable, or even worth fighting at all.

Institutional cooperation was critical to America's renewed investment in law enforcement wiretapping in the early 1980s. After 1982, the DEA solved its Title III staffing problem by sharing concurrent jurisdiction with the FBI, where electronic surveillance had already ascended to the level of investigative art. Cementing the link between the two agencies was Francis "Bud" Mullen Jr., a career FBI official nominated to lead the DEA in 1982. Mullen was a proponent of using telephone taps to fight the drug war at any cost.

"There's no way we would be successful against drug trafficking without wiretapping," Mullen told the Los Angeles Times after settling into his new administrative role. "I don't think we've reached the optimum on using electronic surveillance. . . . It will continue to increase until the drug problem abates."[36]

U.S. Attorney General William French Smith supported Mullen's agenda by promising to exploit Title III wiretapping "to almost the fullest degree required and necessary" in drug investigations. He claimed that combining the resources of the FBI and the DEA to step up Title III work was the "single most important step taken in the drug enforcement area."[37] The soaring number of wiretap investigations in the ensuing years bore vivid witness to the nation's aggressive approach to technology-driven narcotics enforcement.

At long last, the telephone tap—a medium of surveillance first employed by crooks and confidence men, later adapted by private detectives and corporate security specialists, and finally perfected by the nation's foreign intelligence apparatus—had arrived on the streets of America's cities. It was a partial realization of the vision of wiretapping for crime control that the conservative coalition behind the Safe Streets Act had concocted out of the tumult of the late 1960s. After a decade of languishing in the depths of America's gambling crusade, electronic surveillance was reborn.

The nation's leading law enforcement trade journal, *Police Magazine*, would extol the wiretap's "comeback" in the War on Drugs in a series of long-form articles published in 1983.[38] At the same time, the Justice Department courted national news outlets to publicize its victories in Title III narcotics work. High-profile heroin busts in New York, New Jersey, and Las Vegas—the products of methodical wiretap operations, press officials always noted—led to lurid "dope on the table" media spectacles.[39] News reports also drew a direct connection between the cocaine scandals that had begun tarnishing the image of American professional sports in the period and increased electronic surveillance. In 1983–1984, for example, dozens of nationally syndicated articles cited the explosive fact that many of the football and baseball stars who were cooperating with narcotics investigations had come to the attention of authorities when their names and voices surfaced on Title III wiretaps.[40]

In a new twist to the media's regular coverage of government surveillance, news reports on Title III drug cases during the 1980s often took pains to enumerate the mechanics and outcomes of individual wiretap operations: the number of agents required; the amount of tape amassed; the total length of phone conversations intercepted, down to the hour and minute. The details seemed to give the practice of wiretapping a tangible human scale, although reporters at times described the players involved in elevated terms. In December 1985, for instance, the *Wall Street Journal* ran a front-page article on a group of expert witnesses who were busy fielding calls from federal prosecutors. Their only job was to "translate" the slang heard on narcotics wiretaps for juries in courtrooms around the country. The article compared the work of America's wiretap translators to that of the military cryptanalysts who managed to decipher Axis codes during World War II.

"It's an inexact science," one member of the group explained. "The one thing I've learned is that none of the code words are constant. It's whatever grabs the person's fancy." The article even included a brief glossary of slang terms for heroin often overheard on the wire: "Among the many terms that have been used to mean drugs or quantities of drugs are *horse, dog, cat, girl, boy, shirt,* and *car.* Words like *picture* and *injection* sometimes refer to samples."[41]

In these and other instances, a familiar politics of race underwrote the government's new electronic approach to fighting the drug war, just as it had in the effort to push Title III through Congress in 1968. As soon as white America came to see inner-city Black communities as ground zero for the nation's drug problem, wiretapping resurfaced as an acceptable tactic in the battle to clean up the streets. The *Wall Street Journal's* wiretap codebreakers were presumably translating slang spoken on the corners of America's inner cities, a world away from the newspaper's elite white readership. Likewise, the wiretaps at the center of the period's earliest pro sports cocaine scandals had mostly incriminated Black athletes: Tony Dorsett, Ron Springs, Larry Bethea, and Harvey Martin of the Dallas Cowboys; Willie Aikens, Vida Blue, and Willie Wilson of the Kansas City Royals—to name just a few. Notably, the mounting price of the drug war's electronic front ($25 million for federal telephone intercepts in 1984 alone) received little pushback from lawmakers during these years.[42] Once again, arguments about the costs of electronic surveillance would disappear when African American voices were caught on the line.

By the end of the decade there was no turning back. In an effort to keep up with drug war best practices, several states that had resisted granting wiretap authority to police began resorting to telephone taps in narcotics investigations. In Michigan, reluctant legislators caved to hardliners like Joel Shere, the assistant U.S. attorney for Michigan's Eastern District, a region that includes cities with sizable Black populations, such as Detroit and Flint. Shere had traveled to the state capitol to testify that wiretaps were an "essential tool" in the drug war. "Almost any major drug case in our state needs it," he asserted. "I think it's just dynamite when the defendants are incriminated by their own lips."[43] In California, where the state legislature finally legalized police wiretapping in 1988, a Los Angeles police official took to the pages of the *Los Angeles Times* to tout his newfound ability to employ electronic surveillance in the poor, crime-ridden neighborhoods of his jurisdiction: "Wiretaps are the answer. Let everyone know that Big Brother, as you call our government, will be watching the major drug dealers in this state. . . . I hope the new law makes every drug dealer paranoid about using his telephone."[44]

Several other states extended police wiretapping laws that were scheduled to sunset in this period. New Jersey, the electronic drug war's pioneer, was one of them.[45] The demand for law enforcement wiretaps continued to reach record heights.

. . .

Telecommunications companies began sounding alarms about increased threats to privacy around the time of wiretapping's vaunted "comeback" in the early 1980s. But the escalation of electronic surveillance in the War on Drugs wasn't what worried corporate insiders. At least not directly.

At firms like AT&T, MCI, and ITT, all of which were eager to gain a foothold in the post-monopoly marketplace, a host of breakthrough innovations promised to transform how Americans communicated with each other. By the time AT&T agreed to dismantle the Bell System in 1982, the future of the industry appeared to reside in new technologies and applications: computers, satellites, and digital switching systems; cellphones, e-mail, and fiber optic networks. The problem—the fear—was that the laws on the books had little control over the technological advances at the center of the coming revolution. A range of third parties, both official and unofficial, were already rushing to exploit their security vulnerabilities. Meanwhile, the industry lacked the legal recourse to push back in the name of customer privacy.

To make matters more urgent, the date that the novelist George Orwell had famously used to predict the arrival of a total surveillance society in *Nineteen Eighty-Four* was fast approaching. According to H. W. William Caming, AT&T's head of corporate security, there was a "compelling need" for the industry to work with Congress to protect electronic privacy as new technologies and services entered the deregulated market.[46] Business depended on ensuring that "1984" was just a metaphor.

Symbols aside, the limited data at the industry's disposal indicated that the relationship between privacy and surveillance was careening out of balance. As early as 1980, polls showed that ordinary Americans still harbored outsized fears about the privacy of their communications and personal data. One internal survey circulated among AT&T operating company managers showed that more than 74 percent of Americans wanted to make privacy a "fundamental right, akin to life, liberty, and the pursuit of happiness," yet doubted that legal regulations could keep pace with ongoing technological developments. The same survey found that 10 percent of Americans believed that an entity associated with the U.S. government had tapped their telephones "at some time." The figures were higher for government employees themselves: police officers (24 percent), regulatory officials (25 percent), and members of Congress (39 percent).[47] Scanning the data, it was clear that companies working to protect privacy would have an advantage in a competitive market for technologies and services. At the very least, corporate brands needed to signal a basic commitment to allaying the fears of the average customer.

The telecommunications industry's growing concern with privacy in the 1980s also reflected new possibilities in the field of surveillance. Even as agencies like the DEA and FBI doubled down on the tried-and-true medium of the telephone tap, electronic surveillance technology was rapidly changing. According to a study conducted by the U.S. government's Office of Technology Assessment (OTA), innovations in communications systems had "greatly increased the technical options for surveillance activities" in the years since the National Wiretap Commission published its 1976 reports.[48] In Washington, the OTA discovered, law enforcement agencies working on drug interdiction and counterterrorism were on record as having intercepted messages transmitted via satellite, microwave, and electronic mail throughout the early 1980s. More than 25 percent of the federal agencies surveyed by the OTA also reported monitoring computerized data systems for the purposes of crime control and national security. The trends were ominous: the return of the wiretap in the drug war was opening the door to more advanced electronic surveillance activities elsewhere. A "virtual revolution" in surveillance capacity was underway, and the state had a strong political mandate to test its limits.[49]

Robert Kastenmeier was keeping a close watch on these transformations from his seat in Congress. A stalwart House Democrat representing Wisconsin's Second District, Kastenmeier believed that the nation stood at a crossroads as 1984 approached. Breakthrough technological change, aggressive market competition, increased government surveillance: in a world beyond the Bell System, such developments were predictable, even manageable, in and of themselves. Taken together, they had produced, in Kastenmeier's words, an "unregulated morass" that both threatened individual privacy and hindered free enterprise.[50]

Heeding a growing chorus of warnings from telecom executives and regulatory officials, Kastenmeier convened a series of hearings for a subcommittee of the House Committee on the Judiciary in November 1983. With the avowed goal of examining "the state of our civil liberties as well as what the future may hold," he invited a wide array of legal experts and industry representatives to testify.[51] Kastenmeier gave the proceedings an attention-grabbing title: "1984—Civil Liberties and the National

Security State." On January 3, 1984, just three days into Orwell's infamous year, he announced to the *Los Angeles Times* that the "exploding technology" of computers and satellites was making the "specter evoked by '1984'... very real."[52] It was time to act.

Urged on by the telecommunications lobby, Kastenmeier's "1984" hearings would continue in the new year as he searched for partners to help draft a bill that would bring the nation's electronic privacy laws in line with new technological developments. Yet the events that set the stage for the unlikely contingent of supporters he found were far more prosaic than anything Orwell's novel could have predicted. They didn't involve computers or satellites, the fantasies of science fiction turned reality. They involved, of all things, the technology of the cordless telephone.

Anyone who has owned a cordless telephone is likely to remember the problem of crosstalk. Because cordless phones work by transforming electrical telephone signals into radio waves, any nearby AM / FM radio or cordless handset operating on the same transmission frequency can tune into a call. Crosstalk was a technical glitch that cordless users came to accept as routine in the 1980s. Manufacturers even warned of the problem in owner's manuals. (I have a particularly vivid memory of a neighbor's conversation blurting into one of my calls when I was 12 or 13—I was chatting on the cordless handset in my parents' kitchen at the time. Everyone I know has a story like this.) But almost as soon as companies like Sony and Radio Shack began selling cordless phones on the commercial market, they started turning up in drug investigations. When they did, it wasn't entirely clear whether the existing legal guidelines enumerated in Title III applied. In the earliest cases, the courts ruled, somewhat counterintuitively, that cordless users weren't entitled to privacy protections at all.

In 1982, an AM / FM radio enthusiast in Kansas happened upon the cordless phone conversations of his neighbors, Timothy Ray Howard and Rosemarie Howard. When he overheard the Howards discussing the sale of cocaine, he contacted the Kansas Bureau of Investigation (KBI), which promptly provided him with materials to record the incriminating conversations from his own radio set while the police tracked the Howards' calls using a pen register. This unusual arrangement led to a search of the Howard residence, and ultimately to the arrest of the Howards themselves. But prior to the couple's trial, a district court judge dismissed the case on the grounds that the KBI had violated Title III's illegal interception clause. In 1984, the Supreme Court of Kansas overruled the lower court's decision. The seven-judge panel reasoned that Title III didn't actually govern the KBI's investigation into the Howards. As owners of the cordless handset in question, they "had been fully advised by the owner's manual as to the nature of the equipment" and thus "had no reasonable expectation of privacy under the circumstances."[53] A few months later, the Supreme Court of Rhode Island handed down a similar ruling in a case involving a young boy who tuned into his neighbor's cordless crosstalk and overheard a series of drug transactions.[54]

The idea that cordless phones lacked Title III protections was symptomatic of a larger, more intractable problem that Kastenmeier had fretted about for years. The laws that Americans had come to rely on to guarantee privacy and limit surveillance were falling behind the times. By the early 1980s, the language of Title III reflected a fast-fading era in the history of communications. As written, the key provisions of the law regulated the "aural acquisition" of "wire and oral" communications transmitted by common carriers, which made sense in an age of single-line telephones, copper wires, and mechanical switching systems.[55] As soon as service providers began transmitting calls through updated networks and platforms, however, those carefully delineated distinctions began to collapse. Wireless radio, microwave, and satellite technologies all fell outside the purview of the statute, limited as it was to protecting electronic communications carried by "wire." So did the technology of the personal computer, which transformed communications data into a stream of 1s and 0s that couldn't reasonably be construed as "oral." Title III's definition of electronic surveillance was similarly anachronistic. Was it possible to consider the interception of digital communications an act of "aural acquisition"? Certainly not.

As the cordless phone cases showed, semantic discrepancies like these had real-world consequences. Patrick Leahy, the Vermont Democrat who emerged as Kastenmeier's legislative partner in the Senate, would illustrate the practical effects of Title III's failings in September 1984, as the momentum behind electronic privacy reform picked up steam:

> Let me just talk about a problem that grows just as we are sitting here. We have phones ringing all over the country. Answer it, and it is not voices you hear but dots and zeroes and blips and beeps that come out of it. And that is the information that is going in digit form and this is everything from interbank orders to private electronic mail hookups. It is nothing remarkable, but it is remarkable

that none of these transmissions are protected from illegal wiretap, because our primary law passed back in 1968 covered only voice transmission; it failed to cover nonaural acquisitions of communications, of which computer to computer transmissions are a good example. . . .

We send sophisticated legal documents, a bid, a love letter. It makes no difference what it might be. . . . [T]he average person assumes he is protected against wiretapping as they would as if they were on the phone, but as a practical matter they are not. And what happens is a case where technology eats away at what we assume are our protections in the Constitution.[56]

For Leahy, as for Kastenmeier, Title III was succumbing to abstract technicalities. The resulting legal outcomes contradicted the basic expectations of privacy that most Americans took for granted. It was reasonable to assume that the law would protect all forms of electronic communication equally, regardless of medium or message. Because it didn't, the law needed to change.

In 1985, Kastenmeier and Leahy began working together to craft a bill intended to bring Title III in line with the new communications environment: the Electronic Communications Privacy Act (ECPA). On its most basic level, the ECPA extended Title III protections to the technologies and services that were beginning to reshape the telecom industry in the early 1980s. Cellphones, pagers, satellites, and personal computers were all included in the bill's purview. So were fiber-optic networks and digital transmissions. In a quirk of legal interpretation, cordless phones remained beyond the scope of the law. But the ECPA also laid claim on an important new frontier for privacy protections: electronic databases. In the process of delivering their services, communications carriers and computer networks were amassing vast troves of customer information, none of it protected under Title III. Kastenmeier and Leahy's bill changed that, making the personal records retained by network service providers subject to "Fourth-Amendment-like" statutory regulations.[57] Title II of the ECPA, better known as the Stored Communications Act, has loomed large in recent years. Its ambiguities have emerged as an explosive site of controversy as government agencies and corporate conglomerates work to exploit the power of digital dataveillance.[58]

The debates over Kastenmeier and Leahy's legislative proposal were staid and efficient, a far cry from the histrionics that Title III produced eighteen years earlier. Both sides of the political spectrum were united behind the ECPA. In large part, this was because corporate interests were at stake. A wide range of technology firms and industry organizations had lobbied for the bill from the outset: from AT&T, IBM, and General Electric to the Electronic Mail Association, the Cellular Telecommunications Association, and the Institute of Electrical and Electronics Engineers.[59] All along, their contention was that protecting privacy was essential to ensuring "innovation" and "growth" in the industry. As time went on, the Orwellian images that Kastenmeier had conjured up to arouse support for reform eventually gave way to a more business-centered narrative—a narrative about the future of communications after the Bell divestiture.

When the proposal for the ECPA finally reached the floor of the House in 1985, Kastenmeier explained that the legislation was a response to "legitimate business concerns," warning that "emerging industries may be stifled" in the absence of updated legal language.[60] He said little about privacy in and of itself. The material drive for growth had subsumed abstract concerns for civil liberties and constitutional rights. The tell was in the ECPA's final list of supporters. Of the dozens of organizations that endorsed Kastenmeier and Leahy's legislation, there was only one civil liberties group to be found among them: the ACLU.[61]

The ECPA went into effect on October 22, 1986. The signing of the law barely received a press notice.[62] Perhaps Kastenmeier's grandstanding about the nightmare of "1984" turned out to be little more than a canny attempt to publicize a legislative cause that mostly boiled down to semantics. Or perhaps electronic privacy was becoming a consensus policy goal in Washington—a goal created by the telecommunications industry's abiding interest in capturing new markets. In the end, the Electronic Communications Privacy Act wasn't entirely about privacy. It was about profit too.

.   .   .

Law enforcement officials stayed unusually quiet during the lead-up to the passage of the ECPA in 1986. Early in the House Judiciary Committee's hearings on Kastenmeier's bill, a Justice Department spokesman cautioned that updating the language of Title III would cause "confusion and uncertainty" among federal agencies using wiretaps to wage the drug war.[63] But the objection was primarily a matter of procedure. In practice the ECPA did little to change when and what the state could tap. From a certain vantage point, the reforms actually made things easier for law enforcement, because they eliminated the legal uncertainties surrounding new electronic surveillance technologies and tactics. The force of the ECPA thus resembled that of Title III, the law it updated. By carving out a limited set of statutory exceptions to communications privacy rights, the 1986 legislation institutionalized wiretapping activities that otherwise fell outside the law's reach. To sweeten the deal for law enforcement, Kastenmeier had even included a provision that expanded the list of crimes for which a judge could approve a wiretap.[64]

Still, the ECPA was a victory for the telecom industry. In name, if not in substance, the law signaled to the American public that the coming revolution in communications technology would prioritize the interests of individual consumers. The slew of network and service advances that rolled out after the passage of the law soon made good on that promise. Fiber-optic networks, digital switching systems, and cellular phones—to name just a few innovations that took off after 1986—made electronic communications more secure and more mobile. They also made surveillance more complicated, interfering with the conventional electronic methods that law enforcement agencies were employing more than ever in the drug war. If in the early 1980s the communications revolution seemed to pose a threat to privacy, by the early 1990s it seemed to pose a threat to surveillance. Technological change was shifting the balance yet again. For the first time in the medium's history, the telephone was difficult to tap.

Experts in the field of communications engineering expressed concerns about the problem of wiretap interference as early as 1976.[65] It wasn't until the late 1980s that the issue came to a head. Modernized network infrastructures were the initial source of law enforcement frustration. Fiber-optic cables, for instance, were billed as "less susceptible to interference, crosstalk, and wiretapping" when AT&T first began testing them in cities like Chicago, New York, and Washington, D.C.[66] As they proliferated, police discovered they were impossible to monitor without extensive centralized assistance from service providers.

Fax machines and pagers likewise stymied customary surveillance methods, particularly in narcotics investigations. "Wiretaps are most often ineffective because [drug traffickers] don't use the phone," an exasperated Justice Department official running drug interdiction in South Florida told *The New York Times* in 1986. "Now they carry beepers and signal each other."[67] New switching features such as call forwarding and call waiting also presented challenges. Armed with a Title III order, an experienced surveillance technician could tap into most any single-line telephone call. But if the user clicked over to receive an incoming call on a second line, the connected conversation was beyond the tap's reach.[68]

Then there was the cellular phone, a newfangled device with seemingly limitless market potential. In 1985 just 203,000 Americans owned a cellphone. By 1990 the number had climbed to 4 million. The exponential growth of cellphone use confounded law enforcement efforts, and nowhere more noticeably than in the War on Drugs. Monitoring cellular calls was a time-consuming matter that involved petitioning a service provider for access to an electronic surveillance "port" in a mobile switching station. Port access was limited; most carriers could accommodate only a few taps at a time. In some jurisdictions, police found themselves waiting in line for surveillance ports to open as the clock on a thirty-day wiretap order ticked away. In New York, traffickers running an outpost of the Colombian Cali drug cartel discovered that they could evade police wiretaps by ditching their cellphones every week. By the time the feds managed to access a surveillance port on one number, the traffickers were using another.[69]

Police and intelligence officials disagreed about the true extent of the wiretap interference problem throughout the late 1980s. Still, the smattering of thwarted Title III cases augured poorly for the prospects of electronic surveillance in the digital age. "If we don't do something we'll be out of the wiretapping business," an agent in the FBI's New York field office told the Bureau's brass.[70] As the setbacks mounted, the FBI would begin to characterize new communications equipment as an existential threat to its most cherished investigative weapon. One internal report predicted that the "continued and future introduction of [new] technologies will bring about a *de facto* repeal of the existing electronic surveillance authority conferred upon . . . criminal law enforcement agencies by the Congress."[71] Whether or not the Bureau was exaggerating the peril of its position, it seemed clear that the state ran the risk of going dark. Only the phone companies were capable of solving the problem.

Alarmed by internal reports of the Bureau's dwindling wiretap capacity, members of the FBI's technical services division began contacting telephone industry officials in the fall of 1990. In a series of confidential memoranda, they described the Bureau's difficulties in the field and argued that electronic surveillance assistance "must logically increase as the complexity of the technology increased."[72] Closed-door negotiations between the FBI and corporate executives proved fruitless in the ensuing months. The phone companies were unyielding. Apart from the obvious political risks, guaranteeing wiretap access was an expensive technical charge—one complicated by the fact that many of the network services fueling the "digital telephony" problem, as it came to be called, were already in widespread use. Making matters worse, lawmakers on Capitol Hill were wary of implementing a compromise between the two sides that appeared to limit innovation or expand surveillance power.

On January 17, 1992, National Security Advisor Brent Scowcroft sent a classified White House memorandum on the FBI's digital telephony negotiations to Secretary of Defense Dick Cheney, Attorney General William Barr, and CIA director Robert Gates. "The best solution to this problem is to obtain legislation which ensures the cooperation of the telephone companies in providing *direct access* to target communications," Scowcroft explained after weeks of deliberation with FBI and Justice Department representatives. "All agencies agree with this legislative approach and that we should do it fairly soon. Preliminary soundings on the Hill suggest there is a reasonable chance of success even though these kinds of issues raise 'civil liberties' issues with the attendant political fireworks. . . . [T]he costs of waiting (loss of access and the cost to recoup) are growing rapidly, and an attempt to fix it now is worth the political risks."[73]

So began the FBI's top-secret campaign to convince the telephone industry—and with it, both Congress and the American public—that the government needed a backdoor to the digital revolution. In internal correspondence, Bureau officials referred to the initiative as Operation Root Canal.[74]

To complete the historical arc we began tracing in Chapter 9, Operation Root Canal in essence sought to turn the principle of limited assistance into a rule of *incorporated assistance*. Rather than rely on service providers to help install wiretaps on a case-by-case basis, the FBI wanted Congress to compel them to manufacture built-in shortcuts, integrating wiretap capacities into the architecture of the networks themselves. Contradicting the principles that motivated Kastenmeier and Leahy to draft the ECPA—and flouting a generation of Americans who had learned that privacy was a constitutional right—the objective was to make surveillance a permanent feature of communications design.[75] As one of Operation Root Canal's fiercest opponents later explained, the FBI's digital telephony proposal was tantamount to "requir[ing] . . . that the telephone network must have vulnerabilities intentionally built into it."[76] And not only that: the vulnerabilities were there to enable the government to eavesdrop more readily on the network's own customers.

In the spring of 1992, the FBI began canvassing members of Congress about the feasibility of its audacious digital telephony plan, but the details of the scheme quickly leaked to the press.[77] The turn of events required the FBI to launch a coordinated public relations offensive. On March 23, 1992, FBI director William Sessions relayed a confidential memorandum to all FBI field offices, outlining a messaging strategy that Bureau administrators were to follow in order to convey the urgency of the wiretap interference problem to skeptical parties.

"The points to emphasize," Sessions underscored, "are that the implementation of digital telephone technology will preclude law enforcement from using court authorized telephone surveillances; that a technical solution will not be attained industry wide or for future generations of technology absent legislation; and that the . . . proposed legislation . . . will not impede advancing technology. This approach relies on the telephone companies to develop the most effective solution at the least cost."[78]

Despite the ensuing firestorm, Sessions hoped to push the digital telephony proposal through Congress as fast as possible, creating a legislative framework that the FBI could depend on to manage other looming crises in the digital communications field. The most important of them—an issue that would turn up time and again in the decades that followed—was encryption.[79]

The response to Operation Root Canal reveals just how much market thinking had come to dominate the public debate over electronic surveillance by the early 1990s. Once leaked, the FBI's digital telephony proposal drew harsh criticism from industry representatives and civil liberties advocates. More often than not, they expressed their objections to the plan in terms of its probable effects on technical innovation and corporate competition. An AT&T spokesman conveyed his "grave concerns" about the FBI's thinking to the *Washington Post* in March of 1992, predicting that surveillance-friendly legislation would impede the introduction of new services and raise billing rates.[80] Later that same month, the *Post*'s editorial board described the idea behind Operation Root Canal as "an assault . . . on the competitive position of American industry" comparable to "requiring Detroit to produce only automobiles that can be overtaken by faster police cars."[81] Incorporated surveillance assistance seemed a genuine threat to free enterprise. The following year, a study conducted by the federal government's General Services Administration concluded that requiring wiretap-ready networks would create security vulnerabilities and kneecap American telecom companies on the global market.[82] Talk about individual privacy was nowhere to be found.

Lost in the backlash to Operation Root Canal, regardless of its basis, was the fact that the FBI had failed to provide solid evidence of the wiretap interference problem from the start. Media outlets tended to take the FBI's narrative at face value, often opening reports on the digital telephony debates by reiterating the claim that law enforcement agencies were losing the ability to conduct surveillance.[83] Yet the data that telephone executives and policymakers received was mostly anecdotal. As negotiations in Congress entered a more public phase in 1993–1994, the FBI's official account of the digital telephony issue began to show confusing statistical inconsistencies. The figures on surveillance interference would fluctuate so widely, and so often, that the Bureau's account came to resemble the notorious McCarthy-era screeds about the number of card-carrying communists working in the U.S. State Department. What was clear was that the Bureau's statistics, much like McCarthy's, were intended to scare policymakers into action.

In March 1994, two years into Operation Root Canal, FBI Director Louis Freeh testified to the joint Committee on the Judiciary that the Bureau had encountered ninety-one instances of wiretap interference in 1993 alone.[84] Ten of those cases, according to Freeh, involved the limited accessibility of surveillance ports at cellular switching stations. Thirty other cases involved difficulties with new calling services like call forwarding, call waiting, and speed dialing. Five months later, Freeh returned to Congress with a different dataset. Now there were 183 cases of thwarted wiretaps—a variation, he explained, that reflected a more thorough nationwide survey.[85] Fifty-four of those cases were related to "cellular port capacity." Sixty-six more involved new calling services impeding an active telephone intercept or hindering a dialed number registry.

In a vacuum, the updated statistics from the FBI might have seemed more trustworthy. But Freeh had been pedaling inflated figures in the interim. In a keynote address to members of the American Law Institute in May 1994, he claimed that new communications technology had prevented the FBI from carrying out "several hundred" court-ordered wiretaps the previous year.[86] He used the vague numbers as a sign of impending turmoil.

"If you think crime is bad now, just wait and see what happens if the FBI one day soon is no longer able to conduct court-approved electronic surveillance," Freeh told the audience in attendance. He went so far as to claim that the decrease in wiretap capacity risked "increased loss of life attributable to law enforcement's inability to prevent terrorist acts," "increased economic harm . . . caused by the growth of undetected and unprosecuted organized crime," and "increased availability of much cheaper illegal drugs." There would be "disastrous consequences" if the phone company failed to rebuild their networks.[87]

Freeh's scare tactics worked. As with earlier attempts to institutionalize an expansion of the state's surveillance powers—the passage of Title III, most notably—law-and-order imperatives ended up overwhelming the opposition of telephone companies and civil liberties organizations. The fluctuating numbers didn't matter.

According to one House staffer who spoke to the Electronic Privacy Information Center on the condition of anonymity, the politics of policing and crime control ensured that the facts about wiretap interference were immaterial by the summer of 1994.[88] President Bill Clinton had run a campaign to get tough on crime, and Freeh had made it clear to the White House that the digital telephony proposal was the "only issue" that mattered to American law enforcement in the new administration.[89] The effectiveness of Freeh's tireless campaign on behalf of Operation Root Canal, conducted with Clinton's blessing, surprised even the most seasoned veterans of Capitol Hill.[90] As both houses of Congress moved to draft legislation, Freeh met with every lawmaker in Washington who expressed public doubts about the need for digital telephony workarounds. Notably, one of the plan's primary sponsors in Congress turned out to be Patrick Leahy, the same Vermont Democrat who had joined with Robert Kastenmeier to draft the ECPA a decade earlier. Leahy's switch in allegiance—from privacy protector to surveillance advocate—was a measure not only of Freeh's influence but of the broader pressure to support punitive forms of law enforcement that policymakers in Washington faced throughout the Clinton years.

The bipartisan bill that finally emerged from Freeh's lobbying crusade was dubbed the Communications Assistance for Law Enforcement Act (CALEA). Clinton signed it into law on October 25, 1994—a little more than a month after authorizing the Violent Crime Control and Law Enforcement Act of 1994, the notorious crime bill that has fed America's limitless appetite for incarceration.[91]

Despite more than two years' worth of adversarial backroom negotiations, politically motivated leaks, and brazen scare tactics, the final version of the digital telephony legislation looked a lot like the incorporated assistance arrangement that the architects of Operation Root Canal first envisioned in 1991. CALEA charged the phone companies to enable the government to intercept "all wire and electronic communications carried by the carrier within a service area," and to provide "features or modifications as are necessary to permit such carriers to comply with [wiretap] capability requirements" in an expeditious manner.[92] In return, Congress offered the phone companies $500 million to help defray the cost of updating existing systems and manufacturing new equipment. CALEA also gave the Federal Communications Commission the power to levy additional subsidies for the phone companies in order to ensure future wiretap capacity compliance. In an irony that outraged the civil liberties organizations that opposed the digital telephony plan from the start, taxpayers ended up bearing the financial costs of a system designed to make electronic surveillance easier.

CALEA wasn't without a few noteworthy political concessions. In a superficial nod to the telephone industry, the law made cordless phone conversations protectable under Title III, closing the nagging privacy loophole that the ECPA kept open in the 1980s. More importantly, CALEA left Internet providers alone; as written, the law only applied to telephone companies. The omission, which mostly escaped notice in the 1990s, would set up a fight over the power of the state to monitor digital traffic online that has raged in the decades since.[93]

. . .

The Communications Assistance for Law Enforcement Act of 1994 was an extraordinary piece of federal legislation, now all but forgotten in the regime of ubiquitous backdoor surveillance under which we live today. For the first time in the nation's history, law enforcement had inserted itself into the conceptual circuitry of communications networks. Before CALEA, the needs of policing and crime control were incidental to the regular workings of the telephone system—a product of the fact that the regular workings of the telephone system were well-suited for wiretapping. After CALEA, telephone networks were devised to facilitate court-ordered surveillance from start to finish. In so doing, the law appeared to create an alarming political precedent that far outstripped its immediate legislative consequences.

"Any bill which mandates that communications providers make technological changes for the sole purpose of making their systems wiretap-ready creates a dangerous and unprecedented presumption that government not only has the power, subject to warrant, to intercept private communications, but that it can require private parties to create special access," ACLU officials Ira Glasser and Laura Murphy Lee wrote in a joint letter of opposition to CALEA. "It is as if the government had required all builders to construct new housing with an internal surveillance camera for government use."[94] Roy Neel, a lobbyist for the U.S. Telephone Association, similarly likened CALEA to "asking the telephone companies . . . to become the local cop."[95]

The gravest projections that telephone industry executives and civil liberties advocates made about CALEA weren't entirely unfounded. The new rule of incorporated assistance would enable law enforcement to seek out expanded surveillance powers during the late 1990s and early 2000s. With the prospect of built-in wiretap access at hand, the FBI pushed the telecommunications industry to its limits in the name of the era's most destructive crime policies. As a case in point, consider CALEA's complicated relationship to the War on Drugs—the same law enforcement mandate that drove the FBI to launch Operation Root Canal to begin with.

In October 1995, one year after the passage of CALEA, the FBI made its first attempt to define the technical surveillance capacities U.S. phone companies were to build into their networks. The self-reinforcing logic of the drug war was written into the FBI's schema. In a public notice filed in the Federal Registry, the Bureau called for separating the nation's telecommunications systems into three classes, each according to demographic size and average wiretap activity. Category I networks were to serve densely populated cities around the country: New York, Chicago, Miami, Los Angeles. These were urban centers with high rates of crime, areas of the country where the "majority of electronic surveillance activity occurs"—a designation that Title III drug enforcement priorities had of course already helped to create. The FBI requested that the phone companies in Category I regions retool their hardware to enable a 1 percent wiretap capacity rate, a scale that would enable police to monitor as many as 1 out of 100 calls at any given time. Networks in Category II, by contrast, served "suburban areas" with lower rates of crime and surveillance. In turn, these areas required a lower wiretap capacity rate of 0.5 percent. All other telephone networks fell into Category III, with a minimum capacity of 0.25 percent.[96]

The scale of the FBI's initial request was staggering. Initial estimates, later confirmed by the *New York Times* and the ACLU, calculated that the Bureau's proposal would have given law enforcement the ability to monitor 50,000 phone lines at once in most major American cities, almost fifty times more than the number of Title III surveillance orders granted nationwide the year CALEA became law.[97] The proposed capacity percentages appeared to amount to a shameless power grab—evidence that the FBI either was tapping more lines than it admitted to on the record or planned an exponential increase in electronic surveillance work in the future.[98]

More importantly, the Bureau's network classifications were a transparent product of the drug war's racial geography. The phone companies were expected to boost wiretap capabilities in "high-crime areas"—a bureaucratic euphemism for inner-city Black neighborhoods, to be sure—while "low-crime" white suburbs remained free of expanded police surveillance capacity by default. FBI assistant director James Kallstrom, a longtime wiretapping advocate, tried to justify the inequities of the FBI's scheme by likening wiretap rates and capacities to the distribution of fire hydrants in U.S. cities. "Fire hydrants are placed in communities," Kallstrom explained, "not because they necessarily represent the number of fires that are expected to occur, but are deployed so that, in the event a fire should occur in a particular location, there is a hydrant available for use by the fire department."[99] The dispassionate air of Kallstrom's metaphor did little to distract from the obvious: certain groups of Americans—poor communities of color, especially—were going to bear the brunt of the FBI's intensified wiretap activity more than others.

Public indignation forced the FBI to revise its initial capacity request in 1996, initiating a seesaw battle over CALEA standards that continued well into the 2000s.[100] In the years that followed, the implementation of the incorporated assistance rule would fall victim to the conflicting priorities of the phone companies and law enforcement. But the ensuing political squabbles, mired in highly technical considerations of engineering and design, tended to obscure the continued creep of electronic surveillance power into the furthest reaches of American life.

Abetted by integrated phone company assistance, Title III wiretap orders continued to surge throughout the late 1990s and early 2000s, particularly in narcotics investigations.[101] At the same time, CALEA helped to revive the nation's struggling surveillance technology industry, which was for decades limited to serving the government market. As soon as Congress mandated wiretap readiness in 1994, technology start-ups began scrambling to provide American phone companies with devices that could help make digital networks CALEA-compliant on the cheap. In the years since the attacks of September 11, 2001, the international footprint of private firms offering digital wiretapping equipment to the telecommunications industry has only grown.[102]

Perhaps the most important effects of CALEA, however, were buried in the depths of the FBI's litany of standards requests to the phone companies. Although the Bureau eventually rescinded its petition to increase wiretap capacity in "high-crime areas" (in 2000, a D.C. Court of Appeals condemned the requirement as a "classic case of arbitrary and capricious agency action"), subsequent CALEA filings included a "punch list" of items that gave law enforcement access to electronic information that would skyrocket in value as the digital revolution progressed.[103] The most significant of the items was, at the time, a seemingly minor ask: easy access to call location data. The demand would survive a series of regulatory challenges throughout the late 1990s, battered but intact.

"This does not turn wireless phones into tracking devices," FCC chairman William E. Kennard explained in a statement on the FBI's location data requests in 1998. "Law enforcement can only secure this information if a court authorizes it. This capability will help law enforcement make our streets safe.'"[104] Of course, telecommunications officials saw the Bureau's interest in location data differently. Ronald Nessen, a spokesman for the Cellular Telecommunications Industry Association, anticipated that satisfying the FBI's demands would "turn all . . . wireless phones into location beacons." Nessen worried that the endless debates over CALEA standards no longer concerned wiretapping, law enforcement's most extreme tactic of electronic investigation. "This is now about determining where you are," he declared.[105]

As the last two decades in the history of communications technology have demonstrated, only one of these two predictions came true. It would have consequences for the balance between privacy and surveillance that no one, not even the industry at the center of the controversy, could have anticipated.

• • •

On April 18, 1990, a team of reporters at the *Baltimore Sun* broke the story of the arrest of Linwood "Rudy" Williams, the ringleader of a violent drug organization that controlled the streets of Baltimore. The ensuing coverage of the case set out the charges leveled against Williams and his associates in predictable fashion. It detailed the number of suspects named in the indictment (30), and the estimated amount of heroin that Williams had smuggled the previous year (110 pounds); the number of search warrants signed (27), and the street value of the drugs seized during their execution ($175,000). It was a big story, worthy of its position on the paper's front page. Williams was alleged to have had a hand in several Baltimore city homicides dating back to 1983.[106]

Two and a half weeks later, the *Sun* published an unusual coda to its coverage of the case: an in-depth exposé of the byzantine police investigation that had led to Williams's arrest. As it turned out, Williams was a master at exploiting the possibilities of new communications technologies to manage his criminal outfit. For an entire calendar year his ingenious use of cellphones and new digital calling services—Caller ID, especially—had stumped a team of more than 200 local, state, and federal agents who wanted nothing more than to capture his voice on a Title III recording. The article, headlined "Caller ID Latest Hit with High-Technology Drug Dealers," depicted Williams's elaborate communications routine as typical of the tactics that large trafficking organizations were using to do business. It also presented the Williams investigation as emblematic of the problems that U.S. law enforcement agencies had begun to encounter in a changing communications environment.

"If it rings, if it beeps, if it's cordless or mobile or equipped with digital display—if it can be used by one person to communicate with another—then rest assured that someone, somewhere will discover a way to sell drugs with it," the *Sun* report opened.

The article went on to rehearse what, for us, should look like a familiar story. Because communications advancements were proving resistant to established electronic surveillance methods—and because drug crews like Williams's were adopting them for the sole purpose of thwarting Title III investigations—the police needed help, either from the phone companies or from Congress. A year before the start of Operation Root Canal, the *Sun* was offering an endorsement of the thinking behind CALEA.

"Law enforcement officials say technical refinements will help," the article continued, "but nothing can change the fact that telephone communications have permanently transformed the drug trade." Something had to give.[107]

The *Baltimore Sun* reporter who uncovered the story behind the Linwood Williams investigation was David Simon, a writer who has since made something of a legendary name for himself in the television industry. A little more than a decade later, Simon used his experience covering cases like Williams's as an inspiration for his celebrated HBO series *The Wire* (2002–2008). As Chris Albrecht, HBO's chairman and chief executive officer, recalled Simon's initial pitch to the network: "David came in and said that he wanted to do the most detailed, most realistic look at a police wiretap investigation that's ever been done."[108] No television program before or since has succeeded in that goal—and so much more—so forcefully.

The Wire is a direct product of the story we've traced in this chapter: a product of the wiretap's reemergence in the War on Drugs in the early 1980s, and a product of the debates over privacy and surveillance that led to the passage of the ECPA and CALEA in the decade that followed. Each of the show's five seasons revolves around an embattled squadron of Baltimore police officers—the Major Crimes Unit, so-called—as it works to wage the drug war in a progressively complex communications ecosystem: from pay phones and pagers (Season 1) to corporate computer logs (Season 2), and from disposable "burner" cellphones (Seasons 3–4) to text messages and digital images (Season 5). The acquisition of an electronic surveillance warrant functions as a crucial narrative goal in the arc of each season. The Major Crimes Unit can't build its cases effectively until a judge signs off on a thirty-day wiretap order—a dynamic foreshadowed in the passing shot of a Title III affidavit that appears in the opening credits to all five seasons of the show. Only on the wire can "good policing" happen. Only on the wire can the Major Crimes Unit move its investigations off of Baltimore's street corners, up through the network of kingpins and politicians who have corrupted the city.

As the lieutenant of the Major Crimes Unit, Cedric Daniels (Lance Reddick) learns late in the show's opening season, "The wire is what gives us . . . the whole crew. Day by day, piece by piece."

And yet *The Wire* doesn't necessarily depict wiretapping as an infallible solution to the problems of the drug war. As Simon's sprawling story arc unfolds, the targets of the Major Crimes Unit's investigations tend to escape the widening net of police surveillance just as often as they are en-trapped by it.[109] "Don't talk in the car, or on the phone, or in any place that ain't ours, and don't say shit to anybody who ain't us," drug dealer D'Angelo (Lawrence Gilliard Jr.) recites to drug enforcer Wee-Bay (Hassan Johnson) midway through the show's first episode. These are "the rules" that every character in *The Wire* must follow—an informal code of en-gagement that recognizes both the ubiquity of electronic surveillance in-vestigations in the modern drug war and their inherent limitations as a check on the powers of the streets. By following the rules, you can stay off the wire. You can stay alive, and make your name beyond the law's reach.

Stringer Bell (Idris Elba), the savvy lieutenant of the drug crew at the center of Seasons 1–3, follows the rules to an almost monastic degree. He refuses to identify his associates, and he habitually changes the microchip in his cellphone to evade potential police taps. Drug suppliers Proposi-tion Joe (Robert F. Chew) and Spiros Vondas (Paul Ben-Victor) follow the rules too, foiling investigations into their dealings by staying off the phone altogether. (A critical scene at the end of Season 2 involves Vondas hurling his mobile into the Patapsco River.) By Season 5, the police violate their own rules of engagement. The Major Crime Unit's overreliance on Title III surveillance ends up railroading a case the city has built against Marlo Stanfield (Jamie Hector), the ruthless kingpin whose character Simon based on Linwood Williams himself.

All the same, *The Wire* suggests that electronic surveillance has the po-tential to bring order to the chaos and illogic of America's drug war. Al-though wiretapping doesn't catch every crook or solve every case, it's what enables us to connect the players in the game: day by day, piece by piece. More importantly, *The Wire* also suggests that it's in the public's interest for telecom carriers to comply with the demands of law enforcement. As communications schemes evolve on the corners of West Baltimore, the cops find themselves turning to the phone companies for assistance. A tense scene late in Season 3 features Lieutenant Daniels and prose-cutor Rhonda Perlman (Deirdre Lovejoy) arguing with the executive of a local service provider about the need to make its cheap mobile phones surveillance-ready (see Figure 10.1). The episode depicts the executive's indifference to the rule of incorporated assistance as out of touch, even reckless. Meanwhile, Daniels and Perlman are able to force the company's hand and make a break in their case. The message is obvious.

Moments like these—there are several in the show's five-season run—lead *The Wire* into tricky political territory. The history of wiretapping

Figure 10.1. Rhonda Perlman (Deirdre Lovejoy) and Cedric Daniels (Lance Reddick) battle with a wireless telephone provider in HBO's *The Wire,* Season 3, Episode 9: "Slapstick." *The Wire,* HBO, 2008.

in America, both before and after David Simon's series, has revealed the dangers of normalizing the medium of the wiretap as an investigative panacea. It has also revealed what happens when we sideline privacy concerns in the interest of profit motives and police imperatives. Notably, Simon's original 1990 report on the Linwood Williams investigation prompted a *Baltimore Sun* editorial that foresaw many of the developments we've witnessed in recent years, as surveillance powers creep further and further into the hardware of our digital lives.

"It is discomfiting, but not really that surprising, that drug dealers have fastened on the Caller ID and cellular telephone services offered by communications utilities," the *Sun*'s editorial board wrote in a comment published on June 19, 1990, coincidentally twenty-two years to the day after Lyndon Johnson signed the Safe Streets Act into law. "The very nature of criminal behavior is the recognition of opportunities to abuse products and services designed for legitimate uses."

Expectations aside, the article concluded on a more uncertain note:

What does come as something of a surprise is the demand that telephone service providers "think about investigations" the way law enforcement officers do, and build into every new service an easy access for surveillance. The interest of federal and local agencies in reaching quickly into the "information loops" used by organized crime is understandable and laudable. Overreaching obtrusion into the communications services used by ordinary citizens is not. So as the proliferating advances of new technology prompt law enforcement to push for new levels of access, the danger of unwarranted intrusions on the liberty of ordinary citizens must also prompt an ongoing civil liberties re-examination. Just how far is government snooping entitled to go in casting the net to catch criminals? It is a troubling question, and it is not just the cops and robbers with interests at stake.[110]

The *Sun*'s editorial comment seems to capture what many Americans would come to realize in the 1990s, as the digital telephony debates unfolded. It also seems to capture what many of us have come to accept today, as we talk and text through the surveillance-friendly networks that CALEA helped to produce. In one way or another, we all must follow the rules.

Epilogue

King's Call, Hoover's Tap

OUR STORY ENDS in 2001, a date that appears to mark a dramatic break from the past we've been tracing.

The terrorist attacks of September 11, 2001, ushered in a new age for electronic surveillance in the United States. By now, its contours are well-established.[1] The passage of the USA PATRIOT Act (2001) dramatically expanded the government's electronic surveillance authority. The law allowed federal agencies to conduct "roving wiretaps," intercepting communications on any device used by the target of a national security investigation: landline, laptop, or cellphone. More importantly, it also gave federal agencies free reign over massive troves of calling records and other forms of stored communications data. The government institution synonymous with the expansion of electronic surveillance after 9 / 11 was the secretive National Security Agency (NSA). As the 2013 Snowden leaks revealed, provisions in the PATRIOT Act and the Foreign Intelligence Surveillance Act (FISA) of 1978 enabled the NSA to launch data-mining programs of incomparable complexity and scale during the early 2000s. Only in recent years have policymakers managed to stem the tide.

At the same time, 2001 heralds a different sort of break from the past: around the turn of the twenty-first century, our analog nation went digital. Since 2001, Americans have witnessed the proliferation of social media platforms and search engines, "cookies" and "apps," smart phones and smart homes. Among other things, these sweeping technological transformations marked the arrival of what the writer and philosopher Shoshana Zuboff has termed "surveillance capitalism"—a new historical epoch in which economic growth and everyday life depend on the ability to monetize the private information we freely surrender in the service of connectivity.[2] The institutional avatar for the new digital economy was (and still is) the search-engine-turned-corporate-behemoth Google, based in Silicon Valley. At Google your data helps to predict outcomes, produce behaviors, and generate profits. The digital records that companies like Google both create and collect, under feeble legal regulations, have also fallen into the hands of law enforcement, creating possibilities for "big data policing" that some scholars see as the future of American crime control.[3]

The NSA and Google are entities that rely on *dataveillance*. For power and profit, they devour the digital traces we invisibly leave behind: likes and dislikes, locations and destinations, purchases and habits, friends and connections. From all that information, they can discern population-wide trends and make predictions at scale. Our voices and conversations—the auditory contents of our communications, flowing through the Internet's architecture—are largely incidental to what the NSA and Google have been doing in recent years. As individual selves, our significance appears to have vanished. The old ways of protecting the privacy of our communications (*don't talk in the car, or on the phone, or in any place that ain't ours, and don't say shit to anybody who ain't us*) appear to have vanished along with it.

The NSA occasionally makes use of eavesdropping techniques like the "roving wiretap" when carrying out foreign intelligence investigations. But the term "wiretapping," in its original sense, encompasses only a fraction of the agency's incomprehensibly vast menu of electronic surveillance activities. Likewise, companies like Google, Amazon, and Apple have built passive listening capabilities into the smart devices they sell on the market. But calling your Alexa or iPhone a wiretap, as some do in jest, is to make an all-too-common category error. Recall the distinction we drew in Chapter 9 between data "at rest" and communications "in motion," first established in the age of the telegraph.[4] In the twenty-first century, state institutions and corporate entities have generally found it more effective to mine the former rather than listen to the latter.

So, has the age of digital dataveillance rendered wiretapping obsolete? Not quite. Or rather: not at all.

Amid the rise of the NSA's data-mining programs and the explosion of Google's big-data economy, the good old-fashioned wiretap continues to thrive in American law enforcement. Today Title III wiretap rates are more than three times higher than in the 1980s and 1990s. In the past ten years for which we have data, state and federal judges approved an average of 3,374 Title III orders annually. At its peak in 2015, Title III accounted for 4,148 separate wiretaps around the country. Seventy-nine percent of them were in drug investigations, most often conducted in the same communities of color that the FBI designated as "high-crime areas" in its CALEA compliance filings of the 1990s. The share of Title III investigations dedicated to narcotics cases is typically much higher in any given year. In 2014, as much as 89 percent of wiretap work in the United States was drug-related. Notably, the overwhelming majority of Title III cases in this period have involved the interception of telephonic communications, not Internet traffic. Our phones, now mobile and smart, are still the target.[5]

Even if more alarming and intrusive forms of electronic surveillance seem to lie under the aegis of FISA, statistics like these would have shocked the electronic privacy advocates of the 1960s and 1970s. They warned of the dangers of wiretapping when Title III rates hovered around a few hundred cases a year. In the meantime, the law enforcement mission creep continues, with electronic surveillance tactics assuming a dizzying array of technological guises. Billion-dollar government investments in drug control and immigration enforcement have allowed ordinary cops on the beat to exploit eavesdropping devices originally designed for military use. A prime example is the IMSI catcher, or "Stingray"—a handheld device that mimics the function of a cellphone tower in order to capture mobile location data and other identifying caller information. Since 2014, Stingrays have turned up in cases involving drug dealers in Baltimore, undocumented immigrants in Detroit, and Black Lives Matter activists in Chicago.[6]

Technically speaking, the Stingray isn't any more of a wiretap than a data-mining program or a search engine. But it's safe to say that the routine use of wiretaps in criminal investigations has paved the way for the adoption of devices like it, usually in the gray areas of the law. Here we return to a relationship that we've encountered time and again in the pages of this book. The wiretap was the first medium of surveillance to become "normal" in the United States, and its normalization has led to the adoption of unregulated eavesdropping tactics and technologies as time moves on. Even in the digital age, wiretapping remains electronic surveillance's bleeding edge.

· · ·

The normalization of the wiretap in America is one of this book's key storylines. Once wiretapping was "dirty business," an extreme tactic of investigation that seemed to represent a criminal threat to civil liberties.

Now wiretapping is "good police work," as the cops in *The Wire* might say. Almost every other form of electronic surveillance imaginable gets measured by its yardstick.

In surveying the wiretap's 160 years of history, it becomes clear that our acceptance of its place in the most banal corners of American life was a long time in coming. Concerns about the dangers of wiretapping were once mainstream in this country. They held political court for so long that it took policymakers more than a century to bring order to the chaos of U.S. wiretap law. Today debates about wiretapping take place on the margins. Every few years the issue resurfaces in scandal—in the form of the Snowden leaks, for example, or the alleged misuses of the FISA process—only to recede into the background once again. The cycle continues.

As we've seen throughout this book, punitive ideas about policing and crime helped drive the normalization of wiretapping in America. The passage of the Omnibus Crime Control and Safe Streets Act in 1968 is perhaps the most convenient date we can point to in tracing the popular shift from "dirty business" to "good police work." But its roots stretch into earlier historical periods, the Prohibition era in particular. From the Safe Streets Act to the Communications Assistance for Law Enforcement Act—from John McClellan and Lyndon Johnson to Louis Freeh and Bill Clinton—the politics of law and order motivated the institutionalization of the state's electronic surveillance authority. This relationship has important consequences for how we understand the development of the U.S. "surveillance state," a formation that we tend to regard as an outgrowth of Cold War anticommunism and antiradicalism. Viewing the normalization of wiretapping as a byproduct of law and order's ascendance puts the imagined imperative to police communities of color at the center of the story of our surveillance society. In the twenty-first century, America's wars on terrorists and immigrants have only reinforced the relationship between institutionalized electronic surveillance and the "surveillance of Blackness."[7]

The normalization of wiretapping in America has taken more subtle forms, too. The wiretap recording itself—as medium, as message—retains a special authority as evidence: the voice of the past, overheard. Its allure remains as strong in the pages of history as it does in criminal investigations and court proceedings. Perhaps nowhere is this dynamic more visible than in the fraught afterlife of the most famous collection of electronic surveillance intercepts ever produced: the FBI's wiretaps of Martin Luther King Jr. It's here that our story comes to a close.

King, as is well known, was under relentless electronic surveillance from 1963 until his death in 1968. At the request of FBI director J. Edgar Hoover, and with the written approval of Attorney General Robert F. Kennedy, the FBI installed wiretaps on King's home and office telephones. A team of Bureau agents also planted bugs in the bedside lamps of his hotel rooms as he crisscrossed the country.

King's friends and associates were suspicious of wiretaps—and rightly so, since the FBI turns out to have tapped many of their telephones too. But King wasn't convinced of the situation's seriousness until November 1964, when the Bureau agents involved in the operation made a craven attempt to weaponize their recordings of his private activities. The FBI's technical surveillance efforts remained something of an open secret during the final years of King's life, only receiving official confirmation in June 1969, a year after his assassination in Memphis. The heavyweight champion Cassius Clay was in the process of appealing a conviction for draft-dodging on the basis of evidence tainted by an illegal wiretap. In the proceedings of the ensuing trial, Clay's attorneys called several FBI agents to the stand. Under oath, an agent in the FBI's Atlanta field office named Robert Nichols was forced to disclose his role in the King surveillance. Nichols also testified to monitoring the telephones of other prominent African American leaders, including Elijah Muhammad. The revelations made national headlines.[8]

The Church Committee intelligence hearings of the mid-1970s revealed the full extent of the FBI's surveillance of King. It remains the nadir of the Bureau's record of abuse under Hoover: the culmination of a long history of antagonism toward African American political leaders and civil rights activists, driven by paranoia, fear, and outright race-hatred. In 1977 a federal judge ordered the FBI's tapes and transcripts sent to the National Archives and sealed for fifty years. Curiosity about their contents didn't disappear, however. In the decades that followed, the King wiretaps would emerge as a kind of "holy grail" of American political history: evidence offering access to the truth of the civil rights leader's life, forever out of reach.

In October 1983, the FBI's electronic surveillance operation resurfaced as the subject of a lurid controversy in the debates over the establishment of Martin Luther King's birthday as a national holiday. Conservative opponents of the effort, led by the notorious segregationist Senator Jesse Helms, wanted to cast doubts on King's life and legacy. Helms pointed to the unknown contents of the FBI recordings—long suspected to contain evidence of King's communist affiliations, as well as proof of extramarital affairs—to convince the nation that King wasn't worth honoring. Helms wanted them unsealed so the public could hear the truth.

"Any residual interest that Dr. King's former colleagues may retain in privacy as to his personal affairs is now far outweighed by the impending legislation," Helms wrote in a last-minute legal plea for the National Archives to release the wiretaps.[9] He was swiftly rebuffed.

Not to be deterred, the Conservative Caucus in Washington circulated a cache of documents to lawmakers—65,000 pages in all, including redacted wiretap transcripts—intended to demonstrate that King "did not deserve the status as a national hero."[10] In private, President Ronald Reagan expressed reservations of his own about the push to declare King's birthday a national holiday. In a letter to one Republican congressman, Reagan mentioned that the FBI wiretaps, still under seal, were crucial to revealing the private realities behind King's public image. He was later forced to apologize for the suggestion.[11]

At the time, most Americans recognized that the effort to use the wiretaps to tarnish King's good name was as abhorrent as the original wiretaps themselves. The ugliness of Helms's campaign may well have increased support for the "MLK Day" legislation among both parties in Congress. In response to Helms's repeated provocations on the Senate floor, Senator Daniel Patrick Moynihan threw a copy of the wiretap transcripts to the ground, denouncing them as a "packet of filth."[12] According to the *New York Times*, the FBI's electronic surveillance operation had "deprived [King] of his civil liberty of privacy long before he lost his life."[13] Keeping the wiretaps under seal seemed the only way to honor King's legacy and correct the FBI's history of surveillance abuse. Some called for the recordings to be destroyed.[14]

The bill to create Martin Luther King Day passed easily, and the controversy over the FBI wiretaps soon faded. But the allure of the holy grail remained. The first to reach it—or at least get within striking distance—was David Garrow.

A political scientist and historian at the City University of New York, Garrow had cobbled together the story of the FBI's surveillance of King during the late 1970s and early 1980s. His meticulous research on the subject was published in 1981 as *The FBI and Martin Luther King, Jr.: From "Solo" to Memphis*—a study that helped to inspire my own work on this book. In 1985, while compiling materials for a follow-up project documenting King's political life, Garrow struck gold. It was the result of a series of creative Freedom of Information Act (FOIA) requests. Instead of working to access the King wiretaps, Garrow managed to reconstruct their contents by mining the telephone conversations of King's friends and lesser-known associates, who were also under FBI surveillance at the time. Garrow's successes with FOIA garnered a flood of media attention. In September 1985, the *New York Times* announced that Garrow had acquired 180,000 pages of classified wiretap transcripts and electronic surveillance reports, with another 150,000 pages still in the processing pipeline.[15] Reports about the unexpected accessibility of the King wiretaps were almost universally laudatory—a far cry from the coverage of Helms's attempts to unseal their contents two years earlier, which had much less noble aims in mind.

Garrow used his FOIA victories as the basis for *Bearing the Cross,* an exhaustive biography of King published in 1986. Reviewers extolled the book for its detailed reconstruction of King's life and voice, even though portions of Garrow's account at times seemed to revel in the details of King's private behavior. Writing in the *Journal of American History,* the historian David Levering Lewis knocked *Bearing the Cross* for setting "a new floor under the standard for retrieval of the biographical past." But for Lewis, Garrow's reliance on the FBI's wiretap transcripts was to be "absolved from the almost certain charges of voyeurism and malign disclosure." In places the wiretaps gave Garrow's story the appearance of "Rankean comprehensiveness—history as it actually was."[16] *Bearing the Cross* went on to win a Pulitzer Prize in 1987.

Despite the monstrosity of their origins—and despite the appalling political uses to which they were put in the early 1980s—the FBI's electronic surveillance transcripts have provided fodder for an entire generation of historians who have worked to understand King's life and legacy. Taylor Branch is perhaps the most celebrated among them. His three-volume biography of King, compiled over the span of twenty-four years and widely considered definitive, makes extensive use of FBI wiretap records.

"Invasive wiretaps are pretty basic primary, biographical material," Branch explained in 1998. "There's nothing better than a verbatim wiretap transcript of somebody's entire telephone life, you know."[17]

In a promotional interview for *At Canaan's Edge* (2006), the final installment of his King trilogy, Branch described the process of sorting through wiretap recordings at the National Archives in vivid and amusing detail: "Sifting through the reams of FBI transcripts was an ordeal. All you have to do is go to the FBI headquarters and be willing to sit in the basement in a windowless room and endure their security procedures, which are pretty rough. If you want to go to the bathroom, you have to ask for a security escort to come and take you to and from to make sure you're not flushing some document down the toilet."[18] Talk to anyone who has ever monitored a live wiretap and they'll describe their work to you in similar terms. Taxing. Monotonous. And more often than not: conducted in a windowless basement.

. . .

I raise the examples of Garrow and Branch, both decorated and reliable historians, to demonstrate just how routine the use of wiretaps as primary evidence has become over time. Electronic surveillance has proven to be as useful to the discipline of academic history as it has to the practice of criminal investigation. Wiretap recordings are standard fare in the footnotes of books like this one. Using them is normal.

For my own part, I have opted to make minimal use of wiretap recordings and transcripts in the research for this project. This was my own choice, made out of considerations that were as much practical as they were ethical. But even in a case as extreme as King's—a case in which the FBI conducted electronic surveillance to destroy a man's life, a case in which politicians exploited electronic surveillance to tarnish his legacy—wiretaps remain a coveted form of historical information. This was a fascinating subtext in Sam Pollard's recent documentary *MLK / FBI* (2021), an exposé of the FBI's harassment of King based on Garrow's scholarly research. Many of the talking heads featured in the film, from former FBI director James Comey to Garrow himself, cast doubt on the status of the King wiretaps as evidence, even as the film still uses their contents to tell the story of the FBI's ill-begotten surveillance operation. Throughout it all, the recordings under seal tantalize us as the voice of the past, overheard: *history as it actually was.*

Whatever else we might say about the stories we've followed in this book, the history of wiretapping in the United States has left behind a wealth of primary source materials for future generations to mine. Perhaps more than we'll ever know.[19] Even if wiretapping remains, at bottom, a "dirty business," wiretap recordings still offer a version of the past whose authority we tend to take for granted. Whether we choose to access that past is ultimately a matter of time, ethics, and resources.

Meanwhile, the National Archives will unseal the King wiretaps in 2027. Regardless of what interested Americans choose to do then, my sense is that we'll be living the history that created them for a long time to come.

NOTES

Introduction

1. "Our Letter from Placerville: Immense Stock Frauds Attempted to Be Perpetrated through the Telegraph," *San Francisco Daily Alta,* August 11, 1864, 1. For more details of the D. C. Williams case, see these articles from the *Sacramento Daily Union:* "Tapping the Wires for Stock Operations," August 12, 1864, 2; "The Telegraph Frauds," August 12, 1864, 3; "The Stock and Telegraph Operator," August 13, 1864, 1; and "Released on Bail," August 22, 1864, 5. Williams's electronic abilities apparently stemmed from his experience working as an operator for the Alta Telegraph Company in California.

2. Michael Lewis, *Flash Boys: A Wall Street Revolt* (New York: W. W. Norton, 2015).

3. Jay David Bolter and Richard Grusin, *Remediation: Understanding New Media* (Cambridge, MA: MIT Press, 1999); Lev Manovich, *The Language of New Media* (Cambridge, MA: MIT Press, 2001); Lisa Gitelman, *Always Already New: Media, History, and the Data of Culture* (Cambridge, MA: MIT Press, 2006). Similar ideas have animated the well-established subfield of media archaeology. See Thomas Elsaesser, "The New Film History as Media Archaeology," *Cinémas* 14, no. 2–3 (2004): 75–117; Erkki Huhtamo and Jussi Parikka, eds., *Media Archaeology: Approaches, Applications, and Implications* (Berkeley: University of California Press, 2011); Jussi Parikka, *What Is Media Archaeology?* (Cambridge: Polity Press, 2012).

4. Quoted in Richard R. John, *Network Nation: Inventing American Communications* (Cambridge, MA: Belknap Press of Harvard University Press, 2010), 27.

5. Josh Lauer, "Surveillance History and the History of New Media: An Evidential Paradigm," *New Media & Society* 14, no. 4 (2012): 566–582.

6. According to a Pew Research Center survey conducted in February 2017, seven out of ten U.S. adults believe that it is at least "somewhat likely" that the government monitors their phone calls and emails, and 37 percent of them believe that such surveillance is "very likely." See "Most Americans Think the Government Could Be Tracking Their Phone Calls and Emails," September 27, 2017, https://www.pewresearch.org/fact-tank/2017/09/27/most-americans-think-the-government-could-be-monitoring-their-phone-calls-and-emails/. The Pew Research Center has also found that opinions are divided as to whether such surveillance is justified. See "Majority Views NSA Phone Tracking as Acceptable Anti-terror Tactic," June 10, 2013, https://www.people-press.org/2013/06/10/majority-views-nsa-phone-tracking-as-acceptable-anti-terror-tactic/; Lee Rainie and Mary Madden, "Americans' Views on Government Surveillance Programs," March 16, 2015, https://www.pewinternet.org/2015/03/16

/americans-views-on-government-surveillance-programs/. On feelings of resignation and apathy surrounding the need to protect privacy in the digital age, see Eszter Hargittai and Alice Marwick, "'What Can I Really Do?': Explaining the Privacy Paradox with Online Apathy," *International Journal of Communication* 10 (2016): 3737–3757; Nora A. Draper and Joseph Turow, "The Corporate Cultivation of Digital Resignation," *New Media & Society* 21, no. 8 (2019): 1824–1839.

7. *Olmstead v. United States,* 277 U.S. 438, 470 (1928).

8. My argument here can be read as a gloss on John Durham Peters's notion that the "modern experience of communication is . . . often marked by felt impasses" and that it is always haunted by failure. John Durham Peters, *Speaking into the Air: A History of the Idea of Communication* (Chicago: University of Chicago Press, 1999), 1. For more philosophical accounts about the relationship between eavesdropping and communication, see John L. Locke, *Eavesdropping: An Intimate History* (New York: Oxford University Press, 2010); Peter Szendy, *All Ears: The Aesthetics of Espionage,* translated by Roland Vésgö (New York: Fordham University Press, 2017).

9. Samuel D. Warren and Louis D. Brandeis, "The Right to Privacy," *Harvard Law Review* 4, no. 5 (December 15, 1890): 193–220.

10. Michel Foucault, "Governmentality," in *The Foucault Effect: Studies in Governmentality,* edited by Graham Burchell, Colin Gordon, and Peter Miller (Chicago: University of Chicago Press, 1991), 87–104.

11. On the origins and politics of "surveillance capitalism," see, among others, Mark Andrejevic, *iSpy: Surveillance and Power in the Interactive Age* (Lawrence: University Press of Kansas, 2007); Josh Lauer, *Creditworthy: A History of Consumer Surveillance and Financial Identity in America* (New York: Columbia University Press, 2017); Shoshana Zuboff, *The Age of Surveillance Capitalism: The Fight for a Human Future at the New Frontier of Power* (New York: Public Affairs, 2019). On the racial biases of surveillance technology, see Simone Browne, *Dark Matters: On the Surveillance of Blackness* (Durham, NC: Duke University Press, 2015); Alvaro M. Bedoya, "Privacy Should Be a Civil Right," *New York Times,* June 7, 2018, A25.

12. On the history and politics of "end of privacy" rhetoric in America, see Sarah E. Igo, "The Beginnings of the End of Privacy," *Hedgehog Review* 17, no. 1 (Spring 2015); Nicholas A. John and Benjamin Peters, "Why Privacy Keeps Dying: The Trouble with the Talk about the End of Privacy," *Information, Communication, and Society* 20, no. 2 (March 2016): 284–298.

13. See, among many others, David Kahn, *The Codebreakers: The Story of Secret Writing* (New York: Scribner, 1996 [1967]); James Bamford, *The Puzzle Palace: A Report on America's Most Secret Agency* (New York: Penguin Books, 1982); James Bamford, *Body of Secrets: Anatomy of the Ultra-Secret National Security Agency* (New York: Anchor Books, 2002); James Bamford, *The Shadow Factory: The Ultra-Secret NSA from 9 / 11 to the Eavesdropping on America* (New York: Anchor Books, 2008); Matthew Aid, *The Secret Sentry: The Untold History of the National Security Agency* (New York: Bloomsbury Press, 2009).

14. Athan Theoharis, *Spying on Americans: Political Surveillance from Hoover to the Huston Plan* (Philadelphia: Temple University Press, 1978); Alexander Charns, *Cloak and Gavel: FBI Wiretaps, Bugs, Informers, and the Supreme Court* (Urbana: University of Illinois Press, 1992); Christian Parenti, *The Soft Cage: Surveillance in America from Slavery to the War on Terror* (New York: Basic Books, 2003); Athan G. Theoharis, *Abuse of Power: How Cold War Surveillance and Secrecy Policy Shaped the Response to 9 / 11* (Philadelphia:

Temple University Press, 2011); Shane Harris, *The Watchers: The Rise of America's Surveillance State* (New York: Penguin Press, 2011).

15. A recent exception to this rule is Sarah E. Igo's magisterial *The Known Citizen: A History of Privacy in Modern America* (Cambridge, MA: Harvard University Press, 2018). On the development of wiretap law and policy in America, see Walter F. Murphy, *Wiretapping on Trial: A Case Study in the Judicial Process* (New York: Random House, 1965); Edith J. Lapidus, *Eavesdropping on Trial* (Rochelle Park, NJ: Hayden Books, 1972); Whitfield Diffie and Susan Landau, *Privacy on the Line: The Politics of Wiretapping and Encryption,* 2nd ed. (Cambridge, MA: MIT Press, 2007); Susan Landau, *Surveillance or Security? The Risks Posed by New Wiretapping Technologies* (Cambridge, MA: MIT Press, 2010); Cyrus Farivar, *Habeas Data: Privacy vs. the Rise of Surveillance Tech* (Brooklyn: Melville House, 2018). Recent journalistic treatments of the subject include Jill Lepore, "The Prism: Privacy in an Age of Publicity," *New Yorker,* June 17, 2013, https://www.newyorker.com/magazine/2013/06/24/the-prism; David Price, "The Social History of Wiretaps," *Counterpunch,* August 9, 2013, https://www.counterpunch.org/2013/08/09/a-social-history-of-wiretaps-2/.

16. See, in particular, Theoharis, *Spying on Americans;* Charns, *Cloak and Gavel;* Theoharis, *Abuse of Power.*

17. See Patrick Radden Keefe, *Chatter: Uncovering the Echelon Surveillance Network and the Secret World of Global Eavesdropping* (New York: Random House, 2006).

18. See, among others, Christopher Andrew and Vasili Mitrokhin, *The Sword and the Shield: The Mitrokhin Archive and the Secret History of the KGB* (New York: Basic Books, 2000); Anna Funder, *Stasiland: Stories from Behind the Berlin Wall* (New York: Harper Perennial, 2011 [2003]); Kristie Makracis, *Seduced by Secrets: Inside the Stasi's Spy-Tech World* (New York: Cambridge University Press, 2008); Gary Bruce, *The Firm: The Inside Story of the Stasi* (New York: Oxford University Press, 2012).

19. U.S. Congress, Law Library of Congress, *Comparative Study on Wiretapping and Electronic Surveillance Laws in Major Foreign Countries,* LL 75–176 (1975). For an account of the political history of wiretapping in the United Kingdom, focusing mostly on the period after 1960, see Patrick Fitzgerald and Mark Leopold, *Stranger on the Line: The Secret History of Phone Tapping* (London: The Bodley Head, 1987).

20. U.S. companies still occupy a leading position in the international market for eavesdropping equipment, although Israeli firms seem to have assumed the vanguard in recent years. The American government also continues to supply eavesdropping equipment to foreign intelligence agencies and law enforcement outfits across Central America and the Middle East in order to assist the global wars on drugs and terrorism. Bamford, *The Shadow Factory,* 161–268.

21. Meyer Berger, "Tapping the Wires," *New Yorker,* June 18, 1938, 41; William J. Mellin, "I Was a Wire Tapper," *Saturday Evening Post,* September 10, 1949, 63.

22. Theoharis, *Abuse of Power,* 46.

23. *Hoffa v. United States* 387 U.S. 231 (1967). See Charns, *Cloak and Gavel,* 79–89.

24. Quoted in Richard Reeves, *President Nixon: Alone in the White House* (New York: Simon and Schuster, 2001), 75–76.

25. Adam Roston, "Sources: Donald Trump Listened In on Phone Lines at Mar-A-Lago," *BuzzFeed News,* June 30, 2016, https://www.buzzfeednews.com/article/aramroston/sources-donald-trump-listened-in-on-phone-lines-at-mar-a-lag; Mark Fisher, "Trump Has a Long

History of Secretly Recording Calls, according to Former Associates," *Washington Post,* May 12, 2017, https://www.washingtonpost.com/politics/trump-has-a-long-history-of-secretly -recording-calls-according-to-former-associates/2017/05/12/b302b038-372d-11e7-b412 -62beef8121f7_story.html.

26. U.S. Congress, House of Representatives, Committee on Expenditures in the Executive Departments, *Wire Tapping in Law Enforcement,* 71st Cong., 3rd sess. (1931), 5–6.

27. U.S. Congress, *Wire Tapping in Law Enforcement,* 5–6.

28. Clifford S. Fishman and Anne T. McKenna, *Wiretapping and Eavesdropping,* 2nd ed. (New York: Clark Boardman Callaghan, 1995), 1.8–1.10, 2.1–2.7.

29. Foreign Intelligence Surveillance Act of 1978, 50 U.S.C. § 1801 (1978). The statute delimits four different kinds of electronic surveillance, of which only the first three can be categorized as forms of "eavesdropping."

1. Stolen Signals and Whispering Wires

1. My account of the workings of the Olmstead organization, both here and in the paragraphs that follow, relies largely on local news clippings collected in "Memorandum on *Olmstead v. United States*: The Case of the Whispering Wires" (n.d.), Box 244, Folder 1, Samuel Dash Papers, Manuscript Division, Library of Congress, Washington, DC. See, in particular, the following articles from the *Seattle Times*: "Raiders Nab Roy Olmstead and Guests," November 18, 1924; "Special U.S. Grand Jury for Olmstead," November 20, 1924; "Crowded Courtroom Hears Statement," February 17, 1926; "Arguments of Defense Attorneys Are Heard," February 19, 1926; and "Olmstead, 20 Others Guilty," February 22, 1926. Further details of Olmstead's bootlegging activities can also be found in Norman Clark, "Roy Olmstead: A Rumrunning King on Puget Sound," *Pacific Northwest Quarterly* 54, no. 3 (July 1963): 89–103; Walter F. Murphy, *Wiretapping on Trial: A Case Study in the Judicial Process* (New York: Random House, 1965), 16–17; Philip Metcalfe, *Whispering Wires: The Tragic Tale of an American Bootlegger* (Portland, OR: Inkwater Press, 2007), 28–37; Karen Abbott, "The Bootlegger, the Wiretap, and the Beginning of Privacy," *New Yorker,* July 5, 2017, https://www.newyorker.com/culture/culture-desk/the-bootlegger-the -wiretap-and-the-beginning-of-privacy.

2. My account of the Seattle Prohibition Bureau's wiretap operation relies on the testimony of Richard Fryant, Clara Whitney, and William Whitney before the District Court of the United States, Western Washington, January 1926. Supreme Court of the United States, *Roy Olmstead, Jerry L. Finch, Clarence G. Healy, et. al. v. United States of America: Transcript of Record, Volume II* (October Term, 1927), 434–517, 518–579, 664–669. See also Metcalfe, *Whispering Wires,* 38–54.

3. U.S. Congress, House of Representatives, Committee on Expenditures in the Executive Departments, *Wire Tapping in Law Enforcement,* 71st Cong., 3rd sess. (1931), 5–6.

4. Quoted in Murphy, *Wiretapping on Trial,* 61.

5. "Wire Tapping," *Washington Post,* February 18, 1928, 6.

6. *Olmstead v. United States,* 277 U.S. 438, 462 (1928).

7. *Olmstead v. United States,* 277 U.S. 438, 464 (1928).

8. Melvin I. Urofsky, *Dissent and the Supreme Court: Its Role in the Court's History and the Nation's Constitutional Dialogue* (New York: Pantheon Books, 2015), 194–208.

9. Samuel D. Warren and Louis D. Brandeis, "The Right to Privacy," *Harvard Law Review* 4, no. 5 (December 15, 1890): 193–220.

10. *Olmstead v. United States,* 277 U.S. 438, 478 (1928). For the telephone industry's *amicus curiae* brief, filed on Olmstead's behalf in 1927, see "*Olmstead v. United States:* Brief in Support of Petitioners' Contention, Amicus Curiae" (October 1927), Southwestern Bell Telephone Company Records, Collection 2: SBC Communications Inc., Record Group 5: Predecessor and Subsidiary Companies, Box 134, Subject Files: "Wiretapping, 1927–1948," AT&T Archives and History Center, San Antonio.

11. *Olmstead v. United States,* 277 U.S. 438, 473 (1928).

12. *Olmstead v. United States,* 277 U.S. 438, 470 (1928).

13. Murphy, *Wiretapping on Trial,* 95–96.

14. "Police Espionage in a Democracy," *Outlook,* May 31, 1916, 235; "Wire Tapping," *Washington Post,* February 18, 1928, 6; "Wire Tapping," *Baltimore Sun,* June 6, 1928, 12; "Government Lawbreakers," *New York Times,* June 6, 1928, 24.

15. Mabel Walker Willebrandt, "Tapped to Liquor Ring: Mrs. Willebrandt Still Opposed to Obtaining Evidence by This Method," *Los Angeles Times,* August 19, 1929, 1.

16. *Olmstead v. United States,* 277 U.S. 438, 468 (1928).

17. "There may be those who think wiretapping is a dirty business," wrote the outspoken New York City district attorney Edward S. Silver in 1964. "But who among us can deny the fact that murderers, narcotic smugglers and peddlers, labor racketeers, corrupters of public officials, bank robbers, burglars, and extortionists are engaged in far dirtier business?" Edward. S. Silver, "Wiretapping and Electronic Surveillance," *Journal of Criminal Law, Criminology, and Police Science* 55, no. 1 (March 1964): 114. For other uses of the phrase "dirty business" in the service of support for government wiretap authority, see "Wiretapping vs. Crime," *New York Times,* January 20, 1958, 22; C. P. Ives, "Dirty Business," *Baltimore Sun,* February 17, 1958, 10.

18. In a letter to his brother, Chief Justice Taft characterized Holmes's *Olmstead* dissent as "nasty" on these grounds (quoted in Urofsky, *Dissent and the Supreme Court,* 202). Pennsylvania Republican James M. Beck similarly criticized Holmes's use of the phrase "dirty business" as an "unjudicial expression" of moral contempt in a 1931 debate on wiretapping held in the U.S. House of Representatives. *Remarks in House Relative to Wire Tapping by Prohibition Agents,* 71st Cong., 1st sess., *Congressional Record* 74 (January 22, 1931): H 2902.

19. On American ideas about privacy prior to *Olmstead v. United States,* see David J. Seipp, *The Right to Privacy in American History* (Cambridge, MA: Program on Information Resources Policy, Harvard University, 1978); Dorothy J. Glancy, "The Invention of the Right to Privacy," *Arizona Law Review* 21, no. 1 (1979): 1–39; "The Right to Privacy in Nineteenth-Century America," *Harvard Law Review* 94, no. 8 (June 1981): 1892–1910; Frederick S. Lane, *American Privacy: The 400-Year History of Our Most Contested Right* (Boston: Beacon Press, 2009), 1–97; and Sarah E. Igo, *The Known Citizen: A History of Privacy in Modern America* (Cambridge, MA: Harvard University Press, 2018), 17–54.

20. On the role of the telegraph in the American Civil War, see Tom Wheeler, *Mr. Lincoln's T-Mails: How Abraham Lincoln Used the Telegraph to Win the Civil War* (New York: Collins, 2006).

21. Wheeler, *Mr. Lincoln's T-Mails,* 77.

22. David Homer Bates, *Lincoln in the Telegraph Office: Recollections of the United States Military Telegraph Corps during the Civil War* (New York: Century Co., 1907), 11.

23. William R. Plum, *The Military Telegraph during the Civil War in the United States, with an Exposition of Ancient and Modern Means of Communications, and of the Federal and Confederate Cipher Systems; also a Running Account of the War between the States* (Chicago: Jansen, McClurg and Co., 1882), 12.

24. Plum, *The Military Telegraph*, 29. See also A. W. Greely, "The Military-Telegraph Service," in *The Photographic History of the Civil War*, vol. 8: *Soldier Life, Secret Service*, edited by Francis Trevelyan Miller and Robert S. Lanier (New York: Review of Reviews Co., 1912), 360–362.

25. Plum, *The Military Telegraph*, 29.

26. "Novelty in Warfare," *Bell's Life in London and Sporting Chronicle*, September 14, 1862, 3; "Tapping the Wire," *Glasgow Herald*, June 10, 1869, 3.

27. In September 1881, British troops stationed in Egypt took possession of several telegraph stations along the Nile River in order to intercept messages sent by the leaders of the Egyptian counterinsurgency. "The War in Egypt," *Western Mail*, September 11, 1882, 1. On "wire milking," see "Tapping Telegraph Wires," *Mechanics' Magazine*, May 12, 1871, 324.

28. "Novelty in Warfare," 3.

29. Greely, "The Military-Telegraph Service," 364.

30. Francis O. J. Smith, *The Secret Corresponding Vocabulary; Adapted for Use to Morse's Electro-Magnetic Telegraph, and Also in Conducting Written Correspondence, Transmitted by the Mails, or Otherwise* (Portland, ME: Thurston, Ilsey and Co., 1845), 3. On the development and deployment of telegraph ciphers, see David Kahn, *The Codebreakers: The Story of Secret Writing* (New York: Scribner, 1996 [1967]), 189–229.

31. Bates, *Lincoln in the Telegraph Office*, 60.

32. "Military Use of the Telegraph—Necessity of Caution," *New York Times*, July 29, 1862, 4.

33. Many wartime telegraph operators were able to root out bogus messages without the use of ciphers because they knew the distinctive signaling patterns of their correspondents. As one Civil War signal serviceman explained, "The characteristics of each Morse operator's sending are just as pronounced and as easily recognized as those of ordinary handwriting. . . . [W]hen a message is transmitted over a wire, the identity of the sender may readily be known to any other operator within hearing who has ever worked with him." Bates, *Lincoln in the Telegraph Office*, 58.

34. Tom Standage, *The Victorian Internet: The Remarkable Story of the Telegraph and the Nineteenth Century's On-Line Pioneers* (New York: Bloomsbury, 1998), 159.

35. Plum, *The Military Telegraph*, 29–30, 264–265; Bates, *Lincoln in the Telegraph Office*, 58–59.

36. Plum, *The Military Telegraph*, 265.

37. Stephen E. Towne and Jay G. Heiser, "'Everything Is Fair in War': The Civil War Memoir of George A. 'Lightning' Ellsworth, Telegraph Operator for John Hunt Morgan," *Register of the Kentucky Historical Society* 108 (Winter / Spring 2010): 33; Kerry Segrave, *Wiretapping and Electronic Surveillance in America, 1862–1920* (Jefferson, NC: McFarland, 2014), 8.

38. "Tapping a Rebel Telegraph Line in Mississippi," *Frank Leslie's Illustrated Newspaper,* March 18, 1865, 404.

39. Plum, *The Military Telegraph,* 352.

40. Towne and Heiser, "'Everything Is Fair in War,'" 17.

41. Towne and Heiser, "'Everything Is Fair in War,'" 6–7.

42. Greely, "The Military Telegraph Service," 342, 360.

43. Plum, *The Military Telegraph,* 352.

44. Greely, "The Military-Telegraph Service," 344.

45. David W. Blight, *Race and Reunion: The Civil War in American Memory* (Cambridge, MA: Belknap Press of Harvard University Press, 2001), 98–170.

46. Plum, *The Military Telegraph,* 144. Plum relates the story, oft-repeated in similar sources, of a Union signal operator named W. K. Smith. Early in the war, Smith tapped a Confederate line near Lexington, West Virginia. When a Confederate operator detected Smith's presence on the line, the two men ended up exchanging friendly messages and riddles in Morse code.

47. "Eavesdropping Extraordinary," *New York Times,* May 18, 1874, 4.

48. Allan Pinkerton, *Professional Thieves and the Detectives: Containing Numerous Detective Sketches Collected from Private Records* (Toronto: Rose, 1880), 191.

49. For an account of the Pinkerton National Detective Agency's attempts to stop criminal wiretappers in the 1860s and 1870s, see "Agency History: Wire Tappers & Swindlers" (ca. 1940), Pinkerton's National Detective Agency Records, 1853–1999, Box 178, Folder 13, Manuscript Division, Library of Congress, Washington, DC.

50. On the influence of the telegraph ticker on financial markets, see David Hochfelder, "'Where the Common People Could Speculate': The Ticker, Bucket Shops, and the Origins of Popular Participation in Financial Markets, 1880–1920," *Journal of American History* 93, no. 2 (September 2006): 335–358; Alex Preda, "Socio-Technical Agency in Financial Markets: The Case of the Stock Ticker," *Social Studies of Science* 36, no. 5 (October 2006): 753–782.

51. David Hochfelder, *The Telegraph in America, 1832–1920* (Baltimore: Johns Hopkins University Press, 2012), 101–137.

52. Hochfelder, *The Telegraph in America,* 102.

53. Pinkerton, *Professional Thieves,* 142–214.

54. "Telegraphic Revelations," *New York Times,* February 26, 1868, 2.

55. "Cotton Men in a Panic: New Orleans Exchange Suspends on Receiving False Quotations," *New York Times,* September 30, 1899, A11.

56. "Cotton Exchanges in Panic," *New York Sun,* September 30, 1899, 2.

57. On the rise and fall of bucket shops in America, see Ann Fabian, *Card Sharps, Dream Books, and Bucket Shops: Gambling in 19th-Century America* (Ithaca, NY: Cornell University Press, 1990), 188–202; Hochfelder, "'Where the Common People Could Speculate.'"

58. Segrave, *Wiretapping and Electronic Surveillance,* 64–81.

59. "Silent Taps on Exchange: Secret Sounder Panacea for Rival Stock Brokers," *Los Angeles Times,* August 8, 1908, I1.

60. Hochfelder, "'Where the Common People Could Speculate,'" 346. For reports of bucketshop proprietors tapping the lines of licensed stock exchanges, see, for instance, "Stealing Quotations," *Chicago Tribune,* June 18, 1884, 5; "Police Raid Wire-Tappers," *Los Angeles Times,* August 27, 1913, I18.

61. "Wire-Tappers Hit Black's Poolroom," *Los Angeles Times,* September 14, 1902, 6.

62. "The Pool Sellers Sold: A Great Swindle Well Planned and Executed," *New York Times,* October 15, 1883, 1.

63. "Tapping News Conduits," *New York Sun,* July 8, 1883, 6, quoted in Segrave, *Wiretapping and Electronic Surveillance,* 24–25.

64. "Electric Crooks," *Los Angeles Times,* September 15, 1902, 5.

65. "Wire Tappers in Court," *New York Times,* January 30, 1893, 8.

66. "Wire Tapping Gang Escapes," *Atlanta Constitution,* December 23, 1900, 12.

67. "Wire-Tappers Hit Black's Poolroom," 6; "Electric Crooks," 5; "Telegrapher Put behind Bars," *Los Angeles Times,* September 18, 1902, A1; "Wire Tappers Fail to Appear," *Washington Post,* November 13, 1902, 12; "Let the Pool Man In," *Washington Post,* August 21, 1904, S2.

68. "Operators Lose All in Poolroom," *Chicago Daily Tribune,* May 8, 1904, 36.

69. Hochfelder, "'Where the Common People Could Speculate,'" 346–348.

70. "Wire-Tapping Swindlers," *Washington Post,* January 15, 1905, T5; Weber S. Loudock, "Wire Tapper Only a Ghost," *Chicago Daily Tribune,* August 11, 1907, E2; "The Ancient and Profitable Swindle of Wireless Wire-Tapping," *Washington Post,* May 28, 1911, MS4; "Wire Game Best Graft," *Washington Post,* October 20, 1912, M2.

71. "Wire Game Best Graft," M2.

72. Arthur Train, *True Stories of Crime from the District Attorney's Office* (New York: Charles Scribner's Sons, 1908), 117. On Larry Summerfield as the inventor of the wireless wiretap con, see also "Wire Game Best Graft," M2; Will Irwin, *Confessions of a Con Man* (New York: B. W. Huebsch, 1913), 79–80. The *New York Times* reported that one of Summerfield's closest associates, Timothy Oakes, had a hand in the original version of the scheme; see "Tim Oakes Dies in Asylum: Famous as Confidence Man and Inventor of 'Wireless Wire-Tapping,'" *New York Times,* September 10, 1907, 9. For a more extensive account of Summerfield and his ilk, see Amy Reading, *The Mark Inside: A Perfect Swindle, a Cunning Revenge, and a Small History of the Big Con* (New York: Vintage Books, 2012), 101–118.

73. Train, *True Stories of Crime,* 117–118.

74. See, for instance, "Bookkeeper Tells of Losing $5,000 in Aiding Scheme of 'Wire Tappers,'" *Chicago Daily Tribune,* July 12, 1902, 5; "They Got His $100,000: Slick Wire-Tapping Fraud," *Baltimore Sun,* February 8, 1905, 9; "Wire Tappers Raided," *New York Times,* August 16, 1905, 1; "Wire Tapping at Jamaica," *New York Times,* October 18, 1905, 12; "After Wire Tappers; Hear of $70,000 Loss," *New York Times,* January 15, 1907, 4; "Wire Tappers' Dupe," *Washington Post,* February 19, 1907, 1; "Swindlers Got $6,500," *Washington Post,* February 21, 1907, 1; "'Wire Tappers' Found Him 'Easy,'" *Atlanta Constitution,* June 16, 1912, B13; "Wire-Tapping Gang Swindling in South," *Atlanta Constitution,* February 6, 1919, 3.

75. Cornelius F. Cahalane, *Police Practice and Procedure* (New York: E. P. Dutton, 1914), 164.

76. "Swindle Charged by 'Wire Tapping': Ancient Game Used to Separate South Carolinian from His Bank Roll," *Atlanta Constitution*, November 20, 1920; "New Victims Found in Gambling Probe," *Atlanta Constitution*, December 30, 1920, 8.

77. Robert MacDougall, "The Wire Devils: Pulp Thrillers, the Telephone, and Action at a Distance in the Wiring of a Nation," *American Quarterly* 58, no. 3 (September 2006): 715–741. See also Robert MacDougall, introduction to Frank L. Packard, *The Wire Devils* (Minneapolis: University of Minnesota Press, 2013 [1918]), vii–xxv.

78. "Fighting Electric Fiends, or, Bob Ferret among the Wire Tappers," *Nick Carter Weekly*, April 2, 1898, 1–31; Arthur Stringer, *The Wire Tappers* (Boston: Little, Brown, 1906); Arthur Stringer, *The Phantom Wires: A Novel* (Boston: Little, Brown, 1907). MacDougall also lists Franklin Pitt's *Brothers of the Thin Wire* (New York: Street and Smith, 1915) as a wire thriller, but in my reading the novella has little in common with the group of fictional works he identifies. MacDougall, "The Wire Devils," 720.

79. Packard, *The Wire Devils*.

80. Stringer, *The Wire Tappers*, 1.

2. Detective Burns Goes to Washington

1. Warren and Brandeis vaguely allude to the existence of "numerous mechanical devices" that "threaten to make good the prediction that 'what is whispered in the closet shall be proclaimed from the house-tops'" and "modern devices[s] for recording or reproducing . . . sounds." But the bulk of their critique of the invasiveness of modern social life lies elsewhere. Samuel D. Warren and Louis D. Brandeis, "The Right to Privacy," *Harvard Law Review* 4, no. 5 (December 15, 1890): 195, 206.

2. Sarah E. Igo, *The Known Citizen: A History of Privacy in Modern America* (Cambridge, MA: Harvard University Press, 2018), 17–54. As many scholars have noted, the right to privacy was, from the beginning, unevenly distributed along lines of race, class, and gender. See, among others, Anita L. Allen and Erin Mack, "How Privacy Got Its Gender," *Northern Illinois Law Review* 10 (1991): 441–478; Eden Osucha, "The Whiteness of Privacy: Race, Media, Law," *Camera Obscura* 24, no. 1 (May 2009): 67–107.

3. See, for instance, Henry Hitchcock, "The Inviolability of Telegrams," *Report of the Second Annual Meeting of the American Bar Association* (Philadelphia: E. C. Markley and Son, 1879), 93–141.

4. The legal scholar Anuj C. Desai has convincingly shown that American ideas about communications privacy "resulted not from principles embedded in the Fourth Amendment or from an originalist interpretation of the Fourth Amendment or even from existing judicial precedents, but rather from policy choices about the post office a century earlier. . . . [A]s a historical matter, it was the post office—not the Fourth Amendment of its own independent force—that originally gave us the notion of communications privacy that we now view as an abstract constitutional principle applicable to telephone conversations, emails, and the like." Desai also shows why legislative attempts to apply mail privacy statutes to telegraph messages failed throughout the 1870s and 1880s. The courts distinguished the postal system and the telegraph system as radically different forms of communication. "Wiretapping before the Wires: The Post Office and the Birth of Communications Privacy," *Stanford Law Review* 60,

no. 2 (November 2007): 557. See also David J. Seipp, *The Right to Privacy in American History* (Cambridge, MA: Program on Information Resources Policy, Harvard University, 1978), 28–46.

5. Quoted in Seipp, *The Right to Privacy,* 34.

6. Herbert N. Casson, *The History of the Telephone* (Freeport, NY: Books for Libraries Press, 1910), 199. On the early history of the telephone see, among others, Claude S. Fischer, *America Calling: A Social History of the Telephone to 1940* (Berkeley: University of California Press, 1992); Richard R. John, *Network Nation: Inventing American Communications* (Cambridge, MA: Belknap Press of Harvard University Press, 2010), 200–268.

7. Quoted in Michèle Martin, *Hello Central?: Gender, Technology, and Culture in the Formation of Telephone Systems* (Montreal: McGill-Queen's University Press, 1991), 152.

8. Martin, *Hello Central?,* 69, 107–108. See also Josh Lauer, "Surveillance History and the History of New Media: An Evidential Paradigm," *New Media & Society* 14, no. 4 (2011): 577; Kenneth Lipartito, "When Women Were Switches: Technology, Work, and Gender in the Telephone Industry, 1890–1920," *American Historical Review* 99, no. 4 (October 1994): 1101.

9. "Police Espionage in a Democracy," *Outlook,* May 31, 1916, 235. Quoted in Seipp, *The Right to Privacy,* 108.

10. Venus Green, *Race on the Line: Gender, Labor, and Technology in the Bell System, 1880–1980* (Durham, NC: Duke University Press, 2001), 86.

11. Cal. Stats. chap. CCLXII, § 1–19 (1862).

12. Cal. Stats. chap. 117, § 1 (1915).

13. "Tapped a Telephone Wire," *The Sun* (February 21, 1889), 1. Quoted in Kerry Segrave, *Wiretapping and Electronic Surveillance in America, 1862–1920,* 110. Often cited in the literature on wiretapping law, the 1889 Connecticut divorce scandal wasn't the first such case to involve a telephone tap. See "How Telephone Lines Are Worked in Scandal Cases," *Electrical Review,* April 13, 1889, 6, which reports on similar cases in Cleveland and Minneapolis around the same time.

14. Conn. Gen. Stat., chap. 30, § 1240 (1915).

15. Margaret Lybolt Rosenzweig, "The Law of Wiretapping, Part II," *Cornell Law Quarterly* 33, no. 1 (1947–1948): 74.

16. This, at least, was the force of an appellate court's ruling in *State of Washington v. Nordskog,* 76 Wash. 472, 136 Pac. 694 (1913), a case involving a private investigator who had been hired by a newspaper to tap the telephone wires of a rival detective agency. Well into the 1940s, state wiretap statutes around the country were ratified or amended in order to discourage unlicensed individuals from attaching external equipment to the network and disrupting service. On the relationship between the development of wiretap law and the integrity of telecommunications infrastructures, see Brian Hochman, "Wiretapping Stuff: Notes on Sound, Sense, and Technical Infrastructure," *Resilience* 5, no. 3 (Fall 2018): 102–103.

17. Mark Clark, "Suppressing Innovation: Bell Laboratories and Magnetic Recording," *Technology and Culture* 34, no. 3 (July 1993): 533–537; David Morton, *Off the Record: The Technology and Culture of Sound Recording in America* (New Brunswick, NJ: Rutgers University Press, 2001), 114–115.

18. *Hush-a-Phone Corporation v. United States,* 238 F.2d 266 (D.C. Cir. 1956). On the corporate interests behind AT&T's legal effort to suppress the Hush-a-Phone, see Tim Wu, *The Master Switch: The Rise and Fall of Information Empires* (New York: Vintage Books, 2011), 101–114.

19. "Seymour Wires Tapped on Order Given by Woods: Lawyer for Telephone Co. Says Police Spied on Firm with Large War Deal," *New York Times,* May 18, 1916, 8. On the NYPD wiretap squad, see Wesley MacNeil Oliver, "America's First Wiretapping Controversy in Context and as Context," *Hamline Law Review* 34 (2011): 235–236.

20. On the repatriation of military surveillance tactics perfected in America's imperial forays, see Alfred W. McCoy, *Policing America's Empire: The United States, the Philippines, and the Rise of the Surveillance State* (Madison: University of Wisconsin Press, 2009).

21. Regin Schmidt, *Red Scare: The FBI and the Origins of Anticommunism in the United States, 1919–1943* (Copenhagen: Museum Tusculanum Press / University of Copenhagen, 2000), 98–100. The Bureau of Investigation also employed wiretaps sparingly in Mann Act investigations during this period. See Herman Schwartz, "Surveillance: Historical Policy Review," U.S. Government, Office of Technology Assessment US 84–10 / 12 (March 15, 1985), 9.

22. Theodore Kornweibel Jr., *"Seeing Red": Federal Campaigns against Black Militancy, 1919–1925* (Bloomington: Indiana University Press, 1998); Kornweibel *"Investigate Everything": Federal Efforts to Compel Black Loyalty during World War I* (Bloomington: Indiana University Press, 2002).

23. "Burns Tells of Work," *Washington Post,* May 16, 1911, 2.

24. For Burns's biography, see Frank Morn, *The Eye That Never Sleeps: A History of the Pinkerton National Detective Agency* (Bloomington: Indiana University Press, 1982), 172–174; William R. Hunt, *Front-Page Detective: William J. Burns and the Detective Profession, 1880–1930* (Bowling Green, OH: Bowling Green State University Popular Press, 1990).

25. On the rise of detective agencies and private eyes for hire, see Robert P. Weiss, "Private Detective Agencies and Labour Discipline in the United States, 1855–1946," *Historical Journal* 29, no. 1 (1986): 87–107; Robert Michael Smith, *From Blackjacks to Briefcases: A History of Commercialized Strikebreaking and Unionbusting in the United States* (Athens: Ohio University Press, 2003); Jennifer Fronc, *New York Undercover: Private Surveillance in the Progressive Era* (Chicago: University of Chicago Press, 2009); John Walton, *The Legendary Detective: The Private Eye in Fact and Fiction* (Chicago: University of Chicago Press, 2015); S. Paul O'Hara, *Inventing the Pinkertons; or, Spies, Sleuths, Mercenaries, and Thugs: Being a Story of the Nation's Most Famous (and Infamous) Detective Agency* (Baltimore: Johns Hopkins University Press, 2016).

26. Quoted in Morn, *Eye That Never Sleeps,* 173.

27. On Burns's uses of the dictograph, see Dimitrios Pavlounis, "Sound Evidence: An Archaeology of Audio Recording and Surveillance in Popular Film and Media" (PhD diss., University of Michigan, 2016), 26–73.

28. "Scientific Eavesdropping," *Literary Digest,* June 15, 1912, 1239. Quoted in Pavlounis, "Sound Evidence," 26.

29. Hunt, *Front-Page Detective,* 63–64, 68.

30. See "Ohio Assemblymen Fast in Bribe Net: Men in Employ of Manufacturers' Body Take 'Dictaphone' Record in Hotel Room," *Chicago Record-Herald,* May 1, 1911, 1; "Talk Heard

by Sleuths over Phone and Dictagraph [*sic*] Entered as Evidence," *Cincinnati Enquirer,* June 23, 1911, 1; "Name Will Be Omitted, but 'A Voice' Will Figure in the Testimony," *Cincinnati Enquirer,* June 25, 1911, 1. Pinkerton's National Detective Agency Records, 1853–1999, Manuscript Division, Library of Congress, Washington, DC [hereafter cited as Pinkerton Records], Box 67, Folder 3.

31. Pavlounis, "Sound Evidence," 48–49.

32. Pavlounis, "Sound Evidence," 86, 107–112.

33. William J. Burns, "The Imaginary Exploits of 'Sherlock Holmes' Outdone," *Atlanta Constitution,* April 14, 1912, 6. Quoted in Pavlounis, "Sound Evidence," 59.

34. "Talk Heard by Sleuths," 1; "Name Will Be Omitted," 1.

35. State Comptroller's Office, New York City, "In the Matter of the Revocation of the Private Detective License of the William J. Burns International Detective Agency, Inc.: Hearing Held before Hon. William Boardman, Deputy State Comptroller" (July 6, 1917), 296, Pinkerton Records, Box 65, Folder 8. For Burns's lone public recognition of his occasional reliance on the "Tel-Tap," the legality of which he strenuously defended, see "Calls on Seymour to Accuse Burns," *New York Times,* May 22, 1916, 1–2. On Burns's legal spat with the inventor of the dictograph, see Pinkerton Records, Box 68, Folder 3; Pavlounis, "Sound Evidence," 44–45.

36. In a 1912 incident that sparked considerable controversy, the *Chicago Tribune* employed a Burns Agency detective to unseat a recently elected U.S. senator. The detective used a dictograph to listen to the senator's phone calls and private conversations. The case led to hearings in Washington. See U.S. Congress, Senate, Committee Pursuant to S. Res. 60, *Election of William Lorimer: Report Pursuant to S. Res. 60, Directing a Committee of the Senate to Investigate whether Corrupt Methods and Practices Were Used or Employed in the Election of William Lorimer as a Senator of the United States from the State of Illinois,* 62nd Cong., 2nd sess. (1912), 11–12, 38.

37. Oliver, "America's First Wiretapping Controversy," 205–261.

38. "Police Head's Testimony: Wire Spying a Necessity to Detect Crime Here, He Says," *New York Times,* May 20, 1916, 1.

39. "Government Plans Balked: Premature Disclosure of Police Evidence Warned Suspected Men," *New York Times,* May 19, 1916, 1.

40. Quoted in Hunt, *Front-Page Detective,* 105.

41. "Seymour to Ignore Office Wire Spying," *New York Times,* April 22, 1916, 8.

42. "Wire-Tappers vs. The Law: Security against Eavesdropping Guaranteed by Law, Editor Shows," *Washington Post,* May 19, 1916, 6.

43. "Why Not Tap the Mails?," *Washington Post,* May 20, 1916, 6.

44. "Labor Gets Godley in Wiretap Inquiry, Burns Will Be Prosecuted for Raid on Seymour & Seymour Office," *New York Times,* June 14, 1916, 24.

45. Officials at the Pinkerton Agency went on record throughout the 1910s and 1920s to express their disdain for wiretapping and electronic eavesdropping. But evidence in the Pinkerton's National Detective Agency Records suggests that the company made exceptions to its internal ban on technical surveillance later on. In October 1934, for instance, Pinkerton agents tapped fifty-three separate telephone calls placed by employees of the Corn Exchange

Bank and Trust in New York. The wiretaps were part of an ongoing investigation of employee theft. "Telephone Conversations" (October 1934), Pinkerton Records, Box 96, Folder 11. A Pinkerton Agency manual from the early 1950s would also give agents the euphemistic suggestion to maintain active "telegraph company informants" and "telephone company informants" in order to assist surveillance operations. "Course of Instruction for Personnel: Operating Methods" (ca. 1950s), Pinkerton Records, Box 54, Folder 7.

46. "W. J. Burns Guilty, Fined $100; Morgan Agent Acquitted," *New York Times*, January 27, 1917, 1.

47. State Comptroller's Office, New York City, "Brief of the Petitioner: In the Matter of the Revocation of the Private Detective License of the William J. Burns International Detective Agency, Inc." (n.d.), 4–5, Pinkerton Records, Box 65, Folder 6.

48. For the existence of a secret Pinkerton Agency's dossier on the Seymour affair, compiled to aid the case against Burns, see Allan Pinkerton, letter to Meier Steinbrink, February 26, 1917, Pinkerton Records, Box 63, Folder 7; anonymous Pinkerton investigator, letter to Meier Steinbrink, July 2, 1917, Pinkerton Records, Box 63, Folder 7; "Seymour & Seymour" (ca. 1917–1919), Pinkerton Records, Box 68, Folder 3.

49. "Police Head's Testimony," 1.

50. State Comptroller's Office, New York City, "In the Matter of the Revocation of the Private Detective License of the William J. Burns International Detective Agency, Inc.: Hearing Held Before Hon. William Boardman, Deputy State Comptroller" (May 27, 1919), 1396, Pinkerton Records, Box 66, Folder 3.

51. Oliver, "America's First Wiretapping Controversy," 251.

52. The tragic story of Sergeant John Kennell of the NYPD wiretap squad deserves a chapter of its own. Here is how the *New York Times* reported on his case, in a front-page article published the day after he tried to take his own life in the wiretap squad's headquarters at 50 Church Street: "Crazed by criticism he brought upon himself as a witness in the wiretapping proceedings, Sergeant John Kennell of the Police Wiretapping Squad tried to kill himself in the office of the bureau at 50 Church Street yesterday. He shot himself directly under the heart with his .38 calibre service revolver, and he is now at the point of death in the Volunteer Hospital. . . . In a statement he made to Assistant District Attorney Joyce of the Homicide Bureau, he said he could not endure any longer the criticism and jeers inspired by his failure to prove the accuracy of phonograms which were obtained by him and used as evidence in the wiretapping hearing before Justice Greenbaum in the Criminal Branch of the Supreme Court. But more than that, he said, he could not stand the resentment of his own co-religionists, who snubbed him because he, a Catholic, had taken the stand as a witness against two Catholic priests." Kennell died shortly thereafter. "Shunned, Wire Spy Tries to End Life," *New York Times*, July 27, 1916, 1.

53. Alan F. Westin, "Wiretapping: The Quiet Revolution," *Commentary*, April 1960, 334.

54. Tim Weiner, *Enemies: A History of the FBI* (New York: Random House, 2013), 55.

55. *The Nation-Wide Spy System Centering in the Department of Justice* (New York: American Civil Liberties Union, 1924), 6. For accounts of the methods of political espionage that Burns's agents employed, borrowing from the private detective industry, see Sydney Coe Howard and Robert Dunn, *The Labor Spy* (New York: Republic, 1924); Leo Huberman, *The Labor Spy Racket* (New York: Modern Age Books, Inc., 1937). The wiretap was one of many weapons employed in the fight against organized labor in the United States.

56. Seipp, *The Right to Privacy,* 108.

57. Quoted in *Prohibition: Wire Tapping Activities,* 71st Cong., 3rd sess., *Congressional Record* 74 (January 20, 1931): H 2688.

58. U.S. Congress, House of Representatives, Committee on Expenditures in the Executive Departments, *Wire Tapping in Law Enforcement,* 71st Cong., 3rd sess. (1931), 2.

59. "Bootleggers Tap Prohibition Wires: Listen In on Telephone Conversations and Learn Official Secrets in Advance," *New York Times,* January 29, 1922, 25.

60. William P. Helm Jr., "Rum Runners Show Skill in Escapes from Dry Fleet," *Washington Post,* September 9, 1924, 8. On codes and codebreaking during Prohibition, see David Kahn, *The Codebreakers: The Story of Secret Writing* (New York: Scribner, 1996 [1967]), 802–817.

61. "Al Capone's Gang Taps Phone Wires, Federal Agents Say," *Chicago Daily Tribune,* August 12, 1931, 4.

62. See, for instance, "Illegal Dry Law Work Repudiated by Mellon," *Washington Post,* January 28, 1927, 1; Mabel Walker Willebrandt, "Tapped to Liquor Ring: Mrs. Willebrandt Still Opposed to Obtaining Evidence by This Method," *Los Angeles Times,* August 19, 1929, 1.

63. "U.S. High Court to Rule on Wire Tapping by Drys: Briefs Call Practice Violation of Constitution," *Chicago Daily Tribune,* January 10, 1928, 12.

64. U.S. Congress, House, Committee, *Wire Tapping in Law Enforcement,* 21.

65. U.S. Congress, House, Committee, *Wire Tapping in Law Enforcement,* 20–21.

66. U.S. Congress, House, Committee, *Wire Tapping in Law Enforcement,* 23.

67. U.S. Congress, House, Committee, *Wire Tapping in Law Enforcement,* 23.

68. U.S. Congress, House, Committee, *Wire Tapping in Law Enforcement,* 18. This is a crucial departure from how police are legally bound to use wiretaps in criminal investigations today. During Prohibition, wiretapping was regarded as necessary precursor to attaining probable cause. Today, by contrast, probable cause is a necessary precursor to attaining a wiretap.

69. "263-S Telephone Supervision: Prospect 7119" (October 13–16, 1931), U.S. Bureau of Prohibition Records, National Archives and Records Administration, Seattle, Washington, Box 5–6, 263-S: Telephone Supervision.

70. A. E. McFatridge, "Case No. 263-S: Investigation Report" (April 12, 1932), U.S. Bureau of Prohibition Records, Box 6, 263-S: "Shorty" Peterson.

71. "263-S Telephone Supervision: Prospect 7119" (October 26–November 10, 1931), U.S. Bureau of Prohibition Records, Box 5–6, 263-S: Telephone Supervision.

72. "263-S Telephone Supervision: Prospect 7119" (November 10–December 8, 1931), U.S. Bureau of Prohibition Records, Box 5–6, 263-S: Telephone Supervision.

73. "263-S Telephone Supervision: Prospect 7119" (November 10–December 8, 1931).

74. "263-S Telephone Supervision: Main 3330" (February 25–March 9, 1932), U.S. Bureau of Prohibition Records, Box 5–6, 263-S: Telephone Supervision.

75. McFatridge, "Case No. 263-S: Investigation Report."

76. A. E. McFatridge, "In re: 263-S: Investigation Report" (April 12, 1932), U.S. Bureau of Prohibition Records, Box 6, 263-S: "Shorty" Peterson.

77. For more on the malleability of privacy rhetoric in the age of Prohibition, see Priscilla M. Regan, *Legislating Privacy: Technology, Social Values, and Public Policy* (Chapel Hill: University of North Carolina Press, 1995), 113–114.

78. *Wire Tapping*, H.R. 5416, 71st Cong., 1st sess., *Congressional Record* 71 (November 22, 1929): H 5968.

79. *Wire Tapping*, H 5968.

80. President Herbert Hoover established the Wickersham Commission in 1929 to study the American criminal justice system under Prohibition. One of the Commission's primary findings was that the passage of the Volstead Act had led to abuses of police power. According to Senator William S. Kenyon, one of the Commission's members, public animosity toward Prohibition was in large part "stimulated by irritating methods of enforcement, such as the abuse of search and seizure processes, invasions of homes, and violation of the fourth amendment to the Constitution, entrapment of witnesses. . . . That there have been abuses of search and seizure processes is without question; likewise as to entrapment of witnesses." National Commission on Law Observance and Enforcement, *Report on the Enforcement of the Prohibition Laws of the United States* (Washington, DC: U.S. Government Printing Office, 1931), 212–213. On Prohibition as an experiment in federal police tactics and state-building, see Lisa McGirr, *The War on Alcohol: Prohibition and the Rise of the American State* (New York: W. W. Norton, 2015).

81. See Roger N. Baldwin, letter to Margery Grant, January 15, 1930, and C. S. Spear, letter to Roger N. Baldwin, April 23, 1930, American Civil Liberties Union Records, Subgroup 1: The Roger Baldwin Years, 1917–1950, Reel 72, Volume 393. The ACLU published a pamphlet titled "Stop Wire Tapping!" in January 1932. Drawing on Brandeis's dissent in *Olmstead,* the ACLU's basic argument was that sanctioning the use of wiretaps in Prohibition enforcement would lead to its proliferation in other arenas of government activity. "Stop Wire Tapping!" (January 1932), American Civil Liberties Union Records, Subgroup 1, Reel 89, Volume 529.

82. *Remarks in Senate Relative to Wire Tapping,* 72nd Cong., 2nd sess., Congressional Record 76 (February 16, 1933), S 4232–4239.

3. To Intercept and Divulge

1. On the creation and influence of the FCC, see Lucas A. Powe Jr., *American Broadcasting and the First Amendment* (Berkeley: University of California Press, 1987); Paul Starr, *The Creation of the Media: Political Origins of Modern Communications* (New York: Basic Books, 2004), 347–384.

2. Federal Communications Act, 47 U.S.C. § 605 (1934).

3. Margaret Lybolt Rosenzweig, "The Law of Wiretapping, Part I," *Cornell Law Quarterly* 32, no. 4 (1946–1947): 533–534.

4. My summary of the drama behind *Nardone v. United States* (1937) derives from the *New York Times*'s extensive coverage of the events surrounding the case in 1936–1937. See, among others, the following *New York Times* articles: "Held in Liquor Smuggling," March 22, 1936, 24; "Suit Seeks Return of Seized Vessel," April 3, 1936, 47; "Seizure of Ships Voided," June 7, 1936, 37; "High Court Bans Testimony Based on Wire-Tapping," December 21, 1937, 1. Relevant clippings are helpfully collected in "Memorandum on *Nardone v. United States*" (n.d.),

Box 243, Folder 5, Samuel Dash Papers, Manuscript Division, Library of Congress, Washington, DC.

5. *Nardone v. United States,* 302 U.S. 379, 383 (1937).

6. *Nardone v. United States,* 302 U.S. 379, 384 (1937).

7. *Nardone v. United States,* 302 U.S. 379, 385 (1937).

8. *Nardone v. United States,* 302 U.S. 379, 387 (1937).

9. *Weiss v. United States,* 308 U.S. 321 (1939). At the time, the interstate-intrastate designation was crucial in the Supreme Court's application of Section 605 in *Nardone I.* The wiretapped calls that prosecutors had introduced into the record in the 1937 case had all crossed state lines, which made them the clear province of federal law. But the issue of wiretapped calls that stayed within the boundaries of state jurisdictions remained unresolved until the *Weiss* decision.

10. *Nardone v. United States* 308 U.S. 338 (1939). For more details on the second *Nardone* trial, see "High Court Widens Wiretapping Ban; Bars Indirect Use," *New York Times,* December 12, 1939, 1. Not to be outdone, federal prosecutors brought yet another trial against Nardone in 1941. The third time was, at long last, the charm: Nardone was convicted and sentenced to eighteen months in prison, and the U.S. Supreme Court declined to review the case the following year. "New Review Denied Nardone," *New York Times,* June 2, 1942, 17.

11. *Nardone v. United States,* 308 U.S. 338, 340 (1939).

12. *Nardone v. United States,* 308 U.S. 338, 341 (1939).

13. Edwin J. Bradley and James E. Hogan, "Wiretapping: From Nardone to Benanti and Rathburn," *Georgetown Law Journal* 46, no. 3 (Spring 1958): 420.

14. "The Debate on Wiretapping: Congress Must Make New Rules for the FBI," *Time,* January 4, 1954, 13; "Exclusion of Evidence Obtained by Wire Tapping: An Illusory Safeguard," *Yale Law Journal* 61, no. 1 (November 1952): 1221–1226; Alan F. Westin, "The Wire-Tapping Problem: An Analysis and a Legislative Proposal," *Columbia Law Review* 52, no. 2 (February 1952): 188.

15. Quoted in John Neary, "The Big Snoop," *Life,* May 20, 1966, 42.

16. *Olmstead v. United States,* 277 U.S. 438 (1928).

17. Frederick F. Greenman, *Wire-Tapping: Its Relation to Civil Liberties* (Stamford, CT: Overbrook Press, 1938), 33.

18. See Westin, "The Wire-Tapping Problem," 177.

19. Peter Winn, "*Katz* and the Origins of the 'Reasonable Expectation of Privacy' Test," *McGeorge Law Review* 40, no. 1 (2009): 2.

20. "Wire Tapping and Law Enforcement," *Harvard Law Review* 53, no. 5 (March 1940): 870.

21. See U.S. Congress, Senate, Committee on Interstate Commerce, *Investigation of Alleged Wire Tapping: Hearings before a Subcommittee of the Committee on Interstate Commerce,* 76th Cong., 3rd sess. (1940).

22. Rosenzweig, "Law of Wiretapping, Part I," 536.

23. William S. Fairfield and Charles Clift, "The Wiretappers, Part I: Who's Listening—And to What?," *The Reporter*, December 23, 1952, 11–12. On wiretapping at the U.S. Treasury Department, see William J. Mellin and Meyer Berger, "I Was a Wire Tapper," *Saturday Evening Post*, September 19, 1949, 19–21, 46, 48, 51, 54, 57–58, 60, 62–63.

24. Edmond F. DeVine, "Evidence: Federal Communications Act: Admissibility of Evidence Which Became Accessible by Wire-Tapping," *Michigan Law Review* 38, no. 7 (May 1940): 1098–1099.

25. For more on the FBI's electronic surveillance policies during the 1940s, 1950s, and 1960s, see Athan Theoharis, *Spying on Americans: Political Surveillance from Hoover to the Huston Plan* (Philadelphia: Temple University Press, 1978), 94–13; Theoharis, ed., *From the Secret Files of J. Edgar Hoover* (Chicago: Ivan R. Dee, 1991), 131–152; Theoharis, *Abuse of Power: How Cold War Surveillance and Secrecy Policy Shaped the Response to 9/11* (Philadelphia: Temple University Press, 2011), 24–67.

26. Theoharis, *From the Secret Files*, 133. Emphasis in original.

27. Theoharis, *From the Secret Files*, 133.

28. Alexander Charns, *Cloak and Gavel: FBI Wiretaps, Bugs, Informers, and the Supreme Court* (Urbana: University of Illinois Press, 1992), 23.

29. "Justice Department Bans Wire Tapping; Jackson Acts on Hoover's Recommendation," *New York Times*, March 18, 1940, 1, 10.

30. Theoharis, *From the Secret Files*, 134.

31. Rhodri Jeffreys-Jones, *The FBI: A History* (New Haven: Yale University Press, 2007), 103–107, 124–125; Tim Weiner, *Enemies: A History of the FBI* (New York: Random House, 2012), 86–89.

32. U.S. Congress, House of Representatives, Committee on the Judiciary, *To Authorize Wire Tapping: Hearings before Subcommittee No. 1 of the Committee on the Judiciary*, 77th Cong., 1st sess. (1941). The Congressional hearings on national security wiretapping of February 1941 amounted to little in terms of concrete policy changes. But for the historian they offer fascinating glimpses of Justice Department officials working not just to blunt the force of Section 605 but to conceal the fact that the president had already offered a confidential directive to do just that. Holtzoff offered copious testimony to the effect that government wiretapping was still legal under the *Nardone* decisions (5–20). Both Hoover and Roosevelt made statements in support of legislative proposals that the presidential memorandum of May 1940 had already made redundant (112, 257).

33. Theoharis, *From the Secret Files*, 131. For more on the FBI's records destruction programs in this period, see Theoharis, *Abuse of Power*, 68–89.

34. Theoharis, *From the Secret Files*, 135.

35. Robert J. Lamphere and Tom Shachtman, *The FBI-KGB War: A Special Agent's Story* (Macon, GA: Mercer University Press, 1986), 99–125. Coplon's entanglement with the Venona Project is also discussed in John Earl Haynes and Harvey Klehr, *Venona: Decoding Soviet Espionage in America* (New Haven: Yale University Press, 1999), 157–160.

36. The definitive account of the Judith Coplon case is Marcia and Thomas Mitchell, *The Spy Who Seduced America: Lies and Betrayal in the Heat of the Cold War, the Judith Coplon Story* (Montpelier, VT: Invisible Cities Press, 2002). See also Curt Gentry, *J. Edgar Hoover: The Man and the Secrets* (New York: W. W. Norton, 1991), 366–375; John Earl

Haynes and Harvey Klehr, *Early Cold War Spies: The Espionage Trials That Shaped American Politics* (New York: Cambridge University Press, 2006), 192–207; John Earl Haynes, Harvey Klehr, and Alexander Vassiliev, *Spies: The Rise and Fall of the KGB in America* (New Haven: Yale University Press, 2009), 287–290. All of these studies deal with the aftermath of the Coplon affair at the FBI, but none attempt to account for its influence on the wiretapping debates of the early 1950s.

37. For Coplon's biography, see Sam Roberts, "Judith Coplon, Haunted by Espionage Case, Dies at 89," *New York Times,* March 2, 2011, A22; Jim Fitzgerald, "Judith Coplon, Accused and Cleared of Being a Soviet Spy, Dies at 89," *Washington Post,* March 4, 2011, B05; Mitchell and Mitchell, *Spy Who Seduced America,* 15–21.

38. Historians have offered conflicting accounts of the timing of Coplon's recruitment. In an internal memorandum to the attorney general, Hoover cited a "highly confidential and reliable informant"—Venona, most likely—to suggest that Coplon began working for the KGB in the fall of 1944. J. Edgar Hoover, Office Memorandum to Thomas C. Clark, Re: Judith Coplon, WAS Espionage (March 7, 1949), Record Group 65, Department of Justice Litigation Case File 146-1-317: *United States v. Judith Coplon,* National Archives and Records Administration, College Park, MD.

39. Quoted in Haynes, Klehr, and Vassiliev, *Spies,* 288–299.

40. Lamphere and Shachtman, *The FBI-KGB War,* 99–125, details the Venona Project's breakthrough discovery of Coplon's activities in 1948. Internal evidence suggests that a "confidential source" also put Coplon on the radar of the House Un-American Activities Committee as early as September 1946. Memorandum to John S. Wood, re: Foreign Agents Registration Section, Department of Justice (September 3, 1946), Department of Justice Litigation Case File 146-1-317: *United States v. Judith Coplon,* National Archives and Records Administration.

41. Quoted in Mitchell and Mitchell, *Spy Who Seduced America,* 80.

42. "The Nation: Verdict on Coplon," *New York Times,* July 3, 1949, E2.

43. Quoted in Mitchell and Mitchell, *Spy Who Seduced America,* 124.

44. Brief of Appellant, *United States v. Judith Coplon,* No. 21790 (2d Cir., October 1950), 19, in Record Group 276, 2nd Circuit U.S. Court of Appeals, *United States v. Judith Coplon,* Docket 21790, National Archives and Records Administration–Northeast Region, New York, NY. For the original request to tap Coplon's telephone, see J. Edgar Hoover, Personal and Confidential Memorandum for the Attorney General, Re: Judith Coplon Espionage, December 31, 1948, Department of Justice Litigation Case File 146-1-317: *United States v. Judith Coplon,* National Archives and Records Administration. This document asks the Attorney General for authorization to "place technical surveillance on [Coplon's] residence at 2634 Tunlaw Road, N.W., Washington, DC, telephone Woodley 5217." Similarly worded requests on January 19 and 24, 1949, authorized the taps on Gubitchev and Coplon's parents.

45. J. Edgar Hoover, Personal and Confidential Memorandum for the Attorney General, Re: Judith Coplon Espionage, January 24, 1949, Department of Justice Litigation Case File 146-1-317: *United States v. Judith Coplon,* National Archives and Records Administration.

46. Brief of Appellant, *United States v. Judith Coplon,* No. 21790 (2d Cir., October 1950), 19–20.

47. For a small sample of the information contained in the FBI's wiretap logs in the Coplon case, see Mitchell and Mitchell, *Spy Who Seduced America,* 342–348.

48. Mitchell and Mitchell, *Spy Who Seduced America*, 346.

49. This crucial bit of information went undisclosed during Coplon's New York hearings. The Freedom of Information Act request I made in September 2016 turned up applications for "technical surveillance" on Bernard S. Morris, a former colleague in the Foreign Agents Registration Section, and Harold Shapiro, Coplon's romantic partner at the time. Attorney General Thomas C. Clark signed off on both documents. J. Edgar Hoover, Personal and Confidential Memorandum to the Attorney General, Re: Judith Coplon (March 7, 1949), Department of Justice Litigation Case File 146-1-317: *United States v. Judith Coplon*, National Archives and Records Administration.

50. Charles Grutzner, "F.B.I. Order Read on Coplon 'Taps,'" *New York Times*, January 13, 1950, 10; "Mystery Deepens in Wiretap Inquiry," *New York Times*, January 14, 1950, 6.

51. The most damaging portion of the TIGER Memo, dated November 7, 1949, reads as follows: "The above named informant [TIGER] has been furnishing information concerning the activities of Coplon since her conviction. *In view of the imminency [sic] of her trial, it is recommended that this informant be discontinued immediately and that all administrative records in the New York office covering the operations of this informant be destroyed.* Pertinent data furnished by the informant has already been furnished in letter form, and having in mind security, now and in the future, it is believed desirable that the indicated records be destroyed." Aside from confirming the illegal destruction of records in the case, the document also provided concrete proof of a dirty FBI tactic that defense attorneys and civil liberties advocates had long suspected was pervasive: the attribution of wiretap evidence to "confidential informants." Quoted in Mitchell and Mitchell, *Spy Who Seduced America*, 228–229; emphasis in original.

52. Brief of Appellant, *United States v. Judith Coplon*, No. 21790 (2d Cir., October 1950), 21.

53. Brief of Appellant, *United States v. Judith Coplon*, 21–22.

54. Brief of Appellant, *United States v. Judith Coplon*, 21.

55. *Olmstead v. United States*, 277 U.S. 438, 470 (1928).

56. Well into the appeals phase of Coplon's legal ordeal, government prosecutors maintained that "it is clear beyond a doubt that, with one possible exception"—the fact that the FBI learned of the timing of Coplon's trips to New York by listening to her phone conversations with her mother—"none of the government's evidence could conceivably have stemmed in the slightest degree, directly or indirectly, from wiretapping." Brief of Appellee, *United States v. Judith Coplon*, No. 21790 (2d Cir., October 1950), 25, in Record Group 276, 2nd Circuit U.S. Court of Appeals, *United States v. Judith Coplon*, Docket 21790, National Archives and Records Administration–Northeast Region, New York, NY.

57. James Lawrence Fly, letter to Franklin Delano Roosevelt, March 27, 1941, James Lawrence Fly Papers, Columbia University Rare Book and Manuscript Library, New York, NY, Box 27. Notably, Roosevelt dismissed Fly's position on emergency wiretapping authorization as overly technical, even obstructionist; see Franklin D. Roosevelt, letter to James Lawrence Fly (n.d.), Box 41: "Fly on Wiretapping." Mickie Edwardson helpfully details Fly's battles with Hoover and the FBI in "James Lawrence Fly, the FBI, and Wiretapping," *Historian* 61, no. 2 (December 1999): 361–382.

58. Edwardson, "James Lawrence Fly," 367–368.

59. James Lawrence Fly, letter to J. Howard McGrath, December 20, 1949, James Lawrence Fly Papers, Box 41: "Fly on Wiretapping."

60. James Lawrence Fly, letter to the *World Telegram* and *The Baltimore Sun,* January 5, 1950, James Lawrence Fly Papers, Box 41: "Fly on Wiretapping."

61. James Lawrence Fly, 'Threat to Liberty, Defiance of Law Seen in FBI Wire-Tapping," *Washington Post,* January 7, 1950, A9.

62. It's important to note here that many officials in Washington were quietly aware of the FBI's wiretapping activities in the early postwar years. In an official comment buried in a 1949 issue of the *Yale Law Journal,* Hoover himself admitted that "it is no secret that the FBI does tap telephones in a very limited type of cases with the express approval in each instance of the Attorney General, but only in cases involving espionage, sabotage, grave risks to internal security or when human lives are in danger." What wasn't well known at the time was the existence of the Roosevelt memorandum, which gave the FBI license to ignore Section 605 and the *Nardone* decisions. J. Edgar Hoover, "A Comment on the Article 'Loyalty among Government Employees,'" *Yale Law Journal* 58, no. 3 (February 1949): 405.

63. "McGrath Upholds F.B.I. Wiretapping: Criticism by Fly Believed Cause of Attorney General's Stand," *New York Times,* January 9, 1950, 15.

64. J. Edgar Hoover, letter to Clifford Forster, January 11, 1950, James Lawrence Fly Papers, Box 41: "Fly on Wiretapping."

65. "McGrath Upholds F.B.I. Wiretapping," 15.

66. "*United States v. Judith Coplon:* Brief of American Civil Liberties Union, Amicus Curiae" (n.d.), James Lawrence Fly Papers, Box 41: "Judy Coplon"

67. James Lawrence Fly, "Petition re: Investigation of FBI" (February 1950), James Lawrence Fly Papers, Box 41: "Fly on Wiretapping."

68. James Lawrence Fly, "In The Matter of *James Lawrence Fly, Complainant v. American Telephone and Telegraph Company, New York Telephone Company, Chesapeake & Potomac Telephone Company, Respondents*" (February 1950), James Lawrence Fly Papers, Box 41: "Fly v. AT&T."

69. James Lawrence Fly, "The Wire-Tapping Outrage," *New Republic,* February 6, 1950, 14–15.

70. James Lawrence Fly, "The Police State" (n.d.), James Lawrence Fly Papers, Box 41: "ACLU Committee on Wiretapping."

71. James Lawrence Fly, letter to Eleanor Roosevelt, February 10, 1950, James Lawrence Fly Papers, Box 41: "Petition re: Investigation of FBI."

72. James Lawrence Fly, letter to James D. Mann, February 9, 1950, James Lawrence Fly Papers, Box 41: Petition: re: Investigation of FBI."

73. *United States v. Judith Coplon,* 185 F.2d 629 (2d Cir. 1950).

74. *United States v. Judith Coplon,* 185 F.2d 629 (2d Cir. 1950).

75. *United States v. Judith Coplon,* 185 F.2d 629 (2d Cir. 1950).

76. *United States v. Judith Coplon,* 185 F.2d 629 (2d Cir. 1950).

77. Historians of Cold War espionage are remarkably inconsistent in their accounts of the resolution of the Coplon case. For example, in *The FBI: A Comprehensive Reference Guide,* ed. Athan G. Theoharis with Tony G. Poveda, Susan Rosenfeld, and Richard Gid Powers (Phoenix, AZ: Oryx Press, 1999), Athan Theoharis mistakenly claims that "the Supreme

Court's 1939 ruling in *Nardone v. U.S.* required dismissal of the case" (26). Theoharis offers a more accurate version of the Coplon appeal in an earlier study, *Spying on Americans*, but both books fail to mention the importance of the FBI's illegal arrest to Hand's decision to overturn the New York conviction (104). By contrast, Haynes and Kehr, *Early Cold War Spies*, discusses the problem of the FBI's illegal arrest at length, but then proceeds to misconstrue Hand's ruling on the Coplon wiretap operation (206).

78. Edward F. Ryan, "Legal Wire Tapping Proposed for U.S. Agents when Needed," *Washington Post*, December 19, 1950, 9.

79. Quoted in Theoharis, *From the Secret Files*, 136.

80. "Evidence by Wiretapping," *Washington Post*, March 26, 1952, 14.

81. U.S. Congress, House of Representatives, Committee on the Judiciary, *Wiretapping for National Security: Hearings before Subcommittee No. 3, Committee on the Judiciary*, 83rd Cong., 1st sess. (1953), 1–4.

82. U.S. Congress, House, Committee, *Wiretapping for National Security*, 4.

83. See "Yesterday's Wiretaps," *Washington Post*, April 11, 1954, B4, for a case against H.R. 5149's retroactive clause.

84. U.S. Congress, House, Committee, *Wiretapping for National Security*, 5–6.

85. U.S. Congress, House, Committee, *Wiretapping for National Security*, 55–67.

86. Milton Magruder, "Wire-Tap Voted with Court Curb," *Washington Post*, April 9, 1954, 1.

87. U.S. Congress, Senate, Committee on the Judiciary, *Wiretapping for National Security: Hearings before a Subcommittee of the Committee on the Judiciary*, 83rd Cong., 2nd sess. (1954), 9, 10.

88. U.S. Congress, Senate, Committee, *Wiretapping for National Security*, 186.

89. U.S. Congress, Senate, Committee, *Wiretapping for National Security* 194–195.

90. U.S. Congress, Senate, Committee, *Wiretapping for National Security*, 194.

91. Sidney E. Zion, "U.S. Drops Charges in Coplon Spy Case," *New York Times*, January 7, 1967, 1.

4. The Wiretapper's Nest

1. "The Hot Wire-Tapping Debate: The Case for, and a Hard-Hitting Argument against It," *Newsweek*, January 11, 1954, 21.

2. Herbert Brownell Jr., "The Public Security and Wire Tapping," *Cornell Law Quarterly* 39, no. 2 (Winter 1954): 201–202, 212.

3. U.S. Congress, House of Representatives, Committee on the Judiciary, *Wiretapping: Hearings before Subcommittee No. 5 of the Committee on the Judiciary*, 84th Cong., 1st sess. (1955), 2–3.

4. Bernard B. Spindel with Bill Davidson, "Who Else Is Listening?," *Collier's*, June 24, 1955, 50. On Spindel's mysterious role in the 55th Street tap-nest affair, see Samuel Dash, Richard F. Schwartz, and Robert E. Knowlton, *The Eavesdroppers* (New Brunswick, NJ: Rutgers University Press, 1959), 89–90; Bernard B. Spindel, *The Ominous Ear* (New York: Award Books, 1968), 95–107.

5. "Phone Tap Center Raided in 54th Street," *New York Times*, February 18, 1955, 1. The *New York Times* initially misreported the location of the wiretap nest.

6. See Peter M. Costello, "The Tetracycline Conspiracy: Structure, Conduct, and Performance in the Drug Industry," *Antitrust Law and Economic Review* 4 (Summer 1968): 13–44; *United States v. Charles Pfizer & Co., Inc., American Cynamid Company, and Bristol-Myers Company*, 426 F.2d 32 (2d Cir. 1970).

7. Dash, Schwartz, and Knowlton, *The Eavesdroppers*, 88.

8. See U.S. Congress, House, Committee, *Wiretapping*, 167–179, 289–297; U.S. Congress, Senate, Committee on the Judiciary, *Wiretapping, Eavesdropping, and the Bill of Rights: Hearings before the Subcommittee on Constitutional Rights of the Committee on the Judiciary, Part 5*, 86th Cong., 1st sess. (1959), 1702.

9. The 55th Street tap-nest case was hardly Broady's first brush with the law. In 1938, the NYPD investigated Broady for establishing a wiretap headquarters with the notorious New York wiretap professional Robert La Borde. Amazingly, the location of Broady's 1938 wiretap operation was alleged to have been 348 E. 55th Street, just a few doors down from the site he would choose for the wiretap nest setup seventeen years later. As in the later case, Broady and La Borde had enlisted two employees of the New York Telephone Company to help set up the mass wiretapping mechanism. All parties involved ended up escaping indictment. At the grand jury hearings that followed the affair, a NYPD captain told a general counsel for the New York Telephone Company that Broady's case demonstrated that "the extent of wire tapping in New York . . . [is] far in excess of any idea your company [has] regarding the situation." U.S. Congress, Senate, Committee on Interstate Commerce, *Investigation of Alleged Wire Tapping: Hearings before a Subcommittee of the Committee on Interstate Commerce*, 76th Cong., 3rd sess. (1940), 457. In the years that followed, Broady frequently employed another renowned New York wiretap professional, Kenneth Ryan, who tapped wires for the NYPD before entering the field of private investigation. Broady and Ryan were indicted for tapping the phones of Manhattan borough president Hugo Rogers in April 1949, but Broady once again avoided jail time. See William S. Fairfield and Charles Clift, "The Wiretappers, Part I: Who's Listening—And to What?," *Reporter*, December 23, 1952, 22.

10. "Broady Sentenced to 2–4 Years, Judge Hits 'Dirty' Wiretapping," *New York Times*, January 14, 1956, 38.

11. "New York Telephone Company Press Release" (February 18, 1955), Michigan Bell Telephone Company Records, Collection 5: Ameritech Corporation, Record Group 5: Predecessor and Subsidiary Companies, Box 181, Subject Files: "Wiretapping, 1955–1966," AT&T Archives, San Antonio.

12. U.S. Congress, House, Committee, *Wiretapping*, 177.

13. Charles Grutzner, "Rise in Wiretapping Brings State Inquiry," *New York Times*, February 27, 1955, E10.

14. See, for instance, Meyer Berger, "Tapping the Wires," *New Yorker*, June 18, 1938, 45.

15. Ray Graves, "Wire for Hire," *Confidential Magazine*, July 1955, 28.

16. "Bulletin 29.7: Issued at the General Offices of the Michigan Bell Telephone Company" (March 11, 1955), Michigan Bell Telephone Company Records, Box 181, Subject Files: "Wiretapping, 1955–1966."

17. H. G. Vogel, letter to Emanuel Celler, May 10, 1955, Michigan Bell Telephone Company Records, Box 181, Subject Files: "Wiretapping, 1955–1966"; "Notes on Secrecy of Communications" (October 17, 1955), Southwestern Bell Telephone Company Records, Collection 2: SBC Communications Inc., Record Group 5: Predecessor and Subsidiary Companies, Box 134, Subject Files: "Wiretapping, 1949–1955," AT&T Archives, San Antonio. In response to the wiretap nest scandal, Bell subsidiaries also began displaying posters on the subject of telecommunications privacy in company buildings and requiring all employees to be tested on a booklet containing "pertinent sections of the State and Federal laws dealing with the secrecy of communications and with the protection of telephone plant and services." U.S. Congress, Senate, Committee, *Wiretapping, Eavesdropping, and the Bill of Rights, Part 5,* 1700.

18. "Eavesdropping and Wiretapping: Report of the New York State Joint Legislative Committee to Study Illegal Interception of Communications, March 1956" (Albany, NY: Williams Press, 1956), in U.S. Congress, Senate, Committee on the Judiciary, *Wiretapping, Eavesdropping, and the Bill of Rights: Hearing before the Subcommittee on Constitutional Rights of the Committee on the Judiciary, Part 2,* 85th Cong., 2nd sess. (1958), 280.

19. "Eavesdropping and Wiretapping: Report," 282.

20. Here, the Savarese Commission's findings built on earlier studies of law enforcement wiretapping in New York state. See, for instance, Margaret Lybolt Rosenzweig, "The Law of Wiretapping, Part II," *Cornell Law Quarterly* 33, no. 1 (1947–1948): 80–90.

21. "Eavesdropping and Wiretapping: Report," 275.

22. "Eavesdropping and Wiretapping: Report," 276.

23. *Trade Catalogs on Audio Equipment: Microphones, Earphones, Speakers, Pickups, Condensers, Transistor Oscillator Coils, Transistor Antennas, Cartridges* (New York: Argonne Electronics, ca. 1950). Trade Literature Collection, Smithsonian Libraries, Washington, DC.

24. Meyer Berger, "Tapping the Wires," *New Yorker,* June 18, 1938, 47.

25. H. G. Vogel, letter to Emanuel Celler, May 10, 1955, Michigan Bell Telephone Company Records, Box 181, Subject Files: "Wiretapping, 1955–1966."

26. William J. Mellin and Meyer Berger, "I Was a Wire Tapper," *Saturday Evening Post,* September 19, 1949, 20.

27. "Eavesdropping and Wiretapping: Report," 282.

28. Alan F. Westin, "Wiretapping: The Quiet Revolution," *Commentary,* April 1960, 334.

29. *People v. Appelbaum* 277 AD 43 (N.Y. App. Div. 1950).

30. *People v. Appelbaum* 277 AD 43 (N.Y. App. Div. 1950).

31. "Eavesdropping and Wiretapping: Report," 273.

32. "Eavesdropping and Wiretapping: Report," 290.

33. U.S. Congress, House, Committee, *Wiretapping,* 174.

34. U.S. Congress, House, Committee, *Wiretapping,* 173.

35. "Eavesdropping and Wiretapping: Report," 287.

36. On adversarial divorce laws and the rise of national divorce rates in the postwar period, see Nancy F. Cott, *Public Vows: A History of Marriage and the Nation* (Cambridge, MA: Harvard University Press, 2000), 194–196.

37. "Broady Is Upheld in Legal Wiretap," *New York Times,* November 29, 1955, 60.

38. "Eavesdropping, Wiretapping, and Private Detectives: Report of the New York State Joint Legislative Committee to Study Illegal Interception of Communications, March 1957," in U.S. Congress, Senate, Committee, *Wiretapping, Eavesdropping, and the Bill of Rights, Part 2,* 351–360.

39. "Eavesdropping, Wiretapping, and Private Detectives: Report," 352.

40. U.S. Congress, Senate, Committee, *Wiretapping, Eavesdropping, and the Bill of Rights, Part 5,* 1702.

41. U.S. Congress, Senate, Committee, *Wiretapping, Eavesdropping, and the Bill of Rights, Part 5,* 1696.

42. See David Kahn, *The Codebreakers: The Story of Secret Writing* (New York: Scribner, 1996 [1967]), 11; Whitfield Diffie and Susan Landau, *Privacy on the Line: The Politics of Wiretapping and Encryption,* 2nd ed. (Cambridge, MA: MIT Press, 2007), 178–179; Susan Landau, *Surveillance or Security? The Risks Posed by New Wiretapping Technologies* (Cambridge, MA: MIT Press, 2010), 67–68.

43. See Orin S. Kerr, "The Fourth Amendment and New Technologies: Constitutional Myths and the Case for Caution," *Michigan Law Review* 102, no. 5 (2004): 845–846; Colin Agur, "Negotiated Order: The Fourth Amendment, Telephone Surveillance, and Social Interactions, 1878–1968," *Information & Culture* 48, no. 4 (2013): 429–430.

44. Charles Einstein, *Wiretap!* (New York: Dell, 1955), 1.

45. Einstein, *Wiretap!,* 16–17.

46. Einstein, *Wiretap!,* 43.

47. Einstein, *Wiretap!,* 47.

48. Einstein, *Wiretap!,* 127.

49. U.S. Congress, House, Committee, *Wiretapping,* 117.

50. U.S. Congress, House, Committee, *Wiretapping,* 87.

51. "The Busy Wiretappers: What They're Doing, What's Being Done about Them," *Newsweek,* March 7, 1955, 34.

5. Eavesdroppers

1. Robert M. Hutchins, letter to Robert T. McCracken, July 12, 1956, Fund for the Republic Records, Box 63, Folder 1, Public Policy Papers, Department of Rare Books and Special Collections, Princeton University Library, Princeton, NJ.

2. "Final Accounting of Pennsylvania Bar Association Endowment to the Fund for the Republic: Investigation of the Right of Privacy as Affected by Current Law Enforcement Practices" (April 18, 1958), Fund for the Republic Records, Box 63, Folder 1.

3. U.S. Congress, Senate, Committee on the Judiciary, *Wiretapping, Eavesdropping, and the Bill of Rights: Hearing before the Subcommittee on Constitutional Rights of the Committee on the Judiciary, Part 3,* 86th Cong., 1st sess. (1959), 512.

4. U.S. Congress, Senate, Committee, *Wiretapping, Eavesdropping, and the Bill of Rights, Part 3,* 511.

5. Alan F. Westin, *Privacy and Freedom* (New York: Atheneum, 1967), 173–174.

6. See *Silverman v. United States,* 365 U.S. 505 (1961); *Berger v. New York,* 388 U.S. 41 (1967).

7. U.S. Congress, Senate, Committee on the Judiciary, *Wiretapping, Eavesdropping, and the Bill of Rights: Hearing before the Subcommittee on Constitutional Rights of the Committee on the Judiciary, Part 5,* 86th Cong., 1st sess. (1959), 1706–1781.

8. Robert M. Brown, *The Electronic Invasion* (New York: John F. Rider, 1967).

9. See Athan Theoharis, *Spying on Americans: Political Surveillance from Hoover to the Huston Plan* (Philadelphia: Temple University Press, 1978); Alexander Charns, *Cloak and Gavel: FBI Wiretaps, Bugs, Informers, and the Supreme Court* (Urbana: University of Illinois Press, 1992); Shane Harris, *The Watchers: The Rise of America's Surveillance State* (New York: Penguin, 2010).

10. Federal Communications Act, U.S. Code 47 (1934), § 605.

11. Samuel Dash, Richard F. Schwartz, and Robert E. Knowlton, *The Eavesdroppers* (New Brunswick, NJ: Rutgers University Press, 1959), 10.

12. Dash, Schwartz, and Knowlton, *The Eavesdroppers,* 3–4.

13. Richardson Dilworth and Samuel Dash, "A Wire Tap Proposal," *Dickinson Law Review* 59, no. 3 (1955): 196.

14. See, for instance, William J. Mellin, "I Was a Wire Tapper," *Saturday Evening Post,* September 19, 1949, 19–21, 46, 48, 51, 54, 57–58, 60, 62–63; William S. Fairfield and Charles Clift, "The Wiretappers, Part I: Who's Listening—And to What?," *The Reporter,* December 23, 1952, 8–22; Fairfield and Clift, "The Wiretappers, Part II: Listening In with Uncle Sam" *The Reporter,* January 6, 1953, 9–20; "The Debate on Wiretapping," *Time,* January 4, 1954, 14–17; "The Busy Wiretappers: What They're Doing, What's Being Done about Them," *Newsweek,* March 7, 1955, 31–34; Bernard B. Spindel with Bill Davidson, "Who Else Is Listening? How to Stop Wiretapping," *Collier's,* June 24, 1955, 49–55.

15. U.S. Congress, House of Representatives, Committee on the Judiciary, *Wiretapping: Hearings before Subcommittee No. 5 of the Committee on the Judiciary,* 84th Cong., 1st sess. (1955), 348–349.

16. U.S. Congress, Senate, Committee, *Wiretapping, Eavesdropping, and the Bill of Rights, Part 3,* 504.

17. Yale Kamisar, "The Wiretapping-Eavesdropping Problem: A Professor's View," *Minnesota Law Review* 44 (April 1960): 895.

18. US. Congress, House, Committee, *Wiretapping,* 36–37.

19. Edward V. Long, *The Intruders: The Invasion of Privacy by Government and Industry* (New York: Praeger, 1966), 140.

20. Dash, Schwartz, and Knowlton, *The Eavesdroppers,* 68–69, 77, 122–123.

21. Dash, Schwartz, and Knowlton, *The Eavesdroppers,* 164–165, 217.

22. Dash, Schwartz, and Knowlton, *The Eavesdroppers,* 82; U.S. Congress, Senate, Committee, *Wiretapping, Eavesdropping, and the Bill of Rights, Part 3,* 528.

23. Burton H. Alden, Byron C. Campbell, et. al., *Competitive Intelligence: Information, Espionage, and Decision-Making* (Watertown, MA: C.I. Associates, 1959), 70.

24. Dash, Schwartz, and Knowlton, *The Eavesdroppers,* 95–96. AT&T would later monitor tens of millions of phone calls in order to detect toll fraud. On the Bell System's attempts to combat "phone phreaking" and other forms of toll fraud, see Phil Lapsey, *Exploding the Phone: The Untold Story of the Teenagers and Outlaws Who Hacked Ma Bell* (New York: Grove Press, 2013).

25. Dash, Schwartz, and Knowlton, *The Eavesdroppers,* 316.

26. Carl Dreher, *"The Eavesdroppers," Nation,* January 16, 1960, 55.

27. Mari MacInnes, "Attacks on Privacy: *The Eavesdroppers," Commentary,* March 1960, 275.

28. Eugene C. Gerhart, *"The Eavesdroppers," American Bar Association Journal* 46, no. 5 (May 1960): 542.

29. MacInnes, "Attacks on Privacy," 274.

30. Thomas C. Hennings Jr., "The Wiretapping-Eavesdropping Problem: A Legislator's View," *Minnesota Law Review* 44 (April 1960): 818.

31. The details of my account of the *Benanti v. United States* case draw on the case summaries included in "High Court Will Rule on Wiretap Evidence," *Washington Post,* October 10, 1957, A22; Edwin J. Bradley and James E. Hogan, "Wiretapping: From Nardone to Benanti and Rathburn," *Georgetown Law Journal* (Spring 1958): 431–434; "Memorandum on *Benanti v. United States,*" Samuel Dash Papers, Box 242, Folder 4, Manuscript Division, Library of Congress, Washington, DC.

32. *Appendix A: United States Court of Appeals for the Second Circuit. No. 266—October Term, 1956, Docket No. 24427: United States of America v. Salvatore Benanti,* 3a–4a.

33. *Benanti v. United States,* 355 U.S. 96, 100, 102 (1957).

34. Handed down just days after *Benanti v. United States, Rathburn v. United States,* 355 U.S. 107 (1957), ruled that third parties were free to use "extension telephones" to eavesdrop on phone conversations if one of the parties on the line gives prior consent. For legal scholars' reactions to the *Benanti* decision, see "The Benanti Case: State Wiretap Evidence and the Federal Exclusionary Rule," *Columbia Law Review* 57, no. 8 (December 1957): 1159–1171; "Recent State Wiretap Statutes: Deficiencies of the Federal Communications Act Corrected," *Yale Law Journal* 67, no. 5 (April 1958): 932–943; Bradley and Hogan, "Wiretapping: From Nardone to Benanti and Rathburn," 431–434.

35. See Anthony P. Savarese Jr., "Eavesdropping and the Law," *American Bar Association Journal* 46 (March 1960): 334.

36. Edward Bennett Williams, letter to Josephine Scheiber, February 25, 1958, Edward Bennett Williams Papers, Box 54, Folder 5, Manuscript Division, Library of Congress, Washington DC.

37. Peter Kihss, "State Justice Rules All Wiretaps Illegal," *New York Times,* January 3, 1958, 14.

38. Anthony Lewis, "Use of Wiretapping Still Hotly Debated," *New York Times,* December 15, 1957, 199.

39. See, for instance, the chapter-length history of wiretapping law in Whitfield Diffie and Susan Landau, *Privacy on the Line: The Politics of Wiretapping and Encryption,* updated

and expanded ed. (Cambridge: MIT Press, 2010), 173–203, which omits *Benanti v. United States* entirely.

40. The "silver platter" doctrine came to an end in *Elkins v. United States*, 364 U.S. 206 (1960), a decision in which the Court cited *Benanti v. United States* as a precedent. But both *Benanti* and *Elkins* failed to overturn a controversial earlier ruling, *Schwartz v. Texas*, 344 U.S. 199 (1952), which permitted the state's use of wiretap evidence illegally obtained by state officials. It wasn't until *Lee v. Florida*, 392 U.S. 378 (1968), another case argued on the foundation of *Benanti*, that the Court was able to root out the silver platter doctrine in all of its forms.

41. "Bill for Wiretaps by States Would Reverse High Court," *New York Times*, January 17, 1958, 1, 13.

42. On allegations that the Kennedy administration made widespread use of wiretaps and bugs, see Robert M. Cipes, "The Wiretap War: Kennedy, Johnson, and the FBI," *New Republic*, December 24, 1966, 16–22.

43. Quoted in U.S. Congress, Senate, Committee, *Wiretapping, Eavesdropping, and the Bill of Rights, Part 3*, 557–558.

44. "Wiretapping vs. Crime," *New York Times*, January 20, 1958, 22.

45. U.S. Congress, Senate, Committee, *Wiretapping, Eavesdropping, and the Bill of Rights, Part 3*, 540. The "peashooter" analogy was one of Silver's favorites; see also Edward S. Silver, "Law Enforcement 'Wire Tapping,'" *Journal of Criminal Law, Criminology, and Police Science* 50, no. 6 (March–April 1960): 577.

46. Edward S. Silver, "The Wiretapping-Eavesdropping Problem: A Prosecutor's View," *Minnesota Law Review* 44 (April 1960): 848.

47. Dash, Schwartz, and Knowlton, *The Eavesdroppers*, 423.

48. U.S. Congress, Senate, Committee, *Wiretapping, Eavesdropping, and the Bill of Rights, Part 3*, 505. This would change sooner than Dash could have anticipated. In 1961 the U.S. Supreme Court's famous decision in *Mapp v. Ohio* 347 U.S. 643 (1961) banned from state criminal prosecutions evidence obtained in violation of the Fourth Amendment.

49. Gerhart, "*The Eavesdroppers*," 543.

50. U.S. Congress, Senate, Committee, *Wiretapping, Eavesdropping, and the Bill of Rights, Part 3*, 538–539.

51. U.S. Congress, Senate, Committee, *Wiretapping, Eavesdropping, and the Bill of Rights, Part 3*, 532.

52. Silver, "Law Enforcement 'Wire Tapping,'" 579.

53. Frank S. Hogan, "'Wire Tapping': An Answer to the Authors," *Journal of Criminal Law, Criminology, and Police Science* 50, no. 6 (March–April 1960): 575.

54. Yale Kamisar to Edward Bennett Williams, November 12, 1959, Edward Bennett Williams Papers, Box 54, Folder 7, Manuscript Division, Library of Congress, Washington, DC. The articles in the April 1960 symposium, published in *Minnesota Law Review* 44 (April 1960), were: Hennings, "The Wiretapping-Eavesdropping Problem: A Legislator's View," 813–834; Silver, "The Wiretapping-Eavesdropping Problem: A Prosecutor's View," 835–854; Edward Bennett Williams, "The Wiretapping-Eavesdropping Problem: A Defense Counsel's View,"

855–871; Harold K. Lipset, "The Wiretapping-Eavesdropping Problem: A Private Investigator's View," 873–889; and Kamisar, "The Wiretapping-Eavesdropping Problem: A Professor's View," 891–940.

55. "The Editors Comment," *Minnesota Law Review* 44 (April 1960): 808.

56. Silver, "The Wiretapping-Eavesdropping Problem," 835, 854.

57. U.S. Congress, Senate, Committee on the Judiciary, *Wiretapping and Eavesdropping Legislation: Hearings before the Subcommittee on Constitutional Rights of the Committee on the Judiciary*, 87th Cong., 1st Sess. (1961), 1–8.

58. Pennsylvania enacted a ban on wiretapping, including wiretapping by law enforcement, in July 1957. Interestingly, the state legislature went ahead with the prohibition without seeing the results of the PBAE investigation.

59. Alan F. Westin, "Wiretapping: The Quiet Revolution," *Commentary*, April 1960, 340.

6. Tapping God's Telephone

1. On the architecture of eavesdropping, see Dörte Zbikowski, "The Listening Ear: Phenomena of Acoustic Surveillance," in *CTRL [SPACE]: Rhetorics of Surveillance from Bentham to Big Brother*, ed. Thomas Y. Levin, Ursula Frohne, and Peter Weibel (Cambridge, MA: MIT Press, 2002), 37–41.

2. *The Detectifone: A Mechanically Perfect Device for Producing the Evidence* (New York: Carl Anderson Electric Corporation, 1917), 1, Trade Literature Collection, Smithsonian Libraries, Washington, DC.

3. *The Detectifone*, 3. Interestingly, the Detectifone won the endorsement of William J. Burns (see Chapter 2). In June 1917 Burns wrote to the Anderson Electric Corporation to report that "we have used your secret service appliances . . . in considerable quantities. We are glad to state that they have given us excellent satisfaction and from our long experience we have not seen anything that equals your product for accuracy and first class results. We have used your instruments in a number of important cases." William J. Burns, letter to Anderson Electric Corporation, June 6, 1917, Pinkerton's National Detective Agency Records, 1853–1999, Box 63, Folder 7, Manuscript Division, Library of Congress, Washington, DC. For more on early eavesdropping devices like the Detectifone, see Kathryn W. Kemp, "'The Dictograph Hears All': An Example of Surveillance Technology in the Progressive Era," *Journal of the Gilded Age and Progressive Era* 6, no. 4 (October 2007): 409–430.

4. *Oxford English Dictionary Online*, s.v. "bug," http://www.oed.com.

5. U.S. Congress, Senate, Committee on the Judiciary, *Wiretapping, Eavesdropping, and the Bill of Rights: Hearing before the Subcommittee on Constitutional Rights of the Committee on the Judiciary, Part 3*, 86th Cong., 1st sess. (1959), 519.

6. Samuel Dash, Richard F. Schwartz, and Robert E. Knowlton, *The Eavesdroppers* (New Brunswick, NJ: Rutgers University Press, 1959), 316–323, 339–343.

7. Dash, Schwartz, and Knowlton, *The Eavesdroppers*, 346–358.

8. U.S. Congress, Senate, Committee on the Judiciary, *Wiretapping, Eavesdropping, and the Bill of Rights: Hearing before the Subcommittee on Constitutional Rights of the Committee on the Judiciary, Part 1*, 85th Cong., 2nd sess. (1958), 8.

9. Dash, Schwartz, and Knowlton, *The Eavesdroppers,* 305.

10. U.S. Congress, Senate, Committee, *Wiretapping, Eavesdropping, and the Bill of Rights, Part 1,* 10.

11. *Goldman v. United States,* 316 U.S. 129 (1942).

12. *On Lee v. United States* 343 U.S. 747 (1952); *Irvine v. California,* 347 U.S. 128 (1954).

13. U.S. Congress, Senate, Committee, *Wiretapping, Eavesdropping, and the Bill of Rights, Part 3,* 519.

14. Edward Bennett Williams, "The Wiretapping-Eavesdropping Problem: A Defense Counsel's View," *Minnesota Law Review* 44 (April 1960): 862.

15. Samuel Dash, *The Intruders: Unreasonable Searches and Seizures from King John to John Ashcroft* (New Brunswick, NJ: Rutgers University Press, 2004), 87.

16. *Silverman v. United States,* 365 U.S. 505 (1961).

17. *Silverman v. United States,* 365 U.S. 505, 509 (1961).

18. Alan F. Westin, "Wiretapping: The Quiet Revolution," *Commentary,* April 1960, 337.

19. U.S. Congress, House of Representatives, Committee on the Judiciary, *Wiretapping: Hearings before Subcommittee No. 5 of the Committee on the Judiciary,* 84th Cong., 1st sess. (1955), 79, 83.

20. Lawrence Laurent, "Electronic 'Bug' Eludes Law's Gasp," *Washington Post,* April 6, 1962, A5.

21. "Electronic Device Firms Foresee No Drop in Business," *Los Angeles Times,* March 3, 1966, WS1.

22. "Bug Thy Neighbor," *Time,* March 6, 1964, 61–63; Don Meilke, "Nosy 'Bug' Is Snug in Rug or Anywhere," *Chicago Tribune,* May 10, 1964, A1; "When Walls Have Ears, Call a Debugging Man," *Business Week,* October 31, 1964, 154–158; Larry Steckler, "Outwit the Electronic Eavesdroppers," *Popular Mechanics,* December 1965, 70–74, 206; John Neary, "The Big Snoop," *Life,* May 20, 1966, 38–47.

23. See, for instance, Ernest Braun and Stuart Macdonald, *Revolution in Miniature: The History and Impact of Semiconductor Electronics Re-Explored,* 2nd ed. (Cambridge: Cambridge University Press, 1982); Michael Riordan and Lillian Hoddeson, *Crystal Fire: The Birth of the Information Age* (New York: W. W. Norton, 1997); Mara Mills, "Hearing Aids and the History of Electronics Miniaturization," *IEEE Annals of the History of Computing* 33, no. 2 (April–June 2011): 24–44.

24. Harold K. Lipset, "The Wiretapping-Eavesdropping Problem: A Private Investigator's View," *Minnesota Law Review* 44 (April 1960): 888.

25. Dash, Schwartz, and Knowlton, *The Eavesdroppers,* 344.

26. Westin, "Wiretapping: The Quiet Revolution," 333.

27. Edward V. Long, *The Intruders: The Invasion of Privacy by Government and Industry* (New York: Praeger, 1966), 158.

28. The bugging of the U.S. Embassy in Moscow from 1945 to 1952, popularly known as the "Great Seal Bug" affair, might well seem to deserve a chapter of its own. But the incident

is far more important as an episode in the history of U.S.–Soviet relations than as an episode in the history of electronic eavesdropping. Accounts of its public significance have been vastly overstated, particularly because bugs were common knowledge in 1960 when Henry Cabot Lodge revealed the existence of the state seal listening device, ominously nicknamed "The Thing," to the United Nations. For the definitive account of the discovery of the bug in the state seal, see George F. Kennan, *Memoirs, 1950–1963* (New York: Pantheon Books), 152–157. Kennan himself noted that the bugging of foreign embassies was "standard practice" by the time he served as U.S. ambassador to the Soviet Union (153). Nonetheless, the listening device he discovered was a feat of electronic ingenuity well ahead of its time. It turned out to have been designed by Léon Theremin, better known as the father of electronic music. See Albert Glinsky, *Theremin: Ether Music and Espionage* (Urbana: University of Illinois Press, 2000), 256–274.

29. Long, *The Intruders*, 17; Robert M. Brown, *The Electronic Invasion* (New York: Rider, 1967), 23. See also Lawrence Stessin, "'I Spy' Becomes Big Business," *New York Times Magazine*, November 28, 1965, 105–106, 108.

30. See "Bug Thy Neighbor," 61–63; Myron Brenton, *The Privacy Invaders* (New York: Coward-McCann, 1964); Vance Packard, *The Naked Society* (New York: David McKay, 1964); Neary, "The Big Snoop"; Long, *The Intruders*; Brown, *The Electronic Invasion*; Bernard B. Spindel, *The Ominous Ear* (New York: Award Books, 1968); John M. Carroll, *The Third Listener: Personal Electronic Espionage* (New York: Dutton, 1969).

31. U.S. Congress, Senate, Committee on the Judiciary, *Invasions of Privacy (Government Agencies): Hearings before the Subcommittee on Administrative Practice and Procedure of the Committee on the Judiciary, Part 1*, 89th Cong., 1st sess. (1965), 13–21.

32. Vaus is cited several times in the footnotes of *The Eavesdroppers*, but never as an on-the-record source. He later discussed cooperating with Dash's investigation in his second published autobiography. See Jim Vaus and Julie Maxey, *The Devil Loves a Shining Mark: The Story of My Life* (Waco, TX: Word Books, 1974), 11–12.

33. Like Vaus, Lipset is cited throughout the text of *The Eavesdroppers*. Dash later pegged Lipset to serve on the Senate Watergate investigation, but public pressure forced him to resign his post when a prior conviction for electronic eavesdropping came to light. See Bob Woodward and Carl Bernstein, "Hill Bug Unit Aide Quits Job: Investigator Guilty in '66 Eavesdropping," *Washington Post*, April 14, 1973, A1, A4.

34. My account of Vaus's biography draws on three main sources: his first autobiography, his wife's book about his life, and his son's book about his life. See Jim Vaus (with D. C. Haskin), *Why I Quit Syndicated Crime: The Wiretapper's Own Story* (Wheaton, IL: Van Kampen Press, 1951); Alice Vaus, as told to Dorothy C. Haskin, *They Called My Husband a Gangster* (Rockville, MD: Wildside Press, 2011 [1952]); Will Vaus, *My Father Was a Gangster: The Jim Vaus Story* (Washington, DC: Believe Books, 2007).

35. Vaus, *Why I Quit Syndicated Crime*, 21.

36. "Why I Quit Syndicated Crime: Testimony of Jim Vaus" (ca. 1951), Papers of Jim Vaus (1950–2001), Accession 09-46, Box 3, Archives of the Billy Graham Center, Wheaton College, Wheaton, Illinois.

37. "Why I Quit Syndicated Crime: Testimony."

38. Vaus, *They Called My Husband a Gangster*, 60.

39. "Credits Conversion: Vaus Admits Perjury in Police Vice Inquiry," *Los Angeles Times,* November 15, 1949, 2.

40. "Wire-Tapping Vaus Hits Sawdust Trail," *Los Angeles Times,* November 8, 1949, 2.

41. On Billy Graham's emergence as a national icon during the late 1940s and early 1950s, see Grant Wacker, *America's Pastor: Billy Graham and the Shaping of a Nation* (Cambridge, MA: Harvard University Press, 2014), 68–101.

42. Vaus, *Why I Quit Syndicated Crime,* 6. Sales figures for *Why I Quit Syndicated Crime* are hard to come by. Vaus's wife remembers her husband, on his own, selling an entire trailer full of books on a cross-country preaching trip in 1952, but the autobiography clearly had a much wider circulation. Vaus, *They Called My Husband a Gangster,* 86.

43. Vaus's son claims that the British edition of *Why I Quit Syndicated Crime* went through ten printings between 1956 and 1973. He also possesses editions in Swedish, German, French, Italian, and Spanish, and indicates that the book may have been pirated and distributed behind the Iron Curtain. Vaus, *My Father Was a Gangster,* 91.

44. "Wire Tapper: Script" (1954), Records of the Billy Graham Evangelical Association: World Wide Pictures, Inc., Collection 214, Box 7, Folder 13, Archives of the Billy Graham Center, Wheaton College, Wheaton, Illinois.

45. "Wiretapper; n.d," Papers of Earl Wesley Schultz Jr., Collection 181, Box 2, File 34, Archives of the Billy Graham Center, Wheaton College, Wheaton, Illinois.

46. "Local YWCA Group to Show 'Wiretapper' Pix in Fund Drive March 14," *Chicago Defender,* March 12, 195, 19.

47. Vaus, *Why I Quit Syndicated Crime,* 29–30.

48. "Wiretapper Tells All in Melodrama," *Los Angeles Times,* December 8, 1955, B13.

49. Vaus, *Why I Quit Syndicated Crime,* 32.

50. "Evangelism Recruit: Cohen Wire Tapper Turns to Saving Souls," *Los Angeles Times,* October 1, 1950, 39.

51. See "Electronics Used by Evangelist in Revivals," *Los Angeles Times,* October 21, 1953, A2; "The Wiretapper," *Time,* April 25, 1955, 55.

52. Jim Vaus, "Interview with Bob Crossley," May 22, 1967, Papers of Jim Vaus, Accession 09-46, Box 4. Emphasis in original.

53. "Electronics Used by Evangelist in Revivals," A2.

54. Orville Allen, "The Millstone," *Genesee County Express,* July 12, 1956, 1.

55. "Tappers Called Ingenious," *Science News Letter,* April 9, 1955, 226.

56. Peter Wyden, "The Busy Wiretappers: What They're Doing, What's Being Done about Them," *Newsweek,* March 7, 1955, 34.

57. Bernard B. Spindel with Bill Davidson, "Who Else Is Listening?," *Collier's,* June 24, 1955, 50.

58. I have drawn on the following sources for the details of Lipset's biography: James E. Bylin, "Super Snooper: Private Eye Hal Lipset Shuns the Rough Stuff, Prospers by Using Wits," *Wall Street Journal,* February 17, 1971, 1, 29; Patricia Holt, *The Bug in the Martini*

Olive, and Other True Cases from the Files of Hal Lipset, Private Eye (Boston: Little, Brown, 1991); Robert McG. Thomas Jr., "Hal Lipset, Private Detective with a Difference, Dies at 78," *New York Times*, December 12, 1997, B15; "Private Detective Harold Lipset Dies," *Washington Post*, December 13, 1997, B6.

59. Holt, *Bug in the Martini Olive*, 25.

60. Quoted in Holt, *Bug in the Martini Olive*, 59. Lipset's views on this matter weren't exceptional. In 1965 the famed anthropologist Margaret Mead made a similar argument about the social value of electronic eavesdropping in "Margaret Mead Re-examines Our Right to Privacy," *Redbook*, April 1965, 15–16.

61. U.S. Congress, Senate, Committee on the Judiciary, *Wiretapping, Eavesdropping, and the Bill of Rights: Hearing before the Subcommittee on Constitutional Rights of the Committee on the Judiciary, Part 5*, 86th Cong., 1st sess. (1959), 1444.

62. U.S. Congress, Senate, Committee, *Wiretapping, Eavesdropping, and the Bill of Rights, Part 5*, 1454.

63. George Dixon, "Washington Scene: Forces Mobilize for Invasion of Our Privacy," *Washington Post*, December 18, 1959, A21.

64. U.S. Congress, Senate, Committee, *Wiretapping, Eavesdropping, and the Bill of Rights, Part 5*, 1467.

65. See Lipset, "The Wiretapping-Eavesdropping Problem," 875, 880.

66. See U.S. Congress, Senate, Committee, *Invasions of Privacy (Government Agencies)*, 1–4.

67. Holt, *Bug in the Martini Olive*, 67.

68. Lawrence Stern, "Don't Talk to a Martini, the Olive May Be Listening" *Washington Post*, February 19, 1965, A1; "Your Olive May Quote You if Snoopers Know You Drink," *Baltimore Sun*, February 19, 1965, 1.

69. "Snoopers Can 'Bug' Olive in Martini, Probers Learn," *Chicago Tribune*, February 19, 1965, 4.

70. Russell Baker, "Treacherous Vegetables," *Miami News*, February 23, 1965, 4.

71. Long, *The Intruders*, 5–6.

72. Neary, "The Big Snoop," 38–39.

73. Holt, *Bug in the Martini Olive*, 67. When I asked Ralph V. Ward, the retired vice president of sales at Mosler Research Products, a defunct eavesdropping device manufacturing company, about the bugged martini, he questioned the viability of Lipset's invention: "How are you going to get a microphone, an amplifier, a transmitter, and an antenna into an olive? That isn't going to work. Not then, and not now. He [Lipset] cheated. . . . It was a fake." Ralph V. Ward, interview with author, December 7, 2018.

74. "FCC Bans Electronic Snooping in Move to Protect 'Little Man,'" *Washington Post*, March 1, 1966, A1; Gina Stevens and Charles Doyle, *Privacy: Wiretapping and Electronic Eavesdropping* (Huntington, NY: Novinka Books, 2002), 32.

75. Quoted in Holt, *Bug in the Martini Olive*, 70.

76. U.S. Congress, Senate, Committee, *Invasions of Privacy (Government Agencies)*, 63.

77. Bernard B. Spindel with Martin Abramson, "Who's Bugging You? A World Authority on Electronic Security Devices Says You Have to Be a James Bond, These Days, to Protect Your Privacy," *Baltimore Sun* (September 24, 1967), TW4.

78. Brown, *The Electronic Invasion,* 118.

79. Schwartz appeared before the U.S. Senate Subcommittee on Constitutional Rights in May 1958. When asked if the basic message of his research for *The Eavesdroppers* was that there was no longer a "foolproof system" for ensuring the privacy of communications, he offered a terrifyingly direct answer: "That is right." U.S. Congress, Senate, Committee, *Wiretapping, Eavesdropping, and the Bill of Rights, Part 1,* 13.

80. Sarah E. Igo, "The Beginnings of the End of Privacy," *Hedgehog Review* 17, no. 1 (Spring 2015). See also Deborah Nelson, *Pursuing Privacy in Cold War America* (New York: Columbia University Press, 2002), 1–27.

81. Packard, *The Naked Society,* 4.

82. Long, *The Intruders,* viii.

83. Alan F. Westin, *Privacy and Freedom* (New York: Atheneum, 1967), 3–4.

84. Long, *The Intruders,* 64.

85. William Sloane, letter to Mrs. Samuel Dash, May 23, 1973, Samuel Dash Papers, Box 240, Folder 7, Manuscript Division, Library of Congress, Washington, DC.

86. Dash, Schwartz, and Knowlton, *The Eavesdroppers,* 385.

7. Title III

1. Nora Sayre, "The Screen: A Grim 'Conversation,'" *New York Times,* April 8, 1974, 44.

2. Joy Gould Boyum, "A Modern Horror Story," *Wall Street Journal,* April 15, 1975, 14.

3. Lawrence Shaffer, "*The Conversation,*" *Film Quarterly* 28, no. 1 (Autumn 1974): 59.

4. Bob Woodward and Carl Bernstein, "Hill Bug Unit Aide Quits Job: Investigator Guilty in '66 Eavesdropping," *Washington Post,* April 14, 1973, A1, A4.

5. Richard M. Cohen, "Nixon Bug System Simple: Expert Calls Devices Hardly Professional," *Washington Post,* May 12, 1974, A6. On Lipset's role in the production of *The Conversation,* see Burt Prelutsky, "Squinting at a Private Eye," *Los Angeles Times,* April 21, 1974, O18; Patricia Holt, *The Bug in the Martini Olive, and Other True Cases from the Files of Hal Lipset, Private Eye* (Boston: Little, Brown, 1991), 49–52.

6. *Watergate: A Brief History with Documents,* 2nd ed., edited by Stanley I. Kutler (New York: Wiley-Blackwell, 2010), 60. Nixon was right in a more limited sense, too. Both of his predecessors used taping systems to record conversations in the Oval Office. For more on this irony, see Bruce J. Schulman, "Taping History," *Journal of American History* 85, no. 3 (September 1998): 571–578.

7. Marshall McLuhan, "At the Moment of Sputnik the Planet Became a Global Theater in Which There Are No Spectators but Only Actors," *Journal of Communications* 24, no. 1 (March 1974): 54.

8. On the origins of the screenplay that would become *The Conversation*, see Brian De Palma, "The Making of *The Conversation*: An Interview with Francis Ford Coppola," *Film-makers Newsletter*, May 1974, 30; Michael Goodwin and Naomi Wise, *On the Edge: The Life and Times of Francis Coppola* (New York: William Morrow, 1989), 140, 144–145.

9. A. H. Weiler, "Gene Hackman, Snoop," *New York Times*, May 7, 1972, D8.

10. Arthur Miller, *The Archbishop's Ceiling and The American Clock: Two Plays, with an Introduction by the Author* (New York: Grove Press, 1989), viii, x.

11. Deborah Nelson, *Pursing Privacy in Cold War America* (New York: Columbia University Press, 2002), xii. See also Sarah E. Igo, *The Known Citizen: A History of Privacy in Modern America* (Cambridge, MA: Harvard University Press, 2018), 144–182, 221–263.

12. *Berger v. New York*, 388 U.S. 41 (1967); *Katz v. United States*, 389 U.S. 347 (1967); *United States v. United States District Court*, 407 U.S. 297 (1972).

13. *Griswold v. Connecticut*, 381 U.S. 479 (1965).

14. *Omnibus Crime Control and Safe Streets Act of 1968*, Public Law 90–351, *U.S. Statutes at Large* 82 (1968), 211.

15. De Palma, "The Making of *The Conversation*," 34.

16. James R. Wyrsch and Anthony P. Nugent Jr., "Missouri's New Wiretap Law," *Journal of the Missouri Bar* 48, no. 1 (January–February 1992): 21–31.

17. Edward V. Long, "Big Brother Is Listening," *Focus Midwest*, August 1962, 8–10; Long "Big Brother Is Watching You," *Frontier Magazine*, June 1964, 9–11; Long, "How to Kill Big Brother: The Senate Investigation That Revealed His Remarkable Strength, and Perhaps His Fatal Weakness" (ca. 1966), Edward V. Long Papers, 1951–1969 (C1268), Box 172: "Administrative Practice and Procedure REF," Manuscript Collections, The State Historical Society of Missouri, Columbia. This last manuscript was the final draft of an article that Long submitted to *Playboy* in late 1966. It eventually appeared as Long, "Big Brother in America," *Playboy Magazine*, January 1967, 127, 255–259. By 1967 Long's editorials were appearing in national outlets like the *Saturday Evening Post* and *Esquire*, as well.

18. Long, "Big Brother in America," 259.

19. Edward V. Long, "Tom Hennings: The Man from Missouri," *Missouri Law Review* 26, no. 4 (November 1961): 418. The Edward V. Long Papers at the State Historical Society of Missouri contain hundreds of letters from constituents addressed to Long. The vast majority of them are in praise of his campaign against government wiretapping. See, in particular, Box 222: "Administrative Practice and Procedures (J-L), 1966," "Administrative Practice and Procedures (S-Z), 1966," "Administrative Practice and Procedures (A-L), 1967," and "Administrative Practice and Procedures (M-Z), 1967"; also Box 248: "Administrative Practice and Procedures (A-F), 1965," "Administrative Practice and Procedures (G-I), 1965," and "Administrative Practice and Procedures (J-M), 1965." My thanks go to the Long family, especially Annie Miller Devoy, for granting me permission to view the Edward V. Long collection.

20. Robert F. Kennedy, "Attorney General's Opinion on Wiretaps," *New York Times*, June 3, 1962, 21.

21. Long, "Big Brother Is Listening," 10. On the Long's Committee's assessment of Kennedy's Organized Crime Drive, see Dick Anderman, letter to Bernard Fensterwald, August 27, 1965, Edward V. Long Papers, Box 123: "Invasion of Privacy."

22. "Bug Thy Neighbor," *Time,* March 6, 1964, 61–63; Edward V. Long, "Government Snooping" (September 28, 1964), Edward V. Long Papers, Box 248: "Administrative Practice and Procedures, Correspondence."

23. U.S. Congress, Senate, Subcommittee on Administrative Practice and Procedure of the Committee on the Judiciary, *Questionnaire Relating to Invasions of Privacy* (Washington, DC: U.S. Government Printing Office, 1964), 1–5.

24. Long, "Government Snooping."

25. Bernard Fensterwald, "Memorandum for the Files" (March 17, 1964), Edward V. Long Papers, Box 248: "Administrative Practice and Procedures, Correspondence."

26. Edward V. Long, "Remarks by Senator Edward V. Long in Senate Summarizing Progress Made in Curbing Invasions of Privacy by Federal Government" (July 30, 1965), Edward V. Long Papers, Box 123: "Invasion of Privacy."

27. U.S. Congress, Senate, Committee on the Judiciary, *Invasions of Privacy (Government Agencies): Hearings before the Subcommittee on Administrative Practice and Procedure, Parts 2–3,* 89th Cong., 1st sess. (1965). For details of the Long Committee's IRS inquiry, in particular, see Sheldon Cohen, letter to Edward V. Long, July 26, 1965, Edward V. Long Papers, Box 248: "Administrative Practice and Procedure, REF: 1965"; Sheldon S. Cohen, letter to Edward V. Long, July 11, 1967, Edward V. Long Papers, Box 123: "Invasions of Privacy."

28. Fred P. Graham, "Wiretap Hearings Hurt Crime Drive, Justice Aides Feel," *New York Times,* July 18, 1965, 1.

29. Charles Bartlett, "Long Probe Worries Feds," *Washington Star,* August 22, 1965, n.p.

30. "Come On—Admit You've Been Using Harassment and Intimidation Methods!," *Kansas City Star,* October 22, 1965.

31. David Kraslow, "FBI Red-Faced on Use of 'Bugs': Assertions of Illegally Planted Spy Devices Peril Rackets Cases," *New York Times,* December 19, 1965, 1, 19–20; "Who Knew about 'Bugging' . . . RFK's Story—and the FBI's," *U.S. News and World Report,* December 26, 1966, 32–34. Although it lies somewhat beyond the scope of my account here, the spat between R. F. Kennedy and Hoover over electronic surveillance authorization is well-trod territory in the historiography of the FBI. For representative studies, which also focus much more extensively on the behind-the-scenes effects of the Fred Black case, see Athan Theoharis, *Spying on Americans: Political Surveillance from Hoover to the Huston Plan* (Philadelphia: Temple University Press, 1978), 111–115; Alexander Charns, *Cloak and Gavel: FBI Wiretaps, Bugs, Informers, and the Supreme Court* (Urbana: University of Illinois Press, 1992), 55–63, 69–89; Tim Weiner, *Enemies: A History of the FBI* (New York: Random House, 2013), 264–273.

32. David Kraslow, "Plans for Hearing on 'Bugging' by FBI Lag," *New York Times,* January 28, 1966, 20.

33. U.S. Congress, Senate, Select Committee to Study Governmental Operations with respect to Intelligence Activities, *Supplementary Detailed Staff Reports on Intelligence Activities and the Rights of Americans, Book III,* 94th Cong., 2nd sess. (1976), 309–310. Quoted in Charns, *Cloak and Gavel,* 41.

34. U.S. Congress, Senate, Select Committee, *Supplementary Detailed Staff Reports,* 286.

35. U.S. Congress, Senate, Select Committee, *Supplementary Detailed Staff Reports,* 286–287. See also C. D. Brennan, memorandum to W. C. Sullivan re: "Technical Surveillances"

(June 23, 1969), Surreptitious Entries (June Mail-Serials X90): Part 12 of 23, File #62-117-166, Online Federal Bureau of Investigation Records: The Vault, https://vault.fbi.gov/: "In view of these [the Long Committee] inquiries, it was necessary to severely restrict and, in many instances, eliminate the Bureau's use of these techniques [wiretapping and bugging]." At least three of the bugs that Hoover discontinued at the time were actively eavesdropping on Martin Luther King Jr. See Theoharis, *Spying on Americans*, 113.

36. Edward V. Long, *The Intruders: The Invasion of Privacy by Government and Industry* (New York: Praeger, 1966), viii.

37. Robert Cahn and Lyn Shepard, "Senate Debates Bugging," *Christian Science Monitor*, February 6, 1967, 1.

38. U.S. Congress, Senate, Committee on the Judiciary, *Right of Privacy Act of 1967: Hearings before the Subcommittee on Administrative Practice and Procedure*, 90th Cong., 2nd sess. (1967), 2.

39. Stanley K. Laughlin Jr., "Memorandum on *Ralph Berger v. New York:* Docket No. 615" (April 18, 1967), Edward V. Long Papers, Box 222: "Administrative Practice and Procedures (S–Z), 1966."

40. *Berger v. New York* 388 U.S. 41, 113 (1967).

41. Louis M. Kohlmeier, "Eavesdropping: The Supreme Court Majority Seems Headed for a Clash with Congress and the President," *Wall Street Journal*, June 29, 1967, 16.

42. Fred P. Graham, "A Sweeping Ban on Wiretapping Set for U.S. Aides," *New York Times*, July 7, 1967, 1.

43. U.S. Congress, Senate, Select Committee, *Supplementary Detailed Staff Reports*, 301. In 1967 the FBI installed 113 wiretaps, the lowest number since the early 1940s. The use of concealed microphones dropped to zero in the same year.

44. Edward V. Long, letter to Lyndon B. Johnson, July 1967, Edward V. Long Papers, Box 123: "Invasion of Privacy."

45. Robert J. Donovan, "GOP Leaders Charge Johnson Racial Failure," *Los Angeles Times*, July 25, 1967, 13.

46. On the politics of urban crime in sixties America and the rise of law-and-order conservatism, see Michael W. Flamm, *Law and Order: Street Crime, Civil Unrest, and the Crisis of Liberalism in the 1960s* (New York: Columbia University Press, 2005); Jonathan Simon, *Governing through Crime: How the War on Crime Transformed American Democracy and Created a Culture of Fear* (New York: Oxford University Press, 2007); Vesla M. Weaver, "Frontlash: Race and the Development of Punitive Crime Policy," *Studies in American Political Development* 21 (Fall 2007): 230–265.

47. On the origins and influence of the Johnson Crime Commission, see Naomi Murakawa, *The First Civil Right: How Liberals Built Prison America* (New York: Oxford University Press, 2014), 69–92; Elizabeth Hinton, *From the War on Poverty to the War on Crime: The Making of Mass Incarceration in America* (Cambridge, MA: Harvard University Press, 2016), 63–133.

48. Nicholas deB. Katzenbach et. al., *The Challenge of Crime in a Free Society: A Report by the President's Commission on Law Enforcement and Administration of Justice* (Washington, DC: U.S. Government Printing Office, 1967), 201, 203, 94.

49. On wiretapping and crime control policy in the late 1960s, see Richard Harris, "Annals of Legislation: The Turning Point," *New Yorker,* December 14, 1968, 68–179; Herman Schwartz, "The Legitimation of Electronic Eavesdropping: The Politics of 'Law and Order,'" *Michigan Law Review* 67, no. 3 (January 1969): 455–510; Edith J. Lapidus, *Eavesdropping on Trial* (Rochelle Park, NJ: Hayden Book, 1975), 38–48.

50. William Lambert, "Ed Long's Help-Hoffa Campaign," *Life,* May 26, 1967, 26–31, 75. The story of Long's effort to bolster Jimmy Hoffa's case had all of the hallmarks of an FBI smear: hints of cash exchanges and personal favors; innuendoes about legitimate business connections mixed up with underworld corruption; the potential for blackmail. Long strenuously maintained his innocence throughout the ordeal. Regardless of what sort of dirt Hoover may or may not have had on the senator from Missouri, it's clear that the motivations behind the Long Committee's investigation had little to do with the Hoffa case. Long took on government wiretapping as his pet political issue well before Hoffa went to trial. See Curt Gentry, *J. Edgar Hoover: The Man and the Secrets* (New York: Norton, 1991), 586–588.

51. For an entertaining inside look at the details of the *Katz* investigation, written by Katz's own defense attorney, see Harvey A. Schneider, "*Katz v. United States:* The Untold Story," *McGeorge Law Review* 40, no. 1 (2009): 13–23.

52. Louis M. Kohlmeier, "Bugging, Crime, and the Court Flip-Flop," *Wall Street Journal,* January 12, 1968, 8.

53. On the fiftieth anniversary of the *Katz* decision, one of the nation's leading experts on Fourth Amendment jurisprudence, Orin S. Kerr, tweeted the following: "Happy Birthday to Katz v. United States, 389 U.S. 347 (1967), decided 50 years ago today and confusing people ever since." Kerr's sentiment reflects the scholarly consensus. Kerr, Twitter post, December 18, 2017, 12:09 p.m.: https://twitter.com/OrinKerr/status/942788856666931200.

54. *Katz v. United States,* 389 U.S. 347, 352 (1967).

55. See Orin S. Kerr, "The Curious History of Fourth Amendment Searches," *Supreme Court Review* 2012, no. 1 (2013): 67–97; David Gray, *The Fourth Amendment in an Age of Surveillance* (New York: Cambridge University Press, 2017), 76–100.

56. *Katz v. United States,* 389 U.S. 347 (1967).

57. Peter Winn, "*Katz* and the Origins of the 'Reasonable Expectation of Privacy' Test," *McGeorge Law Review* 40, no. 1 (2009): 3–4.

58. Footnote 23 of the *Katz* ruling reads: "Whether safeguards other than prior authorization by a magistrate would satisfy the Fourth Amendment in a situation involving the national security is a question not presented by this case." *Katz v. United States,* 389 U.S. 347, 358 (1967).

59. Justice Hugo Black's famous dissent to the *Katz* decision makes this point explicit, stating up front that the ruling "removes the doubts about state power in this field [electronic surveillance] and abates to a large extent the confusion and near-paralyzing effect of the *Berger* holding." *Katz v. United States,* 389 U.S. 347, 364 (1967).

60. Kohlmeier, "Bugging, Crime, and the Court Flip-Flop," 8. See also Kenneth Ira Solomon, "The Short Happy Life of *Berger v. New York,*" *Chicago-Kent Law Review* 45, no. 2 (Fall–Winter, 1968–1969): 123–142.

61. John Herbers, "Wiretap Clause Put in Crime Bill: Senate Panel Votes Change That Johnson Opposes," *New York Times,* October 31, 1967, 29.

62. Harris, "Annals of Legislation," 70.

63. For an early distillation of McClellan's ideas about criminal justice in America, particularly as they related to the problem of organized crime, see John L. McClellan, *Crime without Punishment* (New York: Duell, Sloan and Pearce, 1962).

64. "Remarks of Senator John L. McClellan, National Council on Crime and Delinquency" (November 14, 1967), John L. McClellan Papers, Box 779, Folder 8A / B, Ouachita Baptist University, Riley-Hickingbotham Library Special Collections and Archives, Arkadelphia, AR.

65. "Notes: Re McClellan's Speech" (ca. 1968), John L. McClellan Papers, Box 777, Folder 5. Later that same year, McClellan made a similar statement in an address to the American Bar Association: "Never since the Civil War has this nation faced such frustration, dissension, strife, turmoil, and violence. During the past decade, we have moved far in the direction of chaos and anarchy. The momentum of this trend is sweeping us rapidly toward the brink of disaster." "Remarks of Senator John L. McClellan, American Bar Association, Philadelphia, PA" (August 5, 1968), John L. McClellan Papers, Box 777, Folder 5.

66. U.S. Congress, Senate, Committee on the Judiciary, *Controlling Crime through More Effective Law Enforcement*, 90th Cong., 1st sess. (1967), 142.

67. Quoted in Lapidus, *Eavesdropping on Trial*, 50.

68. G. Robert Blakey, a law professor at the University of Notre Dame, was instrumental in the drafting of the Senate's initial wiretap authorization provisions. In January 1968 Blakey wrote that *Katz* offered a "constitutional blueprint which will permit us to draw up a fair, effective and comprehensive system of court order wiretapping and bugging. All that remains now is the question of legislative will." He later worked closely with McClellan to finish the job. G. Robert Blakey, "Electronic Surveillance and the Courts," *Washington Post,* January 4, 1968, A16. On McClellan's efforts to engineer a crime control bill more favorable to the law-and-order camp, see Barry Mahoney, "The Politics of the Safe Streets Act, 1965–1973: A Case Study in Evolving Federalism and the National Legislative Process" (PhD diss., Columbia University, 1976), 148–178.

69. Here is McClellan on the floor of the Senate in November 1967, comparing the political tactics of civil rights organizations to those of the Ku Klux Klan and claiming that social protest movements were leading the country toward ruin: "Certainly with crime up 62 percent in 6 years, with rebellious students taking over college campuses, with antiwar demonstrators marching on the Pentagon and harassing high public officials, with national leaders being hooted and shouted down, with minority leaders advocating violence and mass civil disobedience, we are constrained to ask, Is America drifting toward anarchy? Yes, Mr. President, when we have reached the point when what the once dreaded Ku Klux Klan did under hoods and bedsheets is now begin done openly, wantonly, and under an aura of privilege and sanction on the pretense of poverty and inequality, then it is indeed timely to inquire if America is on the verge of anarchy." *Omnibus Crime Control and Safe Streets Act of 1967*, 90th Cong., 1st sess., *Congressional Record* 114 (November 28, 1967): S 17288.

70. Building on Robert F. Kennedy's proposals of the early 1960s, *The Challenge of Crime in a Free Society* only addressed the issue of electronic surveillance in a brief, three-page section appended to a chapter on organized crime in America. Katzenbach et. al., *The Challenge of Crime*, 200–203.

71. *Omnibus Crime Control and Safe Streets Act of 1967*, S. 917, 90th Cong., 2nd sess., *Congressional Record* 114 (May 23, 1968): S 14702–14703. McClellan harbored the racist belief that there were direct connections between organized crime and race riots—that the

pervasiveness of syndicated criminal organizations in American cities had led to a nationwide uptick in urban violence, and that the violence in Newark, Detroit, and elsewhere was itself the product of an organized movement of lawbreakers. In 1968 he wrote that "there is . . . a close, although not often noticed, relationship between organized crime and street crime. . . . [O]rganized crime breeds street crime." "Notes: Re McClellan's Speech" (ca. 1968), John L. McClellan Papers, Box 777, Folder 5. He also appears to have espoused similar views in meetings with members of the Kerner Commission on Civil Disorders, which had been formed to investigate the causes of the urban uprisings across America in the summer of 1967. See Donald F. O'Donnell, memorandum to John L. McClellan (October 5, 1967), John L. McClellan Papers, Box 388, Folder 3.

72. Edward V. Long, "Private and Governmental Use of Surveillance Devices," *Combating Crime: The Forensic Quarterly* 41 (August 1967): 311.

73. Mahoney, "The Politics of the Safe Streets Act," 173.

74. *Omnibus Crime Control and Safe Streets Act of 1968,* 211.

75. *Omnibus Crime Control and Safe Streets Act of 1967,* S. 917, 90th Cong., 2nd sess., *Congressional Record* 114 (May 23, 1968): S 14701.

76. *Omnibus Crime Control and Safe Streets Act of 1968,* 217.

77. See http://www.uscourts.gov/statistics-reports/analysis-reports/wiretap-reports for electronic editions of the U.S. government's annual *Wiretap Report.*

78. *Omnibus Crime Control and Safe Streets Act of 1968,* 220, 214.

79. For accounts of Hoover's influence on the final shape of Title III, see Charns, *Cloak and Gavel,* 91; Jeff A. Hale, "Wiretapping and National Security: Nixon, the Mitchell Doctrine, and the White Panthers" (PhD diss., Louisiana State University, 1995), 132.

80. Quoted in Flamm, *Law and Order,* 137.

81. *Omnibus Crime Control and Safe Streets Act of 1967,* S 14710.

82. Oswald Johnston, "Anti-Crime Bill Signed by Johnson," *Baltimore Sun,* June 20, 1968, A1.

8. Big Brother, Where Art Thou?

1. William H. Erickson et al., *Electronic Surveillance: Report of the National Commission for the Review of Federal and State Laws Relating to Wiretapping and Electronic Surveillance* (Washington, DC: U.S. Government Printing Office, 1976), 43.

2. In 1971 an investigative reporter for the *New York Times* alleged that Richard Nixon's attorney general, John Mitchell, had a habit of deactivating scores of warranted wiretaps at the end of each year, offering deflated numbers to Congress, and then reactivating them after boasting of a decreased use of electronic surveillance. Fred P. Graham, "Wiretapping: A Numbers Game," *New York Times,* May 5, 1971, 33. For an example of Mitchell's duplicity, see "Mitchell Reports Fewer Wiretaps: Says Eavesdropping by the Government Has Been Reduced under Nixon," *New York Times,* July 15, 1969, 1. In a secret deposition to Watergate prosecutors, President Nixon later admitted that his administration learned this tactic from J. Edgar Hoover, who did it for more than fifty years. See Tim Weiner, *Enemies: A History of the FBI* (New York: Random House, 2013), 278–279.

3. Edward V. Long, "Big Brother Is Listening," *Focus Midwest,* August 1962, 8.

4. Looking back on the tumult of the late 1960s, James Baldwin memorably diagnosed the "irresponsible ferocity" of the Safe Streets Act as the direct product "some pale, compelling nightmare—an overwhelming collection of private nightmares." James Baldwin, *No Name in the Street* (New York: Vintage International, 2007 [1972]), 130.

5. For a vivid contemporary account that ponders the public apathy that followed the passage of the Safe Streets Act, see Richard Harris, "Annals of Legislation: The Turning Point," *New Yorker,* December 14, 1968, 68–70, 179.

6. Sarah E. Igo, *The Known Citizen: A History of Privacy in Modern America* (Cambridge, MA: Harvard University Press, 2018), 248.

7. U.S. Congress, House of Representatives, Select Committee on Crime, *Crime in America: Heroin Importation, Distribution, Packaging, and Paraphernalia,* 91st Cong., 2nd sess. (1970), 62.

8. John L. McClellan, "Is Government Electronic Surveillance Necessary?," *American Legion Magazine,* April 1970, 21.

9. Erickson et al., *Electronic Surveillance,* xiii.

10. Louis Harris, "Most Want Wiretap Safeguards," *Chicago Tribune,* September 4, 1974, 16; George Gallup, "Gallup Poll: Public Evenly Divided on Wiretapping Issue," *Los Angeles Times,* August 21, 1969, A21.

11. Kevin Krajick, "Electronic Surveillance Makes a Comeback," *Police Magazine,* March 1983, 12.

12. Herman Schwartz, *Taps, Bugs, and Fooling the People* (New York: Field Foundation, 1977), 15–16.

13. Erickson et al., *Electronic Surveillance,* 73–74.

14. Bernard B. Spindel, *The Ominous Ear* (New York: Award Books, 1968), 254.

15. Robert M. Brown, *The Electronic Invasion* (New York: Rider, 1967), 26.

16. "Bug Thy Neighbor," *Time,* March 6, 1964, 61.

17. "Private eyes and industry are worse offenders than either the Federal Government or local law enforcement groups," Long wrote in 1966. "[T]hey have no legitimate excuse for invading privacy." Edward V. Long, "Snooping Hearings in San Francisco" (February 2, 1966), Edward V. Long Papers, Box 123: "Invasions of Privacy," Manuscript Collections, State Historical Society of Missouri, Columbia.

18. Edward V. Long, *The Intruders: The Invasion of Privacy by Government and Industry* (New York: Praeger, 1966), 17.

19. Brown, *The Electronic Invasion,* 23; Burton H. Alden, Byron C. Campbell, et al., *Competitive Intelligence: Information, Espionage, and Decision-Making* (Watertown, MA: C.I. Associates, 1959), 5.

20. Brown, *The Electronic Invasion,* 23.

21. John P. MacKenzie, "Private 'Bugging' Seen Cut by Law," *Washington Post,* July 31, 1970, A2.

22. Erickson et al., *Electronic Surveillance,* xviii.

23. "Bernard Spindel, Wiretapper, Dies," *New York Times,* February 4, 1971, 39. See also "Bernard B. Spindel, 48, Expert on Wiretapping," *Washington Post,* February 6, 1971, B5.

24. "William Mellin, Wiretapper, 83: Treasury Investigator Who Pioneered in Field Dies," *New York Times,* October 31, 1971, 82.

25. Ronald Kessler, "Wiretap Prosecutions Infrequent, Data Show," *Washington Post,* June 24, 1975, A1.

26. Erickson et al., *Electronic Surveillance,* 160.

27. Erickson et al., *Electronic Surveillance,* xvii.

28. John V. Tunney et al., *Surveillance Technology—Policy and Implications: A Staff Report of the Subcommittee on Constitutional Rights of the Committee on the Judiciary, United States Senate* (Washington, DC: U.S. Government Printing Office, 1976), 7, 11.

29. John Twohey, "The Public: Bugged by Bugs," *Washington Post,* June 24, 1973, C2.

30. "A View from the Basement," *Civil Liberties Review* 2, no. 3 (January 1975): 51.

31. Anthony Pellicano, a Westchester, Illinois, private eye who later gained notoriety for wiretapping celebrities in Hollywood, told the *Chicago Tribune* in 1975 that "bugging and wiretap equipment manufacturers are doing at least 50 per cent more business since Watergate erupted." Clarence Page, "Bug-Finding: Lucrative Business," *Chicago Tribune,* February 16, 1975, 12.

32. Erickson et al., *Electronic Surveillance,* xviii.

33. Erickson et al., *Electronic Surveillance,* 23; Tunney et al., *Surveillance Technology,* 11.

34. John S. VanDewerker, *State of the Art of Electronic Surveillance* (Townsend, WA: Loompanics Unlimited, 1983 [1976]), 23.

35. VanDewerker, *State of the Art,* 23.

36. "Edmund Scientific: 1991 Annual Catalog for the Technical Hobbyist & Science Educator" (Barrington, NJ: Edmund Scientific Co., 1991), Trade Literature Collection, Smithsonian Libraries, Washington, DC.

37. U.S. Congress, Senate, Committee on the Judiciary, *Invasions of Privacy (Government Agencies): Hearings before the Subcommittee on Administrative Practice and Procedure of the Committee on the Judiciary, Part 1,* 89th Cong., 1st sess. (1965), 27–63. Interestingly, a Fargo Corporation salesman named Leo Jones is also listed alongside Harold Lipset as a "technical advisor" in the credits to Francis Ford Coppola's *The Conversation.*

38. Ronald Kessler, "Private Wiretapping: How Widespread?," *Washington Post,* February 8, 1971, A12. Emphasis mine.

39. Tunney et al., *Surveillance Technology,* 29.

40. U.S. Congress, Senate, Committee on Government Operations, *Transfer of Technology to the Soviet Union and Eastern Europe,* 93rd Cong., 2nd sess. (1974), 38, 51.

41. U.S. Congress, Senate, Committee, *Transfer of Technology,* 48.

42. U.S. Congress, Senate, Committee, *Transfer of Technology,* 6.

43. Sam Jaffe, "Russians Invite U.S. Firms to Police Trade Show," *Chicago Tribune,* July 7, 1974, 29.

44. U.S. Congress, Senate, Committee, *Transfer of Technology,* 2.

45. Tunney et al., *Surveillance Technology,* 3.

46. Richard T. Cooper, "U.S. Concerns Seek to Sell Surveillance Devices to Russians," *Los Angeles Times,* July 18, 1974, 21.

47. "Russia to Open Crime Control Exhibit Today: American Participation Limited after Congressional Criticism of Spy Devices," *Los Angeles Times,* August 14, 1974, A15. Emphasis mine.

48. Tunney et al., *Surveillance Technology,* 17.

49. Ralph V. Ward, interview with author, December 7, 2018.

50. Nelson Blackstock, *COINTELPRO: The FBI's Secret War on Political Freedom* (New York: Vintage Books, 1976); Athan Theoharis, *Spying on Americans: Political Surveillance from Hoover to the Huston Plan* (Philadelphia: Temple University Press, 1978), 110–155; Frank J. Donner, *The Age of Surveillance: The Aims and Methods of America's Political Intelligence System* (New York: Vintage Books, 1981); Alexander Charns, *Cloak and Gavel: FBI Wiretaps, Bugs, Informers, and the Supreme Court* (Urbana: University of Illinois Press, 1992); Jeff A. Hale, "Wiretapping and National Security: Nixon, the Mitchell Doctrine, and the White Panthers" (PhD diss., Louisiana State University, 1995); Seth Rosenfeld, *Subversives: The FBI's War on Student Radicals, and Reagan's Rise to Power* (New York: Farrar, Straus and Giroux, 2012).

51. David J. Garrow, *The FBI and Martin Luther King, Jr.: From "Solo" to Memphis* (New York: W. W. Norton, 1981); Kenneth O'Reilly, *"Racial Matters": The FBI's Secret File on Black America* (New York: Free Press, 1989); Kenneth O'Reilly, "Hoover's FBI and Black America," in *Black Americans: The FBI Files,* edited by David Gallen (New York: Carroll and Graf, 1994), 7–65; William J. Maxwell, *F.B. Eyes: How J. Edgar Hoover's Ghostreaders Framed African American Literature* (Princeton, NJ: Princeton University Press, 2015).

52. James Bamford, *The Puzzle Palace: Inside the National Security Agency, America's Most Secret Intelligence Organization* (New York: Penguin Books, 1983), 302–355; Matthew M. Aid, *The Secret Sentry: The Untold History of the National Security Agency* (New York: Bloomsbury, 2009), 128–170.

53. On the cultivated mystique of secrecy at the NSA, see Friedrich Kittler, "NSA: No Such Agency" (1986), translated by Paul Feigelfeld and Jussi Parikka, *Theory, Culture, & Society* (February 12, 2014): http://theoryculturesociety.org/kittler-on-the-nsa/; Patrick Radden Keefe, *Chatter: Uncovering the Echelon Surveillance Network and the Secret World of Global Eavesdropping* (New York: Random House, 2006).

54. Authorities caught the Watergate burglars with woefully outmoded eavesdropping devices in hand, a detail that made them the laughingstock of the professional eavesdropping industry in 1972. "This is fantastic," one freelance wiretapper told the *Los Angeles Times* two days after the news of the botched break-in made headlines. "That kind of equipment went out with high-button shoes. These guys have got to be circus bums." Another manufacturer of electronic surveillance devices called Watergate the "most amateurish job I ever saw." "Attempt to Bug Party Offices Bemuses Pros," *Los Angeles Times,* June 19, 1972, A20; Ronald Kessler, "Experts Heap Scorn on Bungled 'Bug' Caper," *Washington Post,* June 19, 1972, A7.

55. For extended accounts of the political origins and influence of the Keith case, see Hale, "Wiretapping and National Security"; Trevor W. Morrison, "The Story of *United States v.*

United States District Court (Keith): The Surveillance Power," *Columbia Public Law & Legal Theory Working Papers* (2008): http://lsr.nellco.org/columbia_pllt/08155.

56. Hale, "Wiretapping and National Security," 149, 167.

57. Americo R. Cinquegrana, "The Walls (and Wires) Have Ears: The Background and First Ten Years of the Foreign Intelligence Surveillance Act of 1978," *University of Pennsylvania Law Review* 137, no. 3 (January 1989): 799.

58. Christopher Lydon, "ACLU Suit Tests Wiretaps by FBI of Groups in U.S.," *New York Times,* June 27, 1969, 1, 17.

59. Walter R. Gordon, "Some Subversion Wiretaps Ruled Illegal without Order," *Baltimore Sun,* April 9, 1971, A6.

60. U.S. Congress, House of Representatives, Committee on the Judiciary, *White House Surveillance Activities and Campaign Activities,* 93rd Cong., 2nd sess. (1974), 8–31; Richard Reeves, *President Nixon: Alone in the White House* (New York: Simon and Schuster, 2001), 75–6.

61. On the radical political group at the center of *United States v. United States District Court,* see Jeff A. Hale, "The White Panthers' 'Total Assault on the Culture,'" in *Imagine Nation: The American Counterculture of the 1960s and '70s,* edited by Peter Braunstein and Michael William Doyle (New York: Routledge, 2002), 125–157.

62. *U.S. v. Sinclair,* 321 F.Supp. 1074 (E.D. Mich 1971).

63. *United States v. United States District Court,* 407 U.S. 297, 314 (1972).

64. John P. MacKenzie, "Court Curbs Wiretapping of Radicals," *Washington Post,* June 20, 1972, A1; Fred P. Graham, "High Court Curbs U.S. Wiretapping Aimed at Radicals," *New York Times,* June 20, 1972, 1, 23; "The Restraint of Law," *New York Times,* June 20, 1972, 38.

65. Morrison, "The Story," 12–17.

66. U.S. Congress, Senate, Committee on the Judiciary, *Nominations of William H. Rehnquist and Lewis F. Powell, Jr.: Hearings before the Committee on the Judiciary,* 92nd Cong., 1st sess. (1971), 208–213.

67. In so doing, the legislation created an important division of labor in the work of government surveillance that stands to this day: national security wiretapping and FISA on one side; police wiretapping and Title III on the other. When pundits and lawmakers debate wiretapping and electronic eavesdropping today, they tend to overlook this crucial distinction, often worrying about the former at the expense of the latter.

68. George Lardner Jr., "Carter Signs Bill Limiting Foreign Intelligence Surveillance," *Washington Post,* October 26, 1978, A2.

69. On the failures of federal intelligence reform after Watergate and the Church Committee, see Loch K. Johnson, *A Season of Inquiry: The Senate Intelligence Investigation* (Lexington: University Press of Kentucky, 1985), 227–27; Kathryn S. Olmsted, *Challenging the Secret Government: The Post-Watergate Investigations of the CIA and FBI* (Chapel Hill: University of North Carolina Press, 1996).

70. "Secret Court Said to Grant Every Request to Bug Spies," *Washington Post,* March 4, 1980, A8; Cinquegrana, "The Walls (and Wires)," 814–820.

71. *United States v. Kahn*, 415 U.S. 143 (1974).

72. "Privacy: A Year-End Report," *Washington Post*, December 26, 1974, A24.

73. U.S. Congress, Senate, Select Committee to Study Governmental Operations with respect to Intelligence Activities, *Intelligence Activities and the Rights of Americans, Book 2*, 94th Cong., 2nd sess. (1976), 5; Erickson et al., *Electronic Surveillance*, 9.

74. "Panel Urges Broader Powers for Wiretapping and Bugging," *New York Times*, April 30, 1976, 20.

75. Kevin Krajick, "Electronic Surveillance Makes a Comeback," *Police Magazine*, March 1983, 25.

9. Limited Assistance Necessary

1. These figures reflect the data provided in the 1976–1980 editions of the U.S. government's annual *Report on Applications for Orders Authorizing or Approving the Interception of Wire or Oral Communications*, published by the Administrative Office of the U.S. Courts.

2. The account that follows is based on a series of telephone conversations and e-mail exchanges I had with Edward J. Tomas, June 2018–October 2018.

3. Administrative Office of the U.S. Courts, *Report on Applications for Orders Authorizing or Approving the Interception of Wire or Oral Communications: For the Period January 1, 1974 to December 31, 1974* (Washington, DC: U.S. Government Printing Office, 1975), v.

4. Administrative Office of the U.S. Courts, *Report on Applications for Orders Authorizing or Approving the Interception of Wire or Oral Communications: For the Period January 1, 1980 to December 31, 1980* (Washington, DC: U.S. Government Printing Office, 1981), 6.

5. See http://www.leacorp.com/pages/Who-We-Are.html.

6. Richard Sasso (former assistant prosecutor, Hunterdon County, New Jersey), phone interview with author, May 2020.

7. Between 1977 and 1982, the New Jersey Division of Criminal Justice obtained authorization for 894 wiretaps in conjunction with 500 separate criminal investigations. Of the 500 investigations that involved electronic surveillance, more than 90 percent resulted in indictments. New Jersey Department of Law and Public Safety, Division of Criminal Justice, Research and Evaluation Section, *Electronic Surveillance in New Jersey, 1977–1982* (New Brunswick: New Jersey Department of Law and Public Safety, 1983), 20.

8. Phone companies also had their own internal investigative units in those days, a matter that would become controversial in the mid-1970s, when reports of warrantless toll-fraud surveillance made national headlines. See Phil Lapsey, *Exploding the Phone: The Untold Story of the Teenagers and Outlaws Who Hacked Ma Bell* (New York: Grove Press, 2013), 249–261.

9. On the New Jersey State Police Electronic Surveillance Unit's use of leased lines, see Kevin Krajick, "Should Police Wiretap? The States Don't Agree," *Police Magazine*, May 1983, 29–30.

10. "Privilege of the Telegraph," *New York Times*, February 8, 1880, 4. Quoted in David J. Seipp, *The Right to Privacy in American History* (Cambridge, MA: Program on Information Resources Policy, Harvard University, 1978), 106.

11. On the practical differences between monitoring data "at rest" and monitoring communications "in motion," see David Kahn, *The Codebreakers: The Story of Secret Writing* (New York: Scribner, 1996 [1967]), 213.

12. Seipp, *Right to Privacy,* 30–31; Anuj C. Desai, "Wiretapping before the Wires: The Post Office and the Birth of Communications Privacy," *Stanford Law Review* 60, no. 2 (November 2007): 578.

13. "Letter of William Orton, President, Western Union Telegraph Company," 44th Cong., 2nd sess., *Congressional Record* 5 (December 21, 1876): H 353. Quoted in Seipp, *Right to Privacy,* 32.

14. "Seymour Wires Tapped on Order Given by Woods," *New York Times,* May 18, 1916, 8.

15. "Seymour Wires Tapped," 8. On the popular Gilded Age caricature of the telephone company as monstrous octopus or spider, see Robert MacDougall, "The Wire Devils: Pulp Thrillers, the Telephone, and Action at a Distance in the Wiring of a Nation," *American Quarterly* 58, no. 3 (September 2006): 715–741.

16. "Seymour Wires Tapped," 8.

17. "Seymour Wires Tapped," 8.

18. In a 1931 congressional hearing on wiretapping in Prohibition enforcement, U.S. Attorney General William D. Mitchell testified that "in some cases where it is obvious that a criminal conspiracy and criminal acts are being conducted through the use of the telephone, showings have been made to the telephone company and their permission obtained to tap the wires. I think there have been other cases where the wires have been tapped by police officials without any consent from the telephone companies." U.S. Congress, House of Representatives, Committee on Expenditures in the Executive Departments, *Wire Tapping in Law Enforcement,* 71st Cong., 3rd sess. (1931), 24.

19. On the Bell System's "public service principle," see Alan Stone, *Wrong Number: The Breakup of AT&T* (New York: Basic Books, 1989), 24–33.

20. "*Olmstead v. United States:* Brief in Support of Petitioners' Contention, Amicus Curiae" (October 1927), Southwestern Bell Telephone Company Records, Collection 2: SBC Communications Inc., Record Group 5: Predecessor and Subsidiary Companies, Box 134, Subject Files: "Wiretapping, 1927–1948," AT&T Archives and History Center, San Antonio.

21. "*Olmstead v. United States:* Brief in Support of Petitioners' Contention."

22. *Olmstead v. United States,* 277 U.S. 438, 479 (1928).

23. Walter S. Gifford, telegram to E. E. Nims, June 7, 1928, Southwestern Bell Telephone Company Records, Box 134, Subject Files: "Wiretapping, 1927–1948."

24. "Telephone Head Says Wires Must Not Be Tapped," *Atlanta Constitution,* June 9, 1928, 20.

25. See "Secrecy of Communications" (1964), Ohio Bell Telephone Company Records, 1920–1983, Collection 5: Ameritech Corporation, Record Group 5: Predecessor and Subsidiary Companies, Box 26, Subject Files: "Secrecy of Communication," AT&T Archives and History Center.

26. "Secrecy of Communication" (c. 1950–1952), Ohio Bell Telephone Company Records, Box 26, Subject Files: "Secrecy of Communication."

27. See "Bell System's Position on Wiretapping," *The Transmitter,* March–April 1960, 45. Michigan Bell Telephone Company Records, Collection 5: Ameritech Corporation, Record Group 5: Predecessor and Subsidiary Companies, Box 181, Subject Files: "Wiretapping, 1955–1966," AT&T Archives and History Center.

28. U.S. Congress, Senate, Committee on the Judiciary, *Invasions of Privacy (Telephone Systems): Hearings before the Subcommittee on Administrative Practice and Procedure, Part 6,* 89th Cong., 2nd sess. (1966), 2602.

29. On the history of telephone industry's attempts to maintain public goodwill, see Claude S. Fischer, *America Calling: A Social History of the Telephone to 1940* (Berkeley: University of California Press, 1992), 60–85.

30. U.S. Congress, Senate, Committee on the Judiciary, *Wiretapping, Eavesdropping, and the Bill of Rights: Hearing before the Subcommittee on Constitutional Rights of the Committee on the Judiciary, Part 3,* 86th Cong., 1st sess. (1959), 523.

31. "Course of Instruction for Personnel: Operating Methods" (ca. 1950s), 104, Pinkerton's National Detective Agency Records, 1853–1999, Box 54, Folder 7, Manuscript Division, Library of Congress, Washington, DC.

32. Donald Janson, "Wiretap Inquiry Finds a Pattern," *New York Times,* October 24, 1965, 50.

33. Timothy S. Robinson, "FBI's Secret Wiretap Room Glimpsed," *Washington Post,* June 27, 1975, I10–12.

34. Ronald Kessler, "FBI Wiretapping: How Widespread?," *Washington Post,* February 7, 1971, A18; Timothy S. Robinson, "C&P Official Tells of FBI Wiretap Use," *Washington Post,* April 6, 1975, A1, A22; "Reveal Bell Telephone Placed Wiretaps for FBI," *Chicago Tribune,* September 28, 1976, 7. Hampton briefly became a household name when he testified in a lawsuit against AT&T filed by Morton Halperin, a former National Security Council advisor on whom Henry Kissinger had ordered a wiretap in 1969.

35. "On Wiretapping and Electronic Surveillance," *AT&T Management Report* 195 (June 19, 1968), 2–3, Ohio Bell Telephone Company Records, Box 88, Subject Files: "Secrecy of Communications, 1967–1979." A month later, AT&T's assistant vice president sent a similar memorandum to the company's public relations corps. Alvin von Auw, letter to All Public Relations Vice Presidents, June 25, 1968, Michigan Bell Telephone Company Records, Box 182, Subject Files: "Wiretapping, 1967–1977."

36. *Application of the United States for Relief,* 427 F2d 639 (9th Cir. 1970).

37. Bruce Schneier and David Banisar, *The Electronic Privacy Papers: Documents on the Battle for Privacy in the Age of Surveillance* (New York: Wiley, 1997), 48.

38. William P. Mullane Jr., letter to All Bell System News Contact Personnel, August 3, 1973, Michigan Bell Telephone Company Records, Box 182, Subject Files: "Wiretapping, 1973–1974." At no point were the political benefits of limited assistance more clear than during the Watergate scandal. Amid the popular panic over the tapping of phones and the bugging of the nation's highest office, the limited assistance policy enabled AT&T to point to the government as the source of any wrongdoing. See "The Watergate Aftermath: Communications Privacy Remains System Policy," *Pacific Telephone Review,* September 10, 1973, 1–2, Pacific Bell Telephone Company Records, Box 91, Subject Files: "Wiretapping, 1965–1978."

39. On the service and toll fraud monitoring controversies of the mid-1970s, see Lapsey, *Exploding the Phone,* 296–310.

40. Schneier and Banisar, *The Electronic Privacy Papers*, 48.

41. William H. Erickson et al., *Electronic Surveillance: Report of the National Commission for the Review of Federal and State Laws Relating to Wiretapping and Electronic Surveillance* (Washington, DC: U.S. Government Printing Office, 1976), 9.

42. *United States v. New York Telephone Co.*, 434 U.S. 159, 177 (1977).

43. "A Bill on Wiretaps Criticized by AT&T," *New York Times*, June 19, 1978, A15.

44. On the long-term influence of *United States v. New York Telephone Co.* (1977), see Cyrus Farivar, *Habeas Data: Privacy vs. the Rise of Surveillance Tech* (Brooklyn: Melville House, 2018), 26–56. In 2016 the FBI notoriously cited *New York Telephone Co.* and the All Writs Act to justify an attempt to force software engineers at Apple, Inc. to reprogram the password-entry limits on the iPhone. The agency wanted to access the data stored on a standard 5C device owned by Syed Rizwan Farook, one of two terrorists killed in the aftermath of the December 2015 mass shooting in San Bernardino, California. At the eleventh hour, the Bureau backed away from its legal claims after finding its own way into Farook's phone.

10. Off the Wire

1. L. Britt Snider, "Unlucky Shamrock: Recollections from the Church Committee's Investigation of the NSA," *Studies in Intelligence* 43, no. 1 (Winter 1999–2000): 45. See also James Bamford, *The Puzzle Palace: A Report on America's Most Secret Agency* (New York: Penguin Books, 1982), 302–355; James Bamford, *The Shadow Factory: The Ultra-Secret NSA from 9 / 11 to the Eavesdropping on America* (New York: Anchor Books, 2008), 161–168.

2. "AT&T; Whistleblower's Evidence," *Wired*, May 17, 2006: http://www.wired.com/science/discoveries/news/2006/05/70908. Bamford, *The Shadow Factory*, 188–196, contains a full account of the Room 641a affair.

3. For an overview of these programs, which have received extensive news coverage since 2013, see "NSA Prism and UPSTREAM Briefing Slides," in *The Snowden Reader*, edited by David P. Fidler (Bloomington: University of Indiana Press, 2015), 96–100.

4. Tim Wu, *The Master Switch: The Rise and Fall of Information Empires* (New York: Vintage Books, 2011), 105.

5. Edward J. Tomas, interview with author, June 9, 2018.

6. *United States v. New York Telephone Co.*, 434 U.S. 159, 162 (1977).

7. William H. Erickson et al., *Electronic Surveillance: Report of the National Commission for the Review of Federal and State Laws Relating to Wiretapping and Electronic Surveillance* (Washington, DC: U.S. Government Printing Office, 1976), 143.

8. Herman Schwartz, *Taps, Bugs, and Fooling the People* (New York: Field Foundation, 1977), 27.

9. Schwartz, *Taps, Bugs, and Fooling*, 2.

10. Erickson et al., *Electronic Surveillance*, 267.

11. Erickson et al., *Electronic Surveillance*, 271.

12. Erickson et al., *Electronic Surveillance*, 51.

13. Tom Wicker, "A 'Dirty Business' Fails," *New York Times*, July 12, 1977, 25.

14. Erickson et al., *Electronic Surveillance,* 267.

15. Herman Schwartz, *The Wiretapping Problem Today: A Report* (New York: American Civil Liberties Union, 1962).

16. Herman Schwartz, interview with author, September 6, 2018.

17. Herman Schwartz, "The 'Dirty' Business Turns Up Little Dirt," *Baltimore Sun,* October 7, 1979, K3.

18. Erickson et al., *Electronic Surveillance,* 51.

19. Schwartz, *Taps, Bugs, and Fooling,* 33, 29.

20. Angel Castillo, "Wiretap Approvals by Courts Decrease: Officials Say Decline in Electronic Surveillance Use Is Linked to Higher Operating Costs," *New York Times,* October 5, 1980, 65.

21. Administrative Office of the U.S. Courts, *Report on Applications for Orders Authorizing or Approving the Interception of Wire or Oral Communications: For the Period January 1, 1979 to December 31, 1979* (Washington, DC: U.S. Government Printing Office, 1980), 6–7.

22. Ben A. Franklin, "Wiretapping Cost a Record in Trial," *New York Times,* April 2, 1983, 5.

23. "Wiretap Seen as Weapon to Battle Drugs," *Los Angeles Times,* April 12, 1971, 6.

24. Erickson et al., *Electronic Surveillance,* 146.

25. Katherine Beckett, *Making Crime Pay: Law and Order in Contemporary American Politics* (New York: Oxford University Press, 1997); Christian Parenti, *Lockdown America: Police and Prisons in the Age of Crisis,* new ed. (New York: Verso, 2008), 3–68; Elizabeth Hinton, *From the War on Poverty to the War on Crime: The Making of Mass Incarceration in America* (Cambridge, MA: Harvard University Press, 2016), 180–217; Max Feller-Kantor, *Policing Los Angeles: Race, Resistance, and the Rise of the LAPD* (Chapel Hill: University of North Carolina Press, 2018).

26. New Jersey Department of Law and Public Safety, Division of Criminal Justice, Research and Evaluation Section, *Electronic Surveillance in New Jersey, 1977–1982* (New Brunswick: New Jersey Department of Law and Public Safety, 1983), 20.

27. Edward J. Tomas, interview with author, June 9, 2018.

28. Kevin Krajick, "Should Police Wiretap? The States Don't Agree," *Police Magazine,* May 1983, 31.

29. Krajick, "Should Police Wiretap?," 31.

30. Mike Gray, *Drug Crazy: How We Got into This Mess and How We Can Get Out* (New York: Random House, 1998), 101.

31. Ronald J. Ostrow, "Wiretapping in U.S. Cases Soars by 30.9%," *Los Angeles Times,* May 4, 1982, B5.

32. Ben A. Franklin, "Report Says Federal Wiretaps Rose 23% in '82," *New York Times,* May 1, 1983, A21; "Authorized Wiretaps Jump 40% Last Year, Report Says," *Washington Post,* May 2, 1985, A10.

33. Parenti, *Lockdown America,* 47.

34. Quoted in Kevin Krajick, "Electronic Surveillance Makes a Comeback," *Police Magazine,* March 1983, 17.

35. On the historical connections between Prohibition enforcement and the War on Drugs, see Lisa McGirr, *The War on Alcohol: Prohibition and the Rise of the American State* (New York: W. W. Norton, 2015).

36. Ronald J. Ostrow, "Electronic Surveillance Hits New Highs in War on Crime," *Los Angeles Times,* December 18, 1983, A4.

37. Jack Nelson, "Planes, Wiretaps Curbing Drug Traffic, Smith Says," *Los Angeles Times,* April 16, 1982, C9.

38. See Krajick, "Electronic Surveillance," 8–19, 24–25; Krajick, "Should Police Wiretap?," 29–32, 36–41.

39. David Bird, "13 Are Charged with Operating Big Heroin Ring," *New York Times,* August 24, 1983, B2; Wallace Turner, "Bugging by U.S. Produces Big Cases against Las Vegas Crime," *New York Times,* November 8, 1983, A1; Ostrow, "Electronic Surveillance Hits New Highs"; Ralph Blumenthal, "New Technology Helps in Efforts to Fight Mafia," *New York Times,* November 24, 1986, B3.

40. The relationship between the DEA's electronic surveillance crusade and the drug scandals that rocked the National Football League and Major League Baseball during the mid-1980s deserves more attention. For a sampling of news coverage that mentions FBI wiretaps in connection with investigations of cocaine use in professional sports, see Jane Gross, "4 Cowboys Named As Cocaine Users," *New York Times,* July 9, 1983, 29–30; Ann LoLordo, "Two Orioles Give Testimony in Drug Probe," *Baltimore Sun,* July 13, 1983, A1; "Wilson, 2 Ex-Royals Sentenced," *Washington Post,* November 18, 1983, D12; Michael Goodwin, "Baseball and Cocaine: A Deepening Problem," *New York Times,* August 19, 1985, A1, C6.

41. Anthony M. DeStefano, "Drug Agents Turn Linguist to Plumb the Criminal Mind," *Wall Street Journal,* December 27, 1985, 1, 4.

42. David Beier and Deborah Leavy, memorandum to Robert W. Kastenmeier (December 10, 1985), Robert Kastenmeier Papers, 1950–1990 (M91-048), Box 57, Folder 15, Wisconsin Historical Society, Division of Library, Archives, and Museum Collections, Madison.

43. "State Officials Debate whether to Allow Wiretapping in Drug Cases," *Detroit Free Press,* November 29, 1985, 17A, Collection 5: Ameritech Corporation, Record Group 5: Predecessor and Subsidiary Companies, Michigan Bell Telephone Company, Box 182, Subject Files: "Wiretapping, 1982–1997," AT&T Archives and History Center, San Antonio.

44. Brad Gates, "Wiretapping in War against Drugs," *Los Angeles Times,* May 8, 1988, A9.

45. Ralph Ginzburg, "An Outcry on Wiretaps Flares Anew," *New York Times,* July 31, 1988, NJ1; "Stronger Wiretap Law Sought," *Philadelphia Tribune,* September 13, 1988, 2A.

46. David Burnham, "Loophole in Law Raises Concern about Privacy in Computer Age," *New York Times,* December 19, 1983, A1. See also U.S. Congress, Senate, Committee on the Judiciary, *Oversight on Communications Privacy: Hearing before the Subcommittee on Patents, Copyrights and Trademarks,* 98th Cong., 2nd sess. (1984), 4.

47. "Privacy: Increasing Priority for Public and Company," *Illinois Bell Background for Managers* 7 (May 1980), 4, Collection 5: Ameritech Corporation, Record Group 5: Predecessor and Subsidiary Companies, Illinois Bell Telephone Company, Box 30, Subject Files: "Privacy, 1966–1980," AT&T Archives and History Center.

48. U.S. Congress, Office of Technology Assessment, *Federal Government Information Technology: Electronic Surveillance and Civil Liberties*, OTA-CIT-293 (Washington, DC: U.S. Government Printing Office, October 1985), 3.

49. U.S. Congress, Office, *Federal Government Information Technology*, 3.

50. Robert W. Kastenmeier, "Electronic Communication Privacy Act of 1985: Extension of Remarks" (n.d.), Robert Kastenmeier Papers, 1950–1990 (M91-048), Box 32, Folder 5.

51. U.S. Congress, House of Representatives, Committee on the Judiciary, *1984—Civil Liberties and the National Security State: Hearings before the Subcommittee on Courts, Civil Liberties, and the Administration of Justice*, 98th Cong., 1st sess. (1983 / 4), 2.

52. "Privacy and Technology," *Los Angeles Times*, January 3, 1984, C4.

53. *Kansas v. Howard*, 679 P.2d 197 (Kan. 1984).

54. *Rhode Island v. Delaurier*, 488 A.2d 688 (R.I. 1985).

55. 18 U.S.C.A § 2510 (1968). For a closer look at Title III's failings in the early age of digital communications, see Priscilla M. Regan, *Legislating Privacy: Technology, Social Values, and Public Policy* (Chapel Hill: University of North Carolina Press, 1995), 129–136.

56. U.S. Congress, Senate, Committee, *Oversight on Communications Privacy*, 1–2.

57. Orin Kerr, "A User's Guide to the Stored Communications Act, and a Legislator's Guide to Amending It," *George Washington Law Review* 72, no. 6 (August 2004): 1212.

58. See Stephanie K. Pell, "Location Tracking," in *The Cambridge Handbook of Surveillance Law*, edited by David Gray and Stephen Henderson (New York: Cambridge University Press, 2017), 44–70; Andrew Guthrie Ferguson, *The Rise of Big Data Policing: Surveillance, Race, and the Future of Law Enforcement* (New York: NYU Press, 2017); Mark Andrejevic, "Surveillance in the Big Data Era," in *Surveillance Studies: A Reader*, edited by Torin Monahan and David Murakami Wood (New York: Oxford University Press, 2018), 257–260.

59. Robert W. Kastenmeier, memorandum to Members of the House Committee on the Judiciary (June 6, 1986), Robert Kastenmeier Papers, 1950–1990 (M91-048), Box 57, Folder 14.

60. Electronic Communication Privacy Act of 1985, H.R. 4952, 99th Cong., 1st sess., *Congressional Record, Extensions of Remarks* (September 19, 1985), E 4128.

61. Regan, *Legislating Privacy*, 35.

62. "Wiretap Modernization Bill Signed," *New York Times*, October 22, 1986, A22.

63. U.S. Congress, House of Representatives, Committee on the Judiciary, *Electronic Communications Privacy Act: Hearings before the Subcommittee on Courts, Civil Liberties, and the Administration of Justice*, 99th Cong., 1st–2nd sess. (1985–1986), 213.

64. Mary Thornton, "House Votes to Revise Wiretap Law to Restrict Electronic Surveillance," *Washington Post*, June 24, 1986, A4.

65. John S. VanDewerker, *State of the Art of Electronic Surveillance* (Townsend, WA: Loompanics, 1983 [1976]), 12.

66. Len Ackland, "Loop to Get Fiber Optic Lines," *Chicago Tribune*, March 1, 1983, C14.

67. Joel Brinkley, "Florida Effort 'Just Can't Stop Drugs,'" *New York Times*, September 4, 1986, B9; Robert F. Howe, "Tab for Wiretaps High, So Are the Results, Report Says," *Washington Post*, September 25, 1990, A21.

68. David Simon, "Caller ID Latest Hit with High-Technology Drug Dealers," *Baltimore Sun*, May 6, 1990, 1A, 12A.

69. Shane Harris, *The Watchers: The Rise of America's Surveillance State* (New York: Penguin Press, 2010), 71–73.

70. Quoted in Harris, *The Watchers*, 73.

71. Bruce Schneier and David Banisar, *The Electronic Privacy Papers: Documents on the Battle for Privacy in the Age of Surveillance* (New York: Wiley, 1997), 182. My account in this section is deeply indebted to this published collection of confidential documents, which Schneier and Banisar acquired by filing hundreds of Freedom of Information Act requests in the wake of the passage of the Communications Assistance for Law Enforcement Act of 1994.

72. Schneier and Banisar, *The Electronic Privacy Papers*, 138.

73. Schneier and Banisar, *The Electronic Privacy Papers*, 161–162. Emphasis mine.

74. Schneier and Banisar, *The Electronic Privacy Papers*, 135.

75. On the detrimental effects of the FBI's requests to the phone companies in the digital telephony debates, see Lillian R. BeVier, "The Communications Assistance for Law Enforcement Act of 1994: A Surprise Sequel to the Break Up of AT&T," *Stanford Law Review* 51, no. 5 (May 1999): 1049–1125; Charlotte Twight, "Conning Congress: Privacy and the 1994 Communications Assistance for Law Enforcement Act," *The Independent Review* 6, no. 2 (Fall 2001): 185–216.

76. Schneier and Bansiar, *The Electronic Privacy Papers*, 253.

77. Ronald J. Ostrow, "FBI Fears Phone Advances Will Hamper Wiretapping," *Los Angeles Times*, March 7, 1992, A17; Anthony Ramirez, "As Technology Makes Wiretaps More Difficult, FBI Seeks Help," *New York Times*, March 8, 1992, 22.

78. Schneier and Banisar, *The Electronic Privacy Papers*, 169.

79. On the debates over encryption standards that dovetailed the FBI's digital telephony proposals of the early 1990s, see Whitfield Diffie and Susan Landau, *Privacy on the Line: The Politics of Wiretapping and Encryption*, 2nd ed. (Cambridge, MA: MIT Press, 2007), 229–248. The so-called Crypto Wars unfortunately fall outside the scope of my account here.

80. "FBI, Phone firms in Tiff over Turning On the Taps," *Washington Post*, March 10, 1992, C1.

81. "Back to Smoke Signals," *Washington Post*, March 26, 1992, A20.

82. Schneier and Banisar, *The Electronic Privacy Papers*, 172–174; Anthony Ramirez, "FBI's Proposal on Wiretaps Criticized by Federal Agency," *New York Times*, January 15, 1993, A12.

83. See Michael Tackett, "FBI and Phone Companies Face Off over Wiretaps," *Detroit Free Press*, May 19, 1992, 3D; Jube Shiver Jr., "Tapping into High-Tech Talk," *Los Angeles Times*, April 17, 1993, D1; Robert Lee Hotz, "Change in Technology May Curtail Wiretaps," *Los Angeles Times*, October 3, 1993, A29; "Clinton Seeking Legislation to Preserve Wiretap Power," *Wall Street Journal*, February 24, 1994, A6; David Gelernter, "Wiretaps for a Wireless Age," *New York Times*, May 8, 1994, E17.

84. U.S. Congress, House of Representatives / Senate, Committee on the Judiciary, *Digital Telephony and Law Enforcement Access to Advanced Telecommunications Technologies and Services,* 103rd Cong., 2nd sess. (1994), 10.

85. U.S. Congress, House / Senate, Committee, *Digital Telephony,* 107, 121.

86. Schneier and Banisar, *The Electronic Privacy Papers,* 125–126.

87. Schneier and Banisar, *The Electronic Privacy Papers,* 125, 130.

88. Schneier and Banisar, *The Electronic Privacy Papers,* 41.

89. Schneier and Banisar, *The Electronic Privacy Papers,* 148.

90. "The guy was all over the Hill, he never let up," one longtime telephone industry lobbyist remembered of Freeh in the summer of 1994. "In all my years here, I have never seen that kind of lobbying by a top administration official. It was astounding." Rogier van Bakel, "How Good People Helped Make a Bad Law," *Wired,* February 1996, 134.

91. On Clinton's 1994 crime bill and the War on Drugs, see Michelle Alexander, *The New Jim Crow: Mass Incarceration in the Age of Colorblindness* (New York: New Press, 2010), 59–96; John F. Pfaff, *Locked In: The True Causes of Mass Incarceration and How to Achieve Real Reform* (New York: Basic Books, 2017), 21–50.

92. *Communications Assistance for Law Enforcement Act of 1994,* Public Law 103–414, *U.S. Statutes at Large* 108 (1994), 4279–4298.

93. Nate Anderson, *The Internet Police: How Crime Went Online, and the Cops Followed* (New York: W. W. Norton, 2014), 93–118.

94. Schneier and Banisar, *The Electronic Privacy Papers,* 248.

95. John Schwartz, "Industry Fights Wiretap Proposal," *Washington Post,* March 12, 1994, C7.

96. Federal Bureau of Investigation, U.S. Department of Justice, "Implementation of the Communications Assistance for Law Enforcement Act: Initial Notice and Request for Comments," *Federal Register* 60, no. 199 (October 16, 1995): 53644.

97. Seth Schiesel, "FBI Reduces Scope of Proposal on Wiretapping Phone Networks," *New York Times,* January 15, 1997, A11; American Civil Liberties Union, "Big Brother in the Wires: Wiretapping in the Digital Age" (March 1998): https://www.aclu.org/other/big-brother-wires-wiretapping-digital-age.

98. John Markoff, "FBI Wants Advanced System to Vastly Increase Wiretapping," *New York Times,* November 2, 1995, A1.

99. John Schwartz, "FBI Lists Technical Needs for Wiretaps in Digital Age," *Washington Post,* January 15, 1997, A2.

100. See Patricia Moloney Figliola, *Digital Surveillance: The Communications Assistance for Law Enforcement Act* (Washington, DC: Congressional Research Service, 2007), for a thorough account of the Federal Communications Commission's efforts to moderate the FBI's capacity requests after CALEA.

101. Jim McGee, "Wiretapping Rises Sharply under Clinton," *Washington Post,* July 7, 1996, A1.

102. See Bamford, *The Shadow Factory,* 234–253.

103. *United States Telecom Association et al. v. Federal Communications Commission,* 99 A.2d 1442 (D.C. 2000).

104. Roberto Suro, "FCC Proposes Rules for Cellular Wiretaps: Guidelines Are First under 1994 Law," *Washington Post,* October 23, 1998, A16.

105. Jim McGee, "Heightened Tensions over Digital Taps," *Washington Post,* October 27, 1996, H12.

106. See David Simon and Karen E. Warmkessel, "Major Baltimore Drug Ring Closed Down, Police Say," *Baltimore Sun,* April 18, 1990, 1A, 10A; Simon and Warmkessel, "Drug Suspect Williams Ordered Held without Bond," *Baltimore Sun,* April 19, 1990, 1B–2B; Simon and Warmkessel, "Drug War Optimism Weakens despite Successes," *Baltimore Sun,* April 22, 1990, 1A, 13A.

107. Simon, "Caller ID Latest Hit," 1A.

108. Jonathan Abrams, *All the Pieces Matter: The Inside Story of* The Wire (New York: Crown Archetype, 2018), 9–10.

109. For more on the fallibility of wiretapping in *The Wire,* see Joseph Christopher Schaub, "*The Wire:* Big Brother Is Not Watching You in Body-more, Murdaland," *Journal of Popular Film and Television* 38, no. 3 (2010): 122–132; Linda Williams, *On the Wire* (Durham, NC: Duke University Press, 2014), 141–145.

110. "Phone Benefits and Abuses," *Baltimore Sun,* June 19, 1990, 8A.

Epilogue

1. Whitfield Diffie and Susan Landau, *Privacy on the Line: The Politics of Wiretapping and Encryption,* 2nd ed. (Cambridge, MA: MIT Press, 2007), 277–312; Susan Landau, *Surveillance or Security? The Risks Posed by New Wiretapping Technologies* (Cambridge, MA: MIT Press, 2010); Torin Monahan, *Surveillance in the Time of Insecurity* (New Brunswick, NJ: Rutgers University Press, 2010); *The Snowden Reader,* edited by David P. Fidler (Bloomington: Indiana University Press, 2015), 1–18; Lisa Parks, *Rethinking Media Coverage: Vertical Mediation and the War on Terror* (New York : Routledge, 2018).

2. Shoshana Zuboff, *The Age of Surveillance Capitalism: The Fight for a Human Future at the New Frontier of Power* (New York: Public Affairs, 2019).

3. See Andrew Guthrie Ferguson, *The Rise of Big Data Policing: Surveillance, Race, and the Future of Law Enforcement* (New York: NYU Press, 2017); Sarah Brayne, *Predict and Surveil: Data, Discretion, and the Future of Policing* (New York: Oxford University Press, 2021).

4. David Kahn, *The Codebreakers: The Story of Secret Writing* (New York: Scribner, 1996 [1967]), 213.

5. For the most recent installments of the U.S. government's annual *Wiretap Report,* see https://www.uscourts.gov/statistics-reports/analysis-reports/wiretap-reports.

6. "Activists Say Chicago Police Used 'Stingray' Eavesdropping Technology during Protests," *CBS Chicago,* December 6, 2014, https://chicago.cbslocal.com/2014/12/06/activists-say-chicago -police-used-stingray-eavesdropping-technology-during-protests/; Jessica Anderson, "*Sun* Investigates: Cellphone Surveillance Seen Years Earlier in 'The Wire,'" *Baltimore Sun,* April 11, 2015, http://www.baltimoresun.com/news/maryland/investigations/bs-md-sun-investigates

-stingray-20150410-story.html; Robert Snell, "Feds Use Anti-Terror Tool to Hunt the Un-documented," *Detroit News,* May 18, 2017, https://www.detroitnews.com/story/news/local/detroit-city/2017/05/18/cell-snooping-fbi-immigrant/; Alvaro M. Bedoya, "Deportation Is Going High-Tech under Trump," *The Atlantic,* June 21, 2017, https://www.theatlantic.com/technology/archive/2017/06/data-driven-deportation/. On the history of Stingray technologies, see Lisa Parks, "Rise of the IMSI Catcher," *Media Fields* 11 (2016): 1–19.

7. Simone Browne, *Dark Matters: On the Surveillance of Blackness* (Durham, NC: Duke University Press, 2015).

8. "Clay Case Testimony Reveals Long-Term Taps," *Baltimore Sun,* June 6, 1969, 7; Martin Waldron, "Wiretaps on Dr. King Made after Johnson Ban," *New York Times,* June 7, 1969, 29.

9. Ernest B. Furgurson, "Jesse Helms, J. Edgar Hoover and Martin Luther King," *Baltimore Sun,* October 16, 1983, K2.

10. "Hearing Delayed on Dr. King Data," *New York Times,* October 15, 1983, 7.

11. "Reagan Phones King's Widow: Assures Her He Meant No Insult," *Los Angeles Times,* October 21, 1983, A1.

12. Helen Dewar, "Delay of King Bill Defeated," *Washington Post,* October 19, 1983, A1.

13. Steven V. Roberts, "Senators Are Firm on King Holiday," *New York Times,* October 19, 1983, A18.

14. "Dr. King's Foes Seek FBI Files on Him," *Baltimore Afro-American,* October 22, 1983, 1–2; William Safire, "Happy King's Birthday," *New York Times,* October 20, 1983, A27.

15. Ben A. Franklin, "Wiretaps Reveal Dr. King Feared Rebuff on Nonviolence," *New York Times,* September 15, 1985, 30.

16. David Levering Lewis, review of *Bearing the Cross: Martin Luther King, Jr., and the Southern Christian Leadership Conference,* by David J. Garrow, *Journal of American History* 74, no. 2 (September 1987): 482, 483.

17. Taylor Branch, interview by Brian Lamb, *BookTV,* C-SPAN2, April 12, 1998, http://www.booknotes.org/Watch/100454-1/Taylor-Branch.

18. Edward Morris, "Taylor Branch: King's Last Years," *BookPage,* February 2006, https://bookpage.com/interviews/8898-taylor-branch#.W31xbn4nZbU.

19. See Peter Galison, "Removing Knowledge," *Critical Inquiry* 31 (Autumn 2004): 229–243.

ACKNOWLEDGMENTS

Writing is a solitary endeavor. But I'm happy to say that I never felt like I had to work on this book alone. I have so many people to thank for that feeling. It kept me going, especially when the going got tough.

This book is the product of fellowship support from the National Endowment for the Humanities Public Scholars Program and the John W. Kluge Center at the Library of Congress. My thanks go to the readers at the NEH for believing in this project, and to the colleagues at the Kluge Center—too numerous to name here—who helped me wrap up the final stage of my research. My involvement with the Galsworthy Criminal Justice Reform Program, under the stalwart direction of Anthony Bradley, pushed my writing in new and unexpected directions at a critical time. Thanks to Christopher Green, Melynda Price, Samantha Sheppard, Tom Sugrue, Christopher Suprenant, and Shatema Threadcraft for motivating me to think differently. I'll never forget our conversations. Finally, I want to thank the team at Harvard University Press—especially Andrew Kinney, an editor of superhuman sensitivity and patience—for taking a chance on this project and bringing it into the world. I have no doubt that the final product is better for the efforts of two anonymous readers for the press, and for the assistance of Wendy Nelson, Mihaela-Andreea Pacurar, and Stephanie Vyce throughout the editorial process. The Introduction builds on ideas I first presented in "Wiretapping Stuff: Notes on Sound, Sense, and Technical Infrastructure," *Resilience* 5, no. 3 (Fall 2018): 96–108. Portions of Chapters 5 and 6 were first published online at *Post45: Peer Reviewed* as "Eavesdropping in the Age of *The Eavesdroppers;* or, The Bug in the Martini Olive" (February 2016). I thank the editors and reviewers of both venues for their kind and careful attention.

At Georgetown, faculty participants in the GU Americas Initiative read almost every part of this book, heroically enduring my penchant for long submissions. To Katie Benton-Cohen, Denise Brennan, Anna Celenza, Kate Chandler, Erick Langer, Bryan McCann, Chandra Manning, Doug Reed, Milena Santoro, and John Tutino: thanks for your advice—and for your forbearance. Members of the Georgetown Center for Privacy and Technology were generous enough to peruse a version of Chapters 7 and 8. Special thanks go to the legal scholar Paul Ohm, who was charitable enough to offer a brilliant formal response, and to my inspiring colleague Meg Leta Jones, who invited me to gatecrash on the group. Caetlin Benson-Allott,

Nathan Hensley, Seth Perlow, Amanda Phillips, and Nicole Rizzuto commented on drafts and traded stories over pizza and wine, while Marcia Chatelain, Soyica Colbert, Ben Harbert, Al Miner, Brian McCabe, Anne O'Neil-Henry, Coilin Parsons, Robert Patterson, Alis Sandosharaj, Erika Seamon, Dan Shore, and Colva Weissenstein all made working on the hilltop rewarding and fun. Special thanks go to Sherry Linkon, Ricardo Ortiz, and Samantha Pinto—my faculty mentors on campus, past and present. Their support has made all the difference.

Despite the passage of time, I still feel like I'm a student. So it's especially fortunate that I have a number of extraordinary teachers who continue to help me find my way in the world of ideas. Werner Sollors always seems like he's hovering over my shoulder, prosecco in hand, even when thousands of miles away: transmitting foreign dispatches, sending overlooked citations, and making sure I approach the academic racket with grace and good humor. Larry Buell and David Rodowick encouraged my earliest efforts on this project and helped turn me into the teacher I wanted to be. Nancy Cott was the first to force me to write in the past tense. I hope she sees that her patience ended up paying off, even if the shortcomings of this book are mine alone. Long before graduate school, Barry O'Connell was there to inspire my intellectual pursuits and give me a model for what true mentorship looks like. Marisa Parham and John Drabinski were there too—and I'm so excited that they've made the leap to the DMV area, right across town. One of my academic heroes, Jeff Ferguson, passed away while I was working on this project. I wish I could share the final product with him. I'd like him to know I'm still thinking on the lowest of frequencies.

I feel fortunate that some of my closest collaborators also happen to be some of my closest friends. Jack Hamilton, Pete L'Official, and Brian McCammack commented on drafts with humbling speed and traveled the far reaches of the globe just to hang out. Meanwhile, Annie Galvin, Liz Munsell, and Laura McCammack offered the trivia, memes, and good sense that has kept all of us reasonably in line. Tim McGrath has helped shape my thinking for more than fifteen years, and on more than one occasion he has dropped everything to come to my rescue—and with Tenley Archer right behind him. Maggie Gram and Jen Tuohy gave me a second home in D.C. when I needed it most. (How on earth can I ever repay you two?) My sbagliato siblings Nick Donofrio and Verena von Pfetten continue to motivate me to live well. Eva Payne and Evan Kingsley have picked me up and offered on-demand wildlife identification. And from Racine to Cambridge to Amsterdam, George Blaustein has inspired so many of our collective pursuits, fueled by a steady diet of omelets and raw herring.

Sources say that the ACBL is still alive. Aside from the ballers listed above, first-team nominations go to the man in the middle, Derek Etkin, and to Euro-step innovator Eli Cook. Much love to the daughters of the ACBL, forever in a league of their own: Bianca, Ella, Gloria, Juniper, Lena, Kate, Mika, Nilah, Rona, and Viva. A quick tally of their names reveals that these kids can someday run fives. Reserve the court!

If my years on the academic circuit have afforded me anything, it has been the chance to connect with scholars who have asked thought-provoking questions and offered insightful suggestions. Some of them I barely know; others are long-time colleagues and friends. Either way, all of them ended up having an influence on the final shape of this project: Dudley Andrew, Simone Browne, Glenda Carpio, Nicholas Forster, Erica Fretwell, Lisa Gitelman, Hannah Gurman, Mark Hanin, Sarah Igo, Richard John, Michael Kimmage, Josh Lauer, Dimitrios Pavlounis, Claire Potter, Palmer Rampell, Yael Schacher, Scott Selisker, Carlene Stephens, Jonathan Sterne, David Suisman, Nicholas Yablon, and Mary Zost. It's likely that the people included in this list don't remember what they said to me. But I do, and I feel it's important to acknowledge the difference their ideas ended up making. I still recall Hua Hsu's words of encouragement when I first told him I wanted to write a book about wiretapping—and his recommendation to check out the Smith Connection's excellent 1972 single "I'm Bugging Your Phone" ("I'm bugging your phone baby, starting an investigation / I'm bugging your phone baby, taping your conversation"). I'm grateful for the intemperate evening I spent at The Raven with Brian Goodman, which resulted in the title I'd been searching for.

This book is better for its involvement with *Eavesdropping,* a multimedia art exhibit and lecture program curated by Joel Stern and James Parker at the Ian Potter Gallery in Melbourne, Australia (July–August 2018). Joel and James were as gracious as they were inspiring—and so was the group of artists and activists who participated in the event, especially Susan Schuppli. Closer to home, I'm grateful for the audiences that listened in on early versions of these chapters at Amherst College, the Hagley Museum and Library, the University of California–Davis, the University of Illinois at Urbana-Champaign, the University of Pennsylvania, and Yale University. I'm also grateful for all of the archivists and librarians who helped guide me in my search for materials over the last six years. I've listed their contributions in the footnotes, which means they're behind every sentence, on every page.

A number of individuals took the time to sit down with me to talk on the record about their knowledge of the stories I tell in this book. They deserve special mention here.

Early on, Hal Wallace introduced me to the history of telegraph tapping and showed off a fascinating stash of eavesdropping devices stored deep in the vaults of the Smithsonian Museum of American History. Herman Schwartz, a legendary professor of law at American University, met with me to discuss his days agitating and litigating against electronic surveillance on behalf of the American Civil Liberties Union. In December 2018, I spent an unforgettable afternoon with Ralph V. Ward, who invited me to his home in McLean, Virginia, to discuss his tenure as vice president of Mosler Research Products, a company that was at one time the federal government's largest supplier of eavesdropping equipment. For my own reference (and entertainment) he even composed a short memoir of his time in the bugging business. A million thanks go to his daughter, the historian Karen Ward Mahar, for making this connection after we randomly struck up a conversation at an academic conference in Delaware. On behalf of the Edward V. Long family estate, Annie Miller Devoy gave me the rare privilege of accessing Senator Long's papers at the State Historical Society of Missouri. She was also kind enough to help fact-check portions of Chapter 7.

Last, but certainly not least, was Edward J. Tomas: a gift from nowhere. Ed and I spent hours talking on the phone about his work as an electronic surveillance technician for the New Jersey State Police, and our conversations eventually formed the basis of what became the opening section of Chapter 9. Whether our calls were recorded, who can really say? My thanks go to his former colleagues Richard Sasso and Nicholas Dotoli for offering corroborating evidence.

Several other individuals chose to speak with me about wiretapping and electronic surveillance but remain anonymous. They know who they are. Saying thanks doesn't quite cut it, but it'll have to do for now.

Friends in D.C., Cleveland, New York, and elsewhere have picked me up and forced me to be my best self over the years: Brett and Amy Brehm, Jeremy and AJ Brown, Peter and Katy Colarulli, Rivka Friedman and Dena Roth, Sara Gebhardt, Marc and Abby Glick, Eric Jacobstein and Tan Ly, Bill Jensen, Tony Jones and Anna Lobonova, Russell Lang and Erin Dittus, Andrew and Shelley Leonard, Rory and Mike Leraris, Jeff and Tara Malbasa, Scott and Mary Malbasa, Spencer Paul, Eddie Pryce, Kevin and Meghan Rosenberg, Jeff and Katie Sunderland, Jill Tyler and Jon Sybert, Jude Volek and Davida Connon, and Matt and Molly Zeiger. The existence of this book is a testament to the support and camaraderie of this group of people. To Melissa Majerol, who has all of my heart: I can't wait to see what happens next.

To Matt and Stephanie Baumoel, Megan and Mike Considine, David and Cassidy Impink, Susan and Robert Keiser, and Peggy and Cliff Stark: thanks for showing me what it means to be a family. My cousin Cassidy remains the strongest, most inspiring, and most resilient person I've known. Next year—and the year after that, and the year after that—we'll be back together on the water slides. My final words of gratitude are for my sister and brother-in-law, Lisa and Seth Wolkoff, and for my parents, Ken and Carol Hochman: I've left you for last because I'm still not sure how I can express my appreciation for all you've done over these last few years. I love you all so much more than I can say. (Also: hi Benji!)

The most cherished hope I have for this book is that it will provide my daughter, Lena Hochman, with her first encounter with something I've written. Fortunately, by the time it appears in print she will have learned how to read. I plan to start Lena with the five little words printed on the dedication page. Turn back and see for yourself: they are the only words that matter.

INDEX